Handbook of Mixture Analysis

Handbook of Mixture Analysis

Edited by:

Olga Moreira

ARCLER
PRESS

www.arclerpress.com

Handbook of Mixture Analysis

Olga Moreira

Arcler Press
224 Shoreacres Road
Burlington, ON L7L 2H2
Canada
www.arclerpress.com
Email: orders@arclereducation.com

ISBN: 978-1-77407-708-5 (Hardcover)

DECLARATION

Some content or chapters in this book are open access copyright free published research work, which is published under Creative Commons License and are indicated with the citation. We are thankful to the publishers and authors of the content and chapters as without them this book wouldn't have been possible.

ABOUT THE EDITOR

Olga Moreira is a Ph.D. in Astrophysics and B.Sc. in Physics and Applied Mathematics. She is an experienced technical writer and researcher which former fellowships include postgraduate positions at two of the most renown European institutions in the fields of Astrophysics and Space Science (the European Southern Observatory, and the European Space Agency). Presently, she is an independent scientist working on projects involving machine learning and neural networks research as well as peer-reviewing and edition of academic books.

TABLE OF CONTENTS

LIST OF CONTRIBUTORS

Branislav Pani´c
Faculty of Mechanical Engineering, University of Ljubljana, Aškerˇceva ulica 6, 1000 Ljubljana, Slovenia

Jernej Klemenc
Faculty of Mechanical Engineering, University of Ljubljana, Aškerˇceva ulica 6, 1000 Ljubljana, Slovenia

Marko Nagode
Faculty of Mechanical Engineering, University of Ljubljana, Aškerˇceva ulica 6, 1000 Ljubljana, Slovenia

Rafael Coimbra Pinto
Instituto de Informática, Universidade Federal do Rio Grande do Sul, Porto Alegre, RS, Brazil

Paulo Martins Engel
Instituto de Informática, Universidade Federal do Rio Grande do Sul, Porto Alegre, RS, Brazil

Yu Zhang
School of Engineering, University of Lincoln, Lincoln LN6 7TS, UK

Chris Bingham
School of Engineering, University of Lincoln, Lincoln LN6 7TS, UK

Miguel Martínez-García
School of Engineering, University of Lincoln, Lincoln LN6 7TS, UK

Darren Cox
Siemens Industrial Turbomachinery Ltd., Lincoln LN5 7FD, UK

Artemy Kolchinsky
Santa Fe Institute, Santa Fe, NM 87501, USA

Brendan D. Tracey
Santa Fe Institute, Santa Fe, NM 87501, USA

Department of Aeronautics and Astronautics, Massachusetts Institute of Technology, Cambridge, MA 02139, USA

Alicia Quiros
Department of Signal Theory and Communications, Universidad Rey Juan Carlos, C° Molino s/n, Fuenlabrada, Madrid, Spain

Simon P Wilson
Department of Statistics, Trinity College Dublin, Dublin 2, Ireland.

Ji Yeoun Lee
MII826, Biomedical Engineering, The Department of Biomedical Engineering, Jungwon University, Donbu-ri #5, Goesan-eup, Goesan-gun, Chungcheongbuk-do 367-805, South Korea

Thomas Layer
Institute of Telecommunications, Vienna University of Technology, Karlsplatz 13, 1040 Wien Vienna, Austria.
Health & Environment Department, Austrian Institute of Technology, Donau-City-Strasse 1/2, 1220 Wien Vienna, Austria.

Matthias Blaickner
Health & Environment Department, Austrian Institute of Technology, Donau-City-Strasse 1/2, 1220 Wien Vienna, Austria.

Barbara Knäusl
Department of Radiation Oncology, Division of Medical Radiation Physics, Christian Doppler Laboratory for Medical Radiation Research for Radiation Oncology, Medical University of Vienna/AKH Vienna, Währinger Gürtel 18-20, 1090 Wien Vienna, Austria.

Dietmar Georg
Department of Radiation Oncology, Division of Medical Radiation Physics, Christian Doppler Laboratory for Medical Radiation Research for Radiation Oncology, Medical University of Vienna/AKH Vienna, Währinger Gürtel 18-20, 1090 Wien Vienna, Austria.

Johannes Neuwirth
Radiation Safety and Applications, Seibersdorf Labor GmbH, 2444 Seibersdorf, Seibersdorf, Austria.

Richard P Baum
THERANOSTICS Center for Molecular Radiotherapy and Molecular Imaging (PET/

CT) ENETS Center of Excellence, Zentralklinik Bad Berka, Robert-Koch-Allee 9, 99437 Bad Berka, Bad Berka, Germany.

Christiane Schuchardt
THERANOSTICS Center for Molecular Radiotherapy and Molecular Imaging (PET/CT) ENETS Center of Excellence, Zentralklinik Bad Berka, Robert-Koch-Allee 9, 99437 Bad Berka, Bad Berka, Germany.

Stefan Wiessalla
THERANOSTICS Center for Molecular Radiotherapy and Molecular Imaging (PET/CT) ENETS Center of Excellence, Zentralklinik Bad Berka, Robert-Koch-Allee 9, 99437 Bad Berka, Bad Berka, Germany.

Gerald Matz
Institute of Telecommunications, Vienna University of Technology, Karlsplatz 13, 1040 Wien Vienna, Austria.

Jinkai Tian
State Key Laboratory of High Performance Computing, National University of Defense Technology, Changsha 410073, China
College of Computer, National University of Defense Technology, Changsha 410073, China

Peifeng Yan
State Key Laboratory of High Performance Computing, National University of Defense Technology, Changsha 410073, China
College of Computer, National University of Defense Technology, Changsha 410073, China

Da Huang
State Key Laboratory of High Performance Computing, National University of Defense Technology, Changsha 410073, China
College of Computer, National University of Defense Technology, Changsha 410073, China

José Alfredo Jiménez
Department of Statistics, Universidad Nacional de Colombia, Carrera 45 No. 26-85, Bogota, Colombia

Viswanathan Arunachalam[1]
Department of Statistics, Universidad Nacional de Colombia, Carrera 45 No. 26-85, Bogota, Colombia

Chipo Mufudza
Statistics Department, Cukurova University, 01330 Adana, Turkey

Hamza Erol
Statistics Department, Cukurova University, 01330 Adana, Turkey

Muhammad Azhar
College of Computer Science and Software Engineering, Shenzhen University, Shenzhen 518060, China

Mark Junjie Li
College of Computer Science and Software Engineering, Shenzhen University, Shenzhen 518060, China

Joshua Zhexue Huang
College of Computer Science and Software Engineering, Shenzhen University, Shenzhen 518060, China

Vincent Vandewalle
Biostatistics Department, Univ. Lille, CHU Lille, ULR 2694—METRICS: Évaluation des Technologies de Santé et des Pratiques MéDicales, F-59000 Lille, France
Inria Lille—Nord Europe, 59650 Villeneuve d'Ascq, France

Taha Hasan
Department of Statistics, Islamabad Model College for Boys, Sector F-10/4, Islamabad, Pakistan

Sajid Ali
School of Business and Economics, Lahore, Pakistan

Munir Ahmed
Department of Mathematics, Islamabad Model College for Boys, Sector F-10/4, Islamabad, Pakistan

V. S. Sundar
Center for Multimodal Imaging and Genetics, University of California, San Diego, La Jolla, CA, United States
Department of Radiology, University of California, San Diego, La Jolla, CA, United States

Chun-Chieh Fan
Center for Multimodal Imaging and Genetics, University of California, San Diego, La Jolla, CA, United States

Department of Cognitive Sciences, University of California, San Diego, La Jolla, CA, United States

Dominic Holland
Center for Multimodal Imaging and Genetics, University of California, San Diego, La Jolla, CA, United States
Department of Neuroscience, University of California, San Diego, La Jolla, CA, United States

Anders M. Dale
Center for Multimodal Imaging and Genetics, University of California, San Diego, La Jolla, CA, United States
Department of Radiology, University of California, San Diego, La Jolla, CA, United States
Department of Neuroscience, University of California, San Diego, La Jolla, CA, United States
Department of Psychiatry, University of California, San Diego, La Jolla, CA, United States

S. Ruhi
Department of Mathematics, Pabna University of Science & Technology, Pabna, Bangladesh.

S. Sarker
Department of Statistics, University of Rajshahi, Rajshahi 6205, Bangladesh

M. R. Karim
Department of Statistics, University of Rajshahi, Rajshahi 6205, Bangladesh

Habtamu K Benecha
National Agricultural Statistics Service, USDA, Washington, 20250 DC USA.

Brian Neelon
Department of Public Health Sciences, Medical University of South Carolina, Charleston, 29425 SC USA.

Kimon Divaris
Departments of Epidemiology and Pediatric Dentistry, University of North Carolina, Chapel Hill, 27599-7450 NC USA.

John S Preisser
Department of Biostatistics, University of North Carolina, Chapel Hill, 27599-7420 NC USA.

LIST OF ABBREVIATIONS

ADCC	Automatic Decision Cluster Classifier
ANNs	artificial neural networks
ART	Adaptive Resonance Theory
AUC	areas under the curve
CIs	confidence intervals
CMB	cosmic microwave background
CME	Chicago Mercantile Exchange
DFT	discrete Fourier transform
EDF	empirical distribution function
EM	Expectation–Maximization
EMGMM	expectation-maximization algorithm for a Gaussian mixture model
GLMs	generalised linear models
GMM-SVM	GMM supervector kernel-support vector machine
HDD	Heating-Degree-Days
HOS	higher-order statistics
HOS	higher-order statistics
IGMN	Incremental Gaussian Mixture Network
LDA	latent Dirichlet allocation
LDA	Linear Discriminant Analysis
MCMC	Markov Chain Monte Carlo
MEEI	Massachusetts Eye & Ear Infirmary
MFCC	Mel-frequency cepstral coefficients
MM	mixture-model
NECR	noise equivalent count rate
OCA	object cluster association
PCA	principal component analysis
PET	positron emission tomography

PVE	partial volume effects
RND	risk-neutral densities
ROC	receiver operating characteristic
SBR	signal-to-background ratio
ZIP	zero-inflated Poisson
ZIPMR	Zero Inflated Poisson mixture regression

PREFACE

Mixture models are a broad category of probabilistic models for inferring the characteristics of sub-populations within an overall population. The detection and identification of sub-populations are based on the pooled population observed data alone, i.e. it does not require any information about the sub-populations. To describe the characteristics of sub-populations, it is often assumed that each sub-population follows a parametric distribution. These sub-populations parametric distributions are commonly referred to as the components of the mixture model. A mixture model is therefore named after the distribution family of its components. For instance, the components of the Gaussian Mixture Model (GMM) are a family of Gaussian distributions while the components of the Poisson Mixture Model (PMM) are a family of Poisson distributions. The overall mixture distribution is a probability distribution whose density function is a weighted sum of the individual densities.

Through an appropriate choice of its components, a mixture model is able to reproduce complex variations that may be present in the observed data of a given population. Such variations are often due to the existence of several sub-populations and, therefore, the observed data can not be adequately modeled by simply using a single parametric distribution model-based approach. Once estimated, the parameters of a mixture model can be used for solving a wide variety of data analysis tasks. Indeed, mixture models have been successfully applied to the modelling and analysis of statistical data in the fields of biology, astronomy, physics, engineering, genetics, medicine, social sciences, and economics.

The first mixture model-based approach was introduced in 1894 by Karl Pearson. It consisted of a mixture of two Gaussian probability density functions with different means and variances. The parameters of the two-component GMM was obtained by using maximum likelihood to fit this model to the overall population observed data. Pearson's mixture model-based approach suggested that there were two subspecies present. Nowadays, mixed model parameters are estimated using different methods such as the popular expectation-maximization, Markov chain Monte Carlo, moment-matching, spectral method and others.

This edited book, composed of 16 peer-reviewed and open-access articles, aims to provide an up-to-date account of the mixture modeling and its applications. Chapters 1 to 7 focus on the GMM and its applications in neural networks, mechanical engineering, information theory, signal processing, cosmology, and image segmentation. Chapters 8 to 15 feature different mixture models and their applications in data mining, computational medicine, signal processing, genetics, and product reliability. Chapter 16 considers the finite mixture model and its applications in statistical analysis and economics.

Chapter 1 consists of solving the estimation problem of Gaussian Mixture Model parameters using the expectation-maximization (EM) method. To overcome initialization problems found in EM parameter estimation, the authors propose the Rough-Enhanced-Bayes mixture estimation (REBMIX) algorithm as a more effective initialization. Chapter 2 proposes an algorithm for decreasing the time complexity of the Incremental Gaussian Mixture Network (IGMN), a supervised learning algorithm based on the EM algorithm for Gaussian mixture models. The faster algorithm works directly with precision matrices in the IGMN algorithm. The new approach consists of avoiding costly matrix inversions and determinant computations by performing rank-one updates. It effectively eliminates any source of cubic complexity for the supervised learning algorithm and reduces computational running time significantly, turning the IGMN into an efficient technique for the analysis of high-dimensional datasets.

Chapter 3 extends the traditional Gaussian mixture model approach used in recognition of operational states and detection of emerging faults for industrial systems. The authors propose a variational Bayesian method that allows a GMM to cluster with its mixture components that can be used for identifying steady-state operational behavior from transient operation. Chapter 4 proposes a family of estimators based on a pairwise distance function between mixture components that can be used for optimization problems that involve the minimization of entropy or mutual information. The authors evaluate these estimators using numerical simulations of mixtures of Gaussian distributions. The use of a pairwise distance function results in a continuous and smooth estimate of parameters of the mixture components.

Chapters 5 to 7 consist of the application of GMMs in cosmology, signal processing, and image segmentation. Chapter 5 is centered on the factor analysis model for source separation of multi-spectral image data. The proposed methodology is motivated by the need of finding an efficient solution to the separation of components in the radiation maps of cosmic microwave background (CMB) data obtained by Planck's space mission. The prior information about the sources is quantified as a multivariate GMM with an unknown number of factors. The authors use a variational Bayes technique for mixed model parameter estimation. Chapter 6 proposes a two-stage approach using GMMs and higher-order statistics for improving the classification performance between normal and pathological voices. Chapter 7, on the other hand, focuses on classification algorithms for positron emission tomography images and the development of an improved statistical approach using a GMM.

Chapters 8 to 11 are centered on the Dirichlet processes Mixture Model (DMM), Generalized Tukey's distribution Mixture Method, Poisson Mixture Regression Model (PMRM), and Hierarchical Gamma Mixture Model (HGMM) and their applications in spectral analysis, statistical inference, computational medicine, and data mining. Chapter 8 includes a kernel analysis based on DMMs. The proposed DMM-based method can be used to overcome the GMMs overfitting or underfitting problems found in kernel spectral analyses. Chapter 9 focuses on the modeling of mixtures of asymmetric distributions. The authors present a mixture of Tukey's g-generalized distributions which depends on asymmetric parameters such as skewness. Chapter 10 proposes the use of model-based clustering techniques based on PMRMs for detecting heart diseases. Chapter 11 proposed an HGMM-based unsupervised method for classifying high-dimensional data with a large number of classes.

Chapters 12 and 13 introduce two particular mixture models: a multi-partition subspace mixture model and a simplex-centroid constructed quadratic mixture model. Chapter 12 proposes a multi-partition model based on clustering It assumes the existence of several latent clustering variables, each one explaining the heterogeneity of the data with respect to some clustering subspace. The mixture model parameters are estimated using an EM algorithm based on a probabilistic reinterpretation of the factorial discriminant analysis. Chapter 13 reviews the simplex-centroid design which one of the simplest and widely used methods for constructing mixture models. The authors propose and discuss the construction of three to four components mixture models based on shrinkage simplex-centroid designs.

Chapters 14 and 15 consider the application of mixture models and their applications in genetics and manufacturing industry. Chapter 14 proposes a mixture model penalty method that can reduce the regression coefficients while it incorporates any prior information about genetic causal variants. Chapter 15 considers a two-fold Weibull mixture model and its applications in the analysis of failure data for manufactured products.

Chapter 16 considers the finite mixture model and its applications in statistical analysis and economics. In Chapter 16, finite mixture distributions are used to model counts obtained from heterogeneous populations. The authors propose a marginalized finite mixture model that can provide a wide range of alternatives to estimate exposure effects on marginal means of the generated counts. Regression parameters are specified in a two-part marginalized model for a direct estimation of these exposure effects using maximum likelihood methods.

CHAPTER

1

Improved Initialization of the EM Algorithm for Mixture Model Parameter Estimation

Branislav Pani´c, Jernej Klemenc and Marko Nagode

Faculty of Mechanical Engineering, University of Ljubljana, Aškerˇceva ulica 6, 1000 Ljubljana, Slovenia

ABSTRACT

A commonly used tool for estimating the parameters of a mixture model is the Expectation–Maximization (EM) algorithm, which is an iterative procedure that can serve as a maximum-likelihood estimator. The EM algorithm has well-documented drawbacks, such as the need for good initial values and the possibility of being trapped in local optima. Nevertheless, because of its appealing properties, EM plays an important role in estimating the parameters of mixture models. To overcome these initialization problems

Citation: Panić, B., Klemenc, J., & Nagode, M. (2020). Improved initialization of the EM algorithm for mixture model parameter estimation. Mathematics, 8(3), 373. DOI: https://doi.org/10.3390/math8030373

with EM, in this paper, we propose the Rough-Enhanced-Bayes mixture estimation (REBMIX) algorithm as a more effective initialization algorithm. Three different strategies are derived for dealing with the unknown number of components in the mixture model. These strategies are thoroughly tested on artificial datasets, density–estimation datasets and image–segmentation problems and compared with state-of-the-art initialization methods for the EM. Our proposal shows promising results in terms of clustering and density-estimation performance as well as in terms of computational efficiency. All the improvements are implemented in the rebmix R package.

Keywords: mixture model; parameter estimation; EM algorithm; REBMIX algorithm; density estimation; clustering; Image segmentation

INTRODUCTION

The Expectation–Maximization (EM) algorithm was introduced in [1]. Today, after a little more than 40 years, it is still one of the most popular algorithms for statistical pattern recognition. Studies range from the theoretical, i.e., convergence of the EM and its variant DA-EM in [2], to modifications of the EM algorithm for different purposes: image matching [3]; parameter estimation [4,5]; malaria diagnoses [6]; mixture simplification [7]; and audio-visual scene analysis [8]. EM's popularity has risen due to its use in estimating mixture-model (MM) parameters [9,10]. The mixture model (MM) is an umbrella term for probabilistic models in which the modeled population contains several sub-populations. To describe the characteristics of the sub-population, it is assumed that each sub-population follows a simple parametric distribution, usually referred to as the component of the MM. The type of the MM is therefore identified by the distribution family of its components, e.g., components of the Gaussian Mixture Model (GMM) have a Gaussian distribution. Once estimated, the parameters of MM can be used multi-purposely. Their use is equally interesting for density-estimation tasks [7,11] and clustering tasks [6,12,13]. In the context of MM parameter estimation, the EM algorithm can be seen as a clustering algorithm that maximizes the missing data log-likelihood function as the objective function. Therefore, the EM for MM can be seen as the maximum-likelihood estimator. EM also has many appealing properties, such as monotonic convergence and probabilistic constraints, which makes it appealing in the context of MM parameter estimation. The EM algorithm also has well-documented drawbacks [14]. It requires some initial parameters for the first iteration. We will refer to this as initialization.

The MMs have many parameters, which makes initialization difficult. In addition, EM is very sensitive to initialization. This is because of the multi-modality of the likelihood function for the MM. EM is strictly a hill-climbing algorithm (monotonic convergence property) and therefore can become trapped in local optima. This puts a lot of emphasis on the initialization of the EM algorithm, and there is much research on its initialization (Section 2). Good initialization of the EM algorithm not only improves the results of the estimation problem, but it can also speed up the estimation. The EM algorithm has a linear convergence rate that is characterized as slow [14,15]. However, if the initial parameters are good and close to a saddle point (local optima) of the likelihood function, fewer of the more expensive EM iterations are needed to complete the process. The initialization of the EM algorithm is a serious problem and should be treated carefully. This is because the initialization creates an additional problem. The number of components in the MM should be known in advance. However, it is usually unknown. Estimating the optimum number of components in the MM is a difficult task [14]. It often fails due to the spurious local optima maximized by the EM algorithm. Estimating the non-optimal number of components in the MM leads to poor performance, either for clustering, density estimation, or any other application of the MM. The best-possible initialization should always be provided in order to avoid such pitfalls. Therefore, in this paper, we deal with the problem of initializing the EM algorithm for the MM parameter estimation. The initialization uses the Rough-enhanced-Bayes mixture estimation (REBMIX) algorithm [16]. REBMIX is a heuristic for estimating the parameters of the MM that provide a balance between the goodness of the MM parameter estimates and the estimation time [17,18]. In most cases, the parameter estimates with REBMIX are not as good as the ones estimated with the EM algorithm, although the parameter estimates seem to not deviate as much from the parameters estimated with the EM. On other hand, in contrast to the EM, the parameters are estimated quickly. This can be seen in either the REBMIX application to the von-Mises MM parameter estimates in [17] or in the Weibull-Normal MM parameter estimates [18]. Because of this property, we think that REBMIX is a good algorithm for the initialization of the EM algorithm. Although REBMIX can estimate the MM parameters by itself, the need for higher accuracy in a lot of MM applications demands the use of the EM algorithm. The main idea of this paper is that the use of the REBMIX algorithm, as an initialization for the EM algorithm, will result in improved parameter estimates and, therefore, improve the performance in various applications of the MM parameters. In addition to that, due to the fact that REBMIX gives close parameter estimates to the ones of the EM, we believe that the number of iterations of the EM algorithm will be reduced. To prove this, we restrict ourselves to the GMM parameter estimation, being the dominant type of MM and the most troublesome [14]. We propose three different strategies for the GMM estimation, suitable for the model-selection procedures when the number of components in the GMM is unknown. Finally, we conduct a thorough empirical evaluation on the artificial datasets, a density estimation, and image-segmentation tasks. The rest of the paper is structured as follows. Section 2 gives an overview of the relevant work on the initialization of the EM algorithm. Section 3 is the theoretical background for the estimation of the GMM parameters, including the EM, REBMIX, and the coupled REBMIX&EM algorithm. Section 4 gives an overview of the datasets and Section 5 is the results and discussion. The paper ends with Section 6 and the concluding remarks.

RELATED WORK

Estimating the MM parameters requires careful initialization of the EM algorithm. In addition, because the EM algorithm needs initialization, two problems emerge. The first is that the complexity parameter of the MM, i.e., the number of components c, must be known (in most cases it is not). Second, even if the number of components c is known, a good set of initial parameters is required to avoid becoming trapped in local optima.

There are two ways to deal with the problem of an unknown number of components c in the MM. Some [13,19,20] initialize the parameters of the MM for the number of components c from the minimum cmin to maximum cmax number of components and run the EM algorithm on each of them. Then, calculate the goodness-of-fit for each estimated MM with a different number of components and choose the MM that yields the best goodness-of-fit. Others, [12,21–23], start by choosing a large number of components c. Then, they initialize the rest of the MM parameters and run the EM algorithm. Finally, they iteratively decrease the number of components c by split&merge and component annihilation techniques until an optimal number of components is found. Although these strategies are sometimes more robust for the initialization of the EM and use fewer EM iterations, they are also lacking. First, they all require criteria for the split&merge and component annihilation techniques and, second, for this strategy to work, the objective function is rewritten to fit the stopping criteria used, such as the, for example, minimum-message-length criterion in [22,23]. This makes any other criteria superfluous. Therefore, we will restrict ourselves to the first strategy, referring to it as model selection. It is worth mentioning that both strategies are deterministic [23]. Stochastic and re-sampling strategies do exist to tackle this problem [9, 24] and can be beneficial. However, they are often stand-alone algorithms that do not require EM and/or are often computationally more costly.

The second problem is the initialization of the remaining MM parameters when the number of components c is known. References [14,25] focus on random initialization and provide a lot of useful strategies for EM algorithm initialization. In particular, the strategy called short EM, which repeats a smaller number of iterations of EM with looser convergence criteria on multiple random initializations, of which the one with the highest likelihood is chosen and used to initialize the EM algorithm with more iterations and tighter convergence criteria. Their short EM is also used for the stand-alone initialization procedure explained in [26]. The procedure in [26] was restricted to GMM; hence, additional efforts were made in [27] for different types of MM, which was again based on the short EM strategy. Random initialization of the EM algorithm seems to be an attractive choice for initialization of the EM, as it is used in other papers such as [28,29], or [30], which deal with the initialization problem. Another favorable method for initializing the EM algorithm, but only for the GMM estimation, is the K-means algorithm [26, 28, 30]. K-means is a clustering algorithm that can be improved with multiple initializations [31] and therefore is useful for initializing the EM for GMM and can give multiple initial parameters. Additionally, K-means initialization tends to provide initial parameters that are closer to final estimates, so fewer EM iterations are needed. Another popular clustering algorithm for initializing the EM for GMM is hierarchical clustering [32]. This initialization technique is purely deterministic. Hence, it cannot be improved by multiple restarts. Although the initializations provided in [32] give close initial parameters to the final estimates and only a few of the EM iterations are needed, the fact that hierarchical clustering is computationally burdensome means that its use for larger datasets is questionable.

Instead of the multiple-repeated random or multiple-repeated k-means initialization, which tends to be used in the literature [14,25,26], here we use the REBMIX algorithm as the initialization technique for the EM algorithm. In contrast to the multiple repeated random or k-means initialization, REBMIX is a purely deterministic algorithm. However, there are other mechanisms that allow the REBMIX algorithm to provide a wide range of initial parameters for the EM algorithm. For this reason, we have tailored three different strategies, which we have called the Exhaustive, the Best and the Single REBMIX&EM Strategy. Each of the three strategies represents how well the parameter space is explored by the REBMIX&EM strategies. The resulting optimized parameter estimates are then used to perform the above-mentioned model-selection procedures and to select the optimum parameters.

ESTIMATION OF THE GAUSSIAN MIXTURE MODEL PARAMETERS

Let Y = $\{y_1, y_2, \ldots, y_n\}$ be an d-dimensional observed dataset of n continuous observations $y_j = \{y_1, y_2, \ldots, y_d\}$. Each observation y_j is assumed to follow the probability density function (PDF):

$$f(y_j|c, w, \Theta) = \sum_{l=1}^{c} w_l f_l(y_j|\Theta_l), \tag{1}$$

where the components have the multivariate Gaussian PDF:

$$f_l(y_j|\Theta_l) = \frac{1}{\sqrt{(2\pi)^d \det(\Sigma_l)}} \exp\left(-\frac{1}{2}(y_j - \mu_l.)\Sigma_l^{-1}(y_j - \mu_l)^T\right). \tag{2}$$

The GMM consists of finite number of components c and every component has weight w_l. The weights of the components w_l have the properties of the convex combination $w_l \geq 0 \wedge \sum_{l=1}^{c} w_l = 1$[9]. Therefore, the estimation of the parameters in the GMM consists of an estimation of the number of components c, an estimation of each component's weight w_l and each component's distribution parameters Θ_l; in the GMM case, the mean vector μ_l and the covariance matrix Σ_l .

Let $L(Y|\Theta)$ be the likelihood function. The log-likelihood function of the GMM is defined as:

$$\log L(Y|\Theta) = \sum_{j=1}^{n} \log\left(\sum_{l=1}^{c} w_l f_l(y_j|\Theta_l)\right), \tag{3}$$

where $f_l(y_j|\Theta_l)$ is defined in Equation (2). The maximum likelihood estimation of the parameters Θ can be defined as:

$$\widehat{\Theta} = \underset{\Theta}{\operatorname{argmax}} \log L(Y|\Theta). \tag{4}$$

Expectation–Maximization Algorithm

The EM algorithm is an iterative algorithm for an estimation of the parameters Θ of GMM (or any other MM). Instead of maximizing the log-likelihood function for dataset Y containing n observations, it maximizes the complete-data log-likelihood. The complete data Y_c is assembled from the observed data Y (sampled dataset) and the missing data (or data labels) Z. Missing data are the labels of observations representing the belongings of the observations in the dataset to the components in the GMM. Since the labels of the observations z_{kj} are unknown (for example, the label of the observation y_j from the dataset Y) the EM algorithm utilizes two steps, i.e., the expectation (E) step and the maximization (M) step. In the E step of the algorithm, the posterior probability τ_{kj} that the observation y_j from Y arose from the k-th component is calculated:

$$\tau_{kj}^{(t+1)} = \frac{w_k^{(t)} f_k(y_j|\mu_k^{(t)}, \Sigma_l^{(t)})}{\sum_{l=1}^{c} w_l^{(t)} f_l(y_j|\mu_l^{(t)}, \Sigma_l^{(t)})}. \tag{5}$$

In the M step, the iteration-wise update equations for the parameters can be derived by maximization of conditional expectation of the complete-data log-likelihood function [9]. The updates of the parameters can be obtained with the following equations. The weights w_k with:

$$w_k^{(t+1)} = \frac{\sum_{j=1}^{n} \tau_{kj}^{(t)}}{n}, \tag{6}$$

the mean vectors μ_k with

$$\mu_k^{(t+1)} = \frac{\sum_{j=1}^{n} \tau_{kj}^{(t)} y_j}{\sum_{j=1}^{n} \tau_{kj}^{(t)}}, \tag{7}$$

and the covariance matrices Σ_k with update equation:

$$\Sigma_k^{(t+1)} = \frac{\sum_{j=1}^{n} \tau_{kj}^{(t)} (y_j - \mu_k^{(t+1)})(y_j - \mu_k^{(t+1)})^T}{\sum_{j=1}^{n} \tau_{kj}^{(t)}}. \quad (8)$$

After every successful iteration of the E and M steps, the value of the likelihood is guaranteed to increase [9]. The expectation and maximization steps are repeated iteratively until the convergence of the log-likelihood is achieved. The convergence criteria are mostly assessed by the change in the log-likelihood value, i.e., when it is smaller than some predefined threshold value ε [9,33].

Model Selection with EM Algorithm

Let Θ(c) denote the GMM parameters with c components. Let $I = \{\Theta_{\text{init}}(c = c_{\min}), \Theta_{\text{init}}(c = c_{\min} + 1), \ldots, \Theta_{\text{init}}(c = c_{\max})\}$ denote the set of initial parameters of the GMM for the different number of components c. The final set S of estimated parameters becomes $S = \{\text{EM}(i), \forall i \in I\}$, where $\text{EM}(i)$ denotes the iterative process of GMM parameters estimation with the EM algorithm. The problem of model selection can then be summarized as: Choose s from S so that measure C(s) is optimal, where C(s) is the cost function for the calculation of the selected measure.

The most commonly used model selection measure, namely information criterion (IC), utilizes the estimated value of likelihood function (or log-likelihood). There is a lot of IC presented in literature, see Chapter 6 of [9]. Out of these, the ones mostly used are definitely AIC and BIC. AIC is defined as:

$$\text{AIC} = -2 \log L(Y|\Theta(c)) + 2M, \quad (9)$$

where M = c − 1 + cd + cd(d + 1)/2 is the number of parameters in d-dimensional GMM for the number of components c. This criteria mostly favor complex models, due to the small penalization term [25]. Although for most cases of datasets with a larger number of observations n it will select the over-fitted GMM (in terms of number of components c), it is still an interesting criterion for datasets with a smaller number of observations n. The second one, BIC, is defined as:

$$\text{BIC} = -2 \log L(Y|\Theta(c)) + M \log n. \quad (10)$$

BIC penalizes more heavily complex models than AIC due to the additional multiplicative term of logn. On the other hand, for the datasets with a smaller number of observations, it can result in GMM with a very small number of components c (under-fit).

REBMIX Algorithm

The Rough-Enhanced-Bayes mixture estimation (REBMIX) algorithm is a heuristic approach to the estimation of the GMM [16]. Here, the algorithm will be briefly described. For the interested readers, we advise the following papers [16,34,35]. The algorithm utilizes five simple steps, illustrated in Figure 1. The first step, the preprocessing of observations, conducts an empirical density estimation of the dataset. The empirical density estimation can be carried out in three different ways within the REBMIX algorithm, see ([16], p. 16). However, here the attention is given only to the histogram estimation.

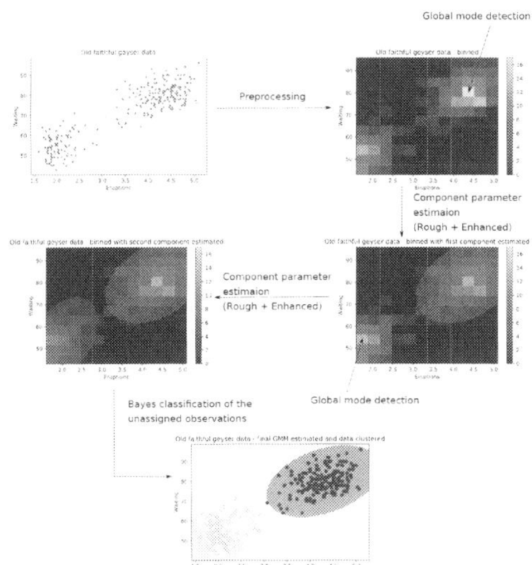

Figure 1: Illustration of the REBMIX algorithm on Old Faithful geyser data.

Let v be the number of bins in histogram. The preprocessing step assigns each bin in the histogram a frequency k_j , which represents the fraction of the observations falling into the volume V_j occupied by the j-th bin. This is used to transform the input dataset into a new shrunken dataset, consisting only of bin origins $\bar{y}_j$ and the corresponding empirical frequency k_j of the bin $j \in \{1, 2 \ldots v^*\}$ (It is worth noting that the v^* is the number of non-empty bins of histogram as they are only used for the estimation. Additionally, for convenience, the number of bins v mentioned refer to a one-dimensional histogram. In the multidimensional setting, the actual number of bins is equal to v^d, where d is the dimension of the problem). This transformed dataset X = (x_1, x_2 . . . x_j , . . . x_{v^*}), where x_j = $(\overline{y}_1, \ldots, \overline{y}_d, k_j)$, will be denoted as the main cluster. The second step in the algorithm is the global mode detection step. The global mode is simply the mode with highest frequency $\overline{\overline{y}}_m$, where m is defined as m = argmax (k_1, k_2, . . . , k_{v^*}). The global mode can be always detected and it serves as the basis for the Rough component parameter estimation (Step 3) [16,36]. Rough component parameters are additionally used to split the main cluster X into two clusters, base cluster X_l , and the residue R, where $X_l = (x_{l1}, \ldots x_{lj}, \ldots, x_{lv}^*) \wedge x_{lj} = (\overline{y}_1, \ldots, \overline{y}_d, k_{lj})$ and $R = (r_1, \ldots, r_j, \ldots, r_v^*) \wedge r_j = (\overline{y}_1, \ldots, \overline{y}_d$. In addition, the equality $k_j = k_{lj} + r_j, \forall j \in \{1, \ldots, v^*\}$ holds. The base cluster holds the observations tied to the currently estimated component of the GMM and residue is the residual cluster. The split process ends when the criteria $D_l \leq D_{min}/w_l$ is met, where the Dl is sum of positive deviations [16], w_l is the component weight, and D_{min} is the internally controlled parameter. After the split process is done, the base cluster is used for the Enhanced component parameter estimation. Enhanced component parameter estimates are an approximation of the Gaussian distribution maximum likelihood estimates for the histogram preprocessing [16,36]. The weight of the currently estimated component l is:

$$w_l = \frac{\sum_{j=1}^{v^*} k_{lj}}{n}, \tag{11}$$

the mean vector:

$$\mu_l = \frac{\sum_{j=1}^{v*} k_{lj}\overline{y}_j}{\sum_{j=1}^{v*} k_{lj}}, \tag{12}$$

and the covariance matrix:

$$\Sigma_l = \frac{\sum_{j=1}^{v^*} k_{lj}(\overline{y}_j - \mu_l)(\overline{y}_j - \mu_l)^T}{\sum_{j=1}^{v*} k_{lj}}. \tag{13}$$

The Enhanced component parameter estimation concludes the l-th component parameter estimation. The residue becomes the main cluster, i.e., X = R, and the Global mode detection, Rough component parameter, and Enhanced component parameter estimation are repeated to estimate the parameters of a newly added (l + 1)-th component. This addition of the components ends when the criteria $n_{res}/n \leq cD_{min}$ is met. The $n_{res} = \sum_{j=1}^{v*} r_j$ is the number of observations in residue and c is the current number of estimated components in GMM. Two intuitive assumptions are made here. The first assumption is the addition of new components should end when the number of observations in the residue is too small to form a new component in the GMM. The control of how small should the residue be is assessed via the parameter D_{min}. The second assumption is that the unassigned observations left in the residue should be redistributed to the already-estimated components and consequently make a certain impact on their parameters. This is made with the Bayes classification of the unassigned observations step [16,36]. The art of the estimation of the parameters of the GMM for the different number of components c lies in updating the parameter D_{min} iteratively after each Bayes classification of the unassigned observation step is conducted. The parameter Dmin is iteratively decreased with the empirical equation $D_{min}^{new} = cD_{min}^{old}/(c+1)$. Parameter D_{min} is decreased until $c > c_{max}$. In doing so, the algorithm is forced to iterate until the final set $\{\Theta_{final}(c = 1), \Theta_{final}(c = 2), \ldots, \Theta_{final}(c = c_{max})\}$ of GMM parameters is obtained.

Initially, value of the parameter D_{min} is taken to be 1. As the weights of the components w_l form a convex combination, the initial value of $D_{min} = 1$ ensures that the first estimated GMM contains only one component. In practice, every decrease in the parameter D_{min} does not guarantee that estimated GMM will have exactly one more component, or generally speaking more components, as the decreasing equation is empirical. Thus, the set of final estimated parameters of the GMM can be, e.g., $\{\Theta_{final}(c = 1), \Theta_{final}(c = 2), \Theta_{final}(c = 2), \Theta_{final}(c = 4) \ldots, \Theta_{final}(c = c_{max})\}$. Still, employing different numbers of bins v eases the burden of the above-mentioned problem. In addition, since there is no specific rule to identify v, the presented algorithm flow in [16] runs over an ordered set of numbers K of bins v.

In [16], there is a suggestion to use the same model-selection procedure as described in Section 3.1.1, to select the optimal GMM parameters. However, it is reported in [17,18] that the model-selection procedure usually makes sub-optimal parameter estimates. This happens because the estimated parameters of the GMM with the REBMIX algorithm can be further optimized to yield higher values of the likelihood function (Equation (3)) and therefore provide a beneficial impact on the values of the commonly used IC, for example, BIC, which is defined in Equation (10).

Combined REBMIX and EM Algorithm

Let R denote the set of solutions (the estimated parameters of the GMM for different numbers of components c) obtained by the REBMIX algorithm. For a certain number of bins v in the histogram preprocessing step and a specified maximum number of components c_{max} and minimum number of components c_{min}, REBMIX will always provide the same set of solutions R. However, there are many different values for the number of bins v that can be chosen for the preprocessing step. The two extreme cases are the case of v = 1 and $v = v_{uni,}$ where v_{uni} is a sufficiently large number of bins for which the $k_j = 1$, □j ∈ {1, 2, . . . , v *} is true. All the values that lie in between these two values can be used for REBMIX to provide different initial parameters for the EM algorithm. As opposed to the stand-alone REBMIX, the use of the EM algorithm ensures that the final estimates yield the maximum likelihood, thus providing more stable grounds for the model-selection procedure. Three different strategies for the coupling of the REBMIX and EM algorithms are proposed. The first strategy, denoted as Exhaustive REBMIX&EM, can be summarized as follows:

1. Select the multiple number of bins v_1, v_2, . . . for histogram preprocessing;
2. For each v from v_1, v_2, . . . , obtain the corresponding set of REBMIX estimates R_{v1}, R_{v2}, . . . ;
3. Merge all solutions from the sets R_{v1}, R_{v2}, . . . into one set R;
4. For each solution r ∈ R, run the EM algorithm and obtain the set S of EM improved estimates of the GMM parameters;
5. Run the model selection procedure on the set S

Although this strategy gives the best potential for the estimation of the parameters of the GMM, it is computationally intensive, due to the many repetitions of the EM algorithm. To ease the computational intensity, the second strategy, denoted as Best REBMIX&EM, is summarized as follows:

1. Select the multiple number of bins v_1, v_2, . . . for histogram preprocessing;
2. For each v from v_1, v_2, . . . , obtain the corresponding set of REBMIX estimates R_{v1} , R_{v2} , . . . ;
3 . Group all solutions from the sets R_{v1} , R_{v2} , . . . by the number of components c, i.e., form the sets $R_c{=}c_{min}$, $R_c{=}_{cmin+1}$, . . . , $R_c{=}c_{max}$, where each set holds the estimated parameters of the GMM with the same number of components c;
4. From each set $R_c{=}j$, choose the solution for which the

likelihood value is the largest r_c=j,best and merge it into new set $R_{\text{best}} = \{r_{c=c_{\min},\text{best}}, r_{c=c_{\min}+1,\text{best}} \cdots r_{c=c\max,\text{best}}\}$;

5. For each solution $r_{c=j,\text{best}} \in R_{\text{best}}$, run the EM algorithm and obtain the set S of EM improved estimates of the GMM parameters;
6. Run the model selection procedure on the set S.

At the end, the least computationally intensive strategy, which additionally minimizes the computation in the REBMIX algorithm, can be derived. This strategy is denoted as the Single REBMIX&EM. If the number of bins v is estimated or intuitively taken as known, the overhead of computing the different parameters Θ of the GMM for the different numbers of bins v can be reduced in the REBMIX algorithm. This strategy can be summarized as follows:

1. Estimate or intuitively select number of bins v;
2. Obtain the corresponding set R of REBMIX estimates for known v;
3. For each solution r ∈ R run EM algorithm and obtain the set S of EM estimates of the GMM parameters;
4. Run the model selection procedure on the set S.

EXPERIMENTS

The proposed method for estimating the parameters of the GMM was tested on the artificially generated datasets, density estimation tasks, and on the color-image segmentation tasks. Additionally for a sanity check, two different strategies were included for the EM algorithm: multiple repeated random initialization with the EM algorithm and multiple repeated k-means initialization with the EM algorithm. The number of repetitions of the k-means and the random initialization was 5. They are referred to as the Random&EM and Kmeans&EM. For the model-selection criterion, the BIC was used.

Artificial Datasets

Artificial datasets were created using the MixSim R package [37]. The utility of this package is in simulating GMMs with various parameters, e.g., with different levels of overlap between the mixture components or number of components (Figure 2). For this research, we have used the dimension of the dataset d, the number of components in the GMM c, and the overlap level a between the components in the GMM.

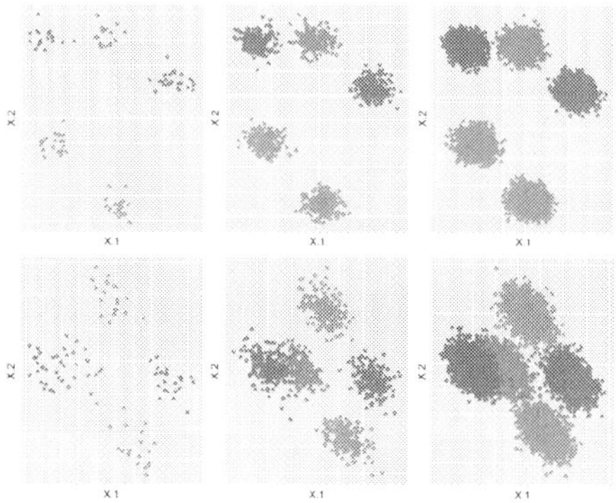

Figure 2. Examples of artificial datasets with two dimensions and five components with a variable number of observations and an overlap level. The upper three graphs had a = 0.001 and the lower three graphs had a = 0.125. From left to right, the number of observations were respectively n = 100, n = 1000, and n = 10,000.

Table 1 summarizes the input parameters. Using all the combinations of input parameters, 18 different parameters of GMMs were simulated. Sampling of the datasets from each simulated GMM was done with n = 100, n = 1000, and n = 10,000 observations, which resulted in 54 different datasets.

Table 1. Input parameters used for simulating different Gaussian mixture models.

Parameter	Values
Dimension (d)	3, 5, 10
Number of components (c)	5, 10, 15
Overlap level (a)	0.001, 0.125

Density Estimation

To study the performance of the proposed strategies for the density estimation, we chose the multivariate distributions presented in [11], for which their implementation performed well. The distributions are: Trips-2d,

Blob 2-d, Bend 2-d, Wave 2-d, Trips 3-d, Blob 3-d, Bend 3-d, Spiral 3-d, and Tetra 3-d, where 2-d and 3-d represent the dimension of the distribution. The datasets were sampled with different number of observations n = 50, 100, 200, 300, 400, 500, 750, 1000, 1500. The number of replications of each dataset was 10. For the performance measure of the density estimates, we have chosen the same measure as in the original paper, the mean integrated square error (MISE) between the true density and estimated density. MISE is defined as:

$$\mathrm{MISE} = \frac{1}{r}\sum_{i=1}^{r}\int (f(y) - \hat{f}(y))^2 dy, \tag{14}$$

where r is the number of replications, $\hat{f}(y)$ estimated density, and f(y) the true density. Additionally, Reference [11] provides an R implementation of their work, so we included the results obtained with their Adaptive density estimation by Bayesian averaging (ADEBA) algorithm.

Image Segmentation

To study the performance of proposed strategies for the clustering, we have selected the digital-image segmentation task. The digital-image-segmentation task can be seen as a real-world clustering problem. To elaborate, segmentation of an image means grouping, i.e., clustering, the image into coherent parts. The pixels in the digital image hold certain values of color intensity (brightness). In general, color digital images are represented by three channels, where each channel holds the intensity of one primary color, such as red, green, and blue (known as RGB images). The clustering problem in this setting can be conceived as color quantification of the digital image or, more simply, grouping the parts of the digital image by the color similarity. Hence, the use of GMMs for this is appropriate. The estimation and the model selection are carried on all the pixel color values in the digital image. Once the GMM parameters are estimated, the clustering scheme can be obtained by assigning each pixel $j \in \{1, \dots, n\}$ in a given image with n pixels to the one of the components in the estimated GMM. The assignment is made with the posterior probabilities τ_{lj}, namely the component for which the posterior probability is highest $l = \mathrm{argmax}(\tau_{1j}, \tau_{2j}, \dots, \tau_{cj})$ [33]. Posterior probabilities τ_{lj} for pixel j are obtained using Equation (5). The number of clusters, therefore, becomes the number of components c in estimated GMM.

The clustering performance is assessed through the Adjusted Rand Index (ARI) measure [38]. To define the ARI, first we need to define the contingency table (Table 2). The contingency table is used to summarize the overlap between the two clustering schemes M = {M_1, M_2, . . . , M_r} and N = {N_1, N_2, . . . , Ns} with different numbers of clusters r and s. Each entry n_{ij} in the contingency table represents the common number of objects in clusters i and j, respectively, from the M and N clustering schemes.

Table 2. Contingency table

M/N	N_1 N_2 ... N_s	Sums
M_1	n_{11} n_{12} ... n_{1s}	a_1
M_2	n_{21} n_{22} ... n_{2s}	a_2
$\vdots$	$\vdots$ $\vdots$ $\ddots$ $\vdots$	$\vdots$
M_r	n_{r1} n_{r2} ... n_{rs}	a_r
Sums	b_1 b_2 ... b_s	

ARI is defined as

$$\mathrm{ARI} = \frac{\sum_i^r \sum_j^s \binom{n_{ij}}{2} - [\sum_i^r \binom{a_i}{2} \sum_j^s \binom{b_j}{2}]/\binom{n}{2}}{\frac{1}{2}[\sum_i^r \binom{a_i}{2} + \sum_j^s \binom{b_j}{2}] - [\sum_i^r \binom{a_i}{2} \sum_j^s \binom{b_j}{2}]/\binom{n}{2}}, \quad (15)$$

where n_{ij}, a_i , b_j can be acquired from the contingency table. The ARI is bounded between −1 and 1, where 1 means that the two clustering schemes have 100% agreement and a negative ARI value means that the agreement is not better than the random. Therefore, the higher the ARI value, the better the match between the two clustering schemes.

For the image-segmentation tasks, five images from the famous Berkeley dataset were used [39], see Figure 3. The images additionally have human segmentation, which can serve as a ground-truth segmentation for the evaluation of the performance of the clustering. As reported in [39], each image in the dataset has multiple human segmentations and they are all considered valid. Therefore, the ARI is calculated for each human segmentation and the maximum value is reported.

Figure 3. Showcase of selected images from [39]. First row: flower image (label 353013; herd image (label 38092); woman image (label 216053). Second row: fisherman image (label 48055); bridge image (label 22093).

RESULTS AND DISCUSSION

Results on Artificially Generated Datasets

First, let us introduce the parameters used for the proposed strategies. The Exhaustive REBMIX&EM and Best REBMIX&EM strategies require an ordered set of the numbers of bins K, and the Single REBMIX&EM strategy requires a single number of bins v as the input. The ordered set of the numbers of bins K had 3–100 bins. For the Single REBMIX&EM strategy, the number of bins v was estimated using the Knuth rule [40]. The maximum number of components c_{max} = 20 and the minimum numbers of components c_{min} = 1 is kept the same for every strategy, along with the Kmeans&EM and the Random&EM. The threshold value for the average log-likelihood value (for the convergence assumption) was 0.0001 and the maximum number of EM iterations allowed was 1000. The choice of convergence of the EM with the average log-likelihood value was made because it removes the impact of larger and smaller datasets (in terms of the number of observations). The average log-likelihood was calculated simply by dividing the log-likelihood value $L(y|\Theta)$ by the number of observations n in the dataset.

Secondly, let us describe the methodology used to compare the strategies here. As explained in Section 3.1.1, the purpose of the model-selection procedure is to choose the appropriate GMM parameters based on the defined measure, which was in our case BIC. Therefore, the quality of the estimated GMM parameters can be expressed by the corresponding value of the BIC.

If the strategy estimated the GMM parameters that gave the best value of the BIC, i.e., the lowest value compared to other strategies used, then we assume that the strategy also estimated the best GMM parameters. This is a reasonable and commonly used method for comparing the quality of the estimated GMM parameters [14, 32]. In addition, the use of the MixSim R package to sample the artificial datasets (Section 4.1) allows the true density estimation and the clustering for each artificial dataset to be determined based on the fact that the GMM from which the data set was sampled is known. This was used to additionally compare the clustering (Section 5.1.1) and density-estimation (Section 5.1.2) performance of the estimated GMM parameters. The general results are presented in Figure 4 which are acquired on all artificial datasets. The first plot represents the percentage of selected GMM with the best BIC value for each REBMIX&EM strategy versus the Kmeans&EM and Random&EM strategy. Exhaustive REBMIX&EM selected better GMM on more datasets than Kmeans&EM and Random&EM strategy, judging by the BIC value. On other hand, strategies Best and Single REBMIX&EM had a smaller number of datasets on which they yielded best GMM in terms of BIC value than Kmeans&EM and equal to the Random&EM strategy. The second plot represents the box-plots of the EM iterations used by each strategy. The Random&EM strategy needed the largest number of EM iterations. Following this is the Exhaustive REBMIX&EM strategy, then the Kmeans&EM strategy, and the smallest number of iterations was needed by the Best and Single REBMIX&EM strategies. This is inferred based on the median value of the box-plots (horizontal line inside the box). In addition, the Best and Single REBMIX&EM strategies had a much smaller number of iterations of the EM algorithm, due to the fact that they had only one set of initial parameters for each number of components $c = 1, 2, 3 \dots 20$, as opposed to, e.g., the Random&EM, which had five sets of initial parameters for each number of components. However, interestingly, the Exhaustive REBMIX&EM strategy had at least 90 different sets of initial parameters for each number of components and still had a smaller number of iterations than the Random&EM and had the same median value as the distribution of the number of iterations for the Kmeans&EM. That does seem to be beneficial to the REBMIX algorithm, which seems to provide very close to the initial parameters of the GMM to the final estimated parameters. Finally, the third plot shows the amount of time spent on estimating the GMM parameters with different strategies. The Best and Single REBMIX&EM strategies had similar computing times and therefore their effectiveness seems to be equivalent. Most of the differences in the performance of these two strategies stem from the fact that the Best REBMIX&EM strategy had a multiple number of bins v (3–100) for the REBMIX algorithm, as opposed to the Single REBMIX&EM strategy, for which we used the Knuth rule to obtain the single number of bins v. Moving forward, the Exhaustive REBMIX&EM strategy gradually increased the required computation time with the increase in the number of observations. This is to be expected, since this strategy required the most EM iterations (second plot) compared to the other two REBMIX&EM strategies. The Random&EM and Kmeans&EM strategy had the worst performance in terms of the computing time. Although the Random&EM strategy needed less time when the number of observations was lower, the Kmeans&EM strategy outperformed it as the number of observations increased. Judging by the actual values, however, the difference between these two strategies was not great.

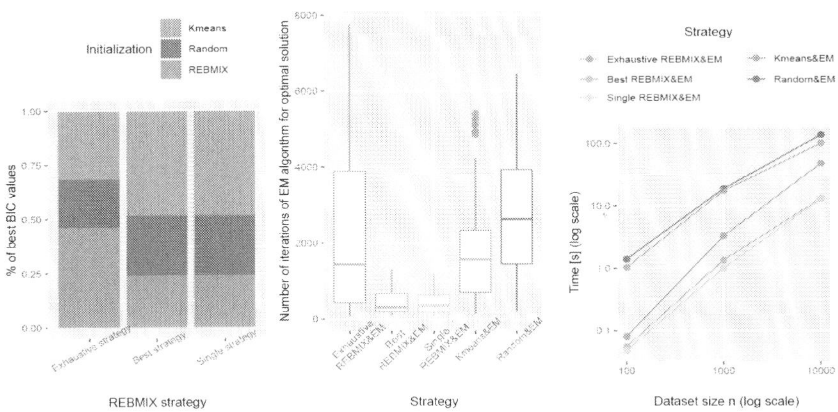

Figure 4. Benchmarking different REBMIX&EM strategies vs. Kmeans&EM and Random&EM.

Application to Clustering

The ARI is used to study the clustering performance. The results are given as box-plots in Figure 5. The grouping in Figure 5 was selected because it reflects the difficulty of the dataset, i.e., a small number of observations in the dataset along with a high degree of overlap between the components can be seen as a difficult clustering problem, while the large number of observations with a small degree of overlap can be seen as a simpler clustering problem.

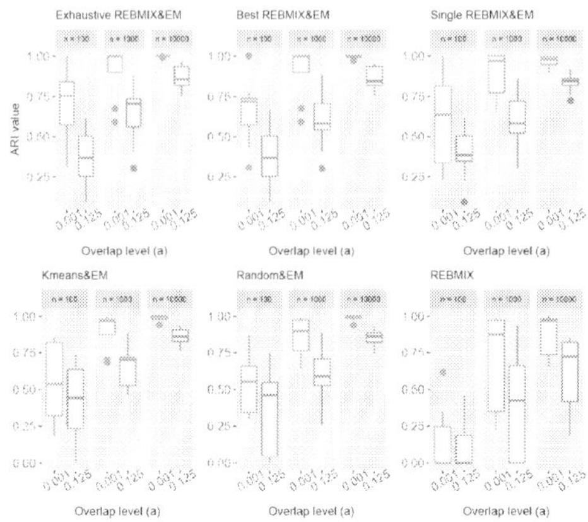

Figure 5. Clustering performance of the Exhaustive REBMIX&EM strategy, Best REBMIX&EM strategy, Single REBMIX&EM strategy, Kmeans&EM, Random&EM, and REBMIX on artificial datasets.

The performance of the Exhaustive REBMIX&EM strategy and the Best REBMIX&EM strategy on smaller datasets (n = 100) with a lower degree of overlap was better than the multiple repeated Kmeans&EM and multiple repeated Random&EM strategy. The Kmeans&EM strategy improved the results for the datasets with a larger number of observations with smaller overlap although the Exhaustive REBMIX&EM strategy and the Best REBMIX&EM strategy performed equivalently or better. The Random&EM only had equivalent results on the largest datasets n = 10,000. The Single REBMIX&EM performed slightly worse than the Exhaustive REBMIX&EM strategy and the Best REBMIX&EM strategy, although equivalently to the Kmeans&EM strategy and mostly better than the Random&EM strategy. On the other hand, the datasets with larger overlap reduced the clustering performance of all the strategies, especially for datasets with n = 100 and n = 1000. In this context, the Kmeans&EM performed better than all the REBMIX&EM strategies for datasets with a small number of observations n = 100. On datasets with a larger number of observations n = 1000 only the Exhaustive REBMIX&EM strategy performed as well as the Kmeans&EM strategy, while only the Random&EM strategy performed poorly. On the datasets with the largest number of observations n = 10,000, the strategies Exhaustive and Best REBMIX&EM slightly outperformed the other strategies. We have also included the results we obtained just by running the REBMIX algorithm. For the K, we chose the 3–100 range—in other words, the same as for the Exhaustive and Best REBMIX&EM strategy. The results for the n = 100 were mainly worse than the results from all the other strategies. As this number increased, i.e., n = 1000 and n = 10,000, the performance improved, although the other strategies still performed better.

Application to the Density Estimation

To study the performance of the estimated GMM with different strategies in terms of density estimation, we used the mean square error (MSE) measure [33]. The MSE is defined as:

$$\mathrm{MSE} = \frac{1}{N}\sum_{j}^{n}(f(y_j|\hat{\Theta}) - f(y_j|\Theta))^2, \tag{16}$$

where $\hat{\Theta}$ are estimated GMM parameters with different strategies and Θ are the true GMM parameters from simulation. The choice of MSE instead of MISE was made here because the distributions used for sampling the datasets had higher dimensions, i.e., 3, 5, and 10, and the creation of the good integration grid requires a lot of computing power in terms of memory. The results are therefore presented as box-plots of MSE values in the upper plot in Figure 6. These box-plots are grouped by the number of observations in the dataset. Again, for the estimation, we have used all the strategies and the stand-alone REBMIX algorithm.

Judging by the values presented, all the strategies and the stand-alone REBMIX gave good results for the datasets with n = 1000 and n = 10000. In contrast, for n = 100, the Kmeans&EM, Random&EM, and REBMIX had much better results than the REBMIX&EM strategies. To study more carefully why this happened, the box-plots for the datasets with n = 100 observations only are given in the lower plot (Figure 6), grouped by the number of dimensions d in the dataset. It is clear that, as the dimensionality of the dataset increased, the results became worse. Although the stand-alone REBMIX algorithm had good results, it is clear that the results of the REBMIX&EM strategies deteriorated. To understand why this happened, we must recall that the log-likelihood function, defined in Equation (3), for the GMM is unbounded. Namely, a singularity in one of the component covariance matrices Σ_l can arise when $\det(\Sigma_l) \rightarrow 0$ [33]. When this problem appears, the EM can be jeopardized and the obtained solution can degenerate. A high-dimensional setting, such as d = 5 or d = 10, emphasizes this problem even more [14]. Because we did not include any of the safeguards for the EM algorithm, in terms of bounding the log-likelihood function, the resulting log-likelihood increased and the model-selection procedure was misled into choosing the degenerated solution.

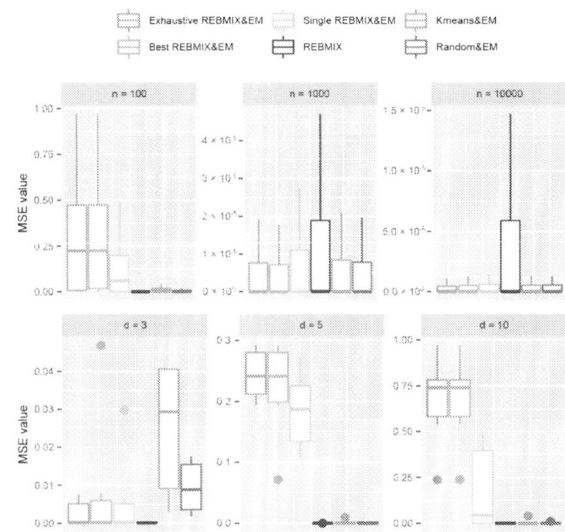

Figure 6. Density-estimation performance evaluation of Exhaustive REBMIX&EM strategy, Best REBMIX&EM strategy, Single REBMIX&EM strategy, Kmeans&EM, Random&EM and REBMIX on the artificial datasets; First row: Plots given are grouped by the number of observations n. Second row: Plots are grouped by the dimension d of the dataset for the n = 100 observations in the dataset.

Density-Estimation Tasks

For the density estimates, the same parameters were used for the three REBMIX&EM strategies as in the case of the Artificial datasets and also the same for the Random&EM and Kmeans&EM strategies. For the ADEBA algorithm, we used the parameters for the ADEBA − α implementation (see [11], Section 3), i.e., the parameter α is chosen with Bayesian averaging and $\beta = 1$ for which we will use the generic term ADEBA.

The acquired values of MISE for each pair of strategy/method and distribution are shown on different plots in Figure 7. In general, the GMM estimates using different strategies yielded better results with an increase in the dataset size n. For the small dataset sizes, e.g., n = 50 and n = 100, ADEBA usually had the best results, although the REBMIX&EM strategies proved to be good competitors, which can be seen from the results on the distributions Bend 2-d, Bend 3-d, Trips-2d, Blob 2-d, Tetra 3-d and Wave 2-d. For the datasets sampled from the distributions Trips-2d and Trips-3d with an increasing size of the dataset, all the strategies used outperformed the ADEBA.

This is expected because the Trips-2d and Trips-3d distributions are actually GMM distributions with c = 3 non-overlapping components. Interestingly, the ADEBA outperformed all the strategies used for all the dataset sizes n on the Blob 2-d distribution, which is also a GMM distribution with c = 8 overlapping components. However, as the dimension of the distribution increased (i.e., the distribution Blob 3-d is only the same distribution as Blob 2-d in three-dimensional space), the results from the ADEBA deteriorated. On the distributions Bend 2-d and Bend 3-d, all the estimators had their performance deteriorated. Kmeans&EM and Random&EM performed worse than the REBMIX&EM strategies and the ADEBA.

On the Spiral 3-d distribution for the n = 50 REBMIX&EM strategies, the Kmeans&EM and Random&EM strategy performed worse than the ADEBA, although their performance improved as the n increased (for the REBMIX&EM strategies again from n = 100 and and for the Kmeans&EM and Random&EM strategies from the n = 300 and n = 500, respectively). For the distributions Tetra 3-d and Wave 2-d, the results are similar to the ones for the Blob 2-d distribution. Between the different REBMIX&EM strategies, the results did not differ greatly. The slight difference can be seen on the smaller dataset sizes n.

However, as n increased, the results were almost identical. We have also evaluated REBMIX, as a stand-alone algorithm, for this experiment. Again, K was 3–100. REBMIX was better when the number of observations in the dataset was small, usually only for n = 50. On other hand, just the increase from the n = 50 to n = 100 improved the results from the REBMIX&EM strategies on many distributions, most notably on the GMM distributions such as Trips 3d.

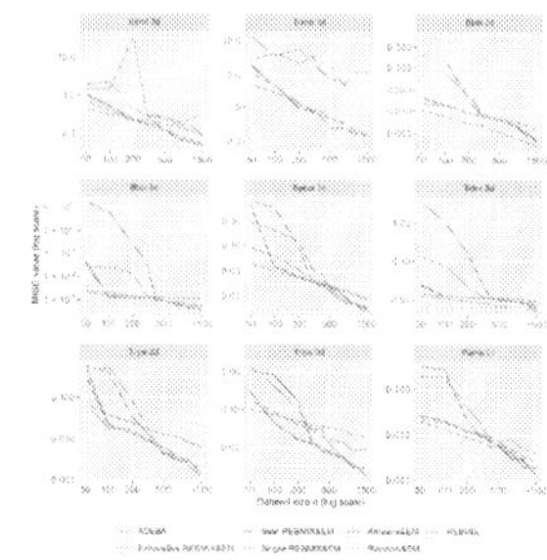

Figure 7. Density-estimation performance evaluation of Exhaustive REBMIX&EM strategy, Best REBMIX&EM strategy, Single REBMIX&EM strategy, Kmeans&EM, Random&EM, and ADEBA on datasets sampled from multivariate distributions in [11].

To study the performance of different EM strategies and the ADEBA, the computation times for each method/strategy were measured for different distributions and different dataset sizes n. The values are grouped by the dataset size n and the method/strategy, and the mean values are shown in Figure 8. The Single and Best REBMIX&EM strategy far outperformed all the other strategies used and the ADEBA algorithm for all the dataset sizes, with the exception of size n = 50, where the ADEBA and REBMIX&EM strategies achieved equivalent results. The Exhaustive REBMIX&EM strategy performed as well as the ADEBA. The Kmeans&EM and Random&EM performed worse on all the dataset sizes, with the exception of the dataset size n = 1500, where they performed slightly better than the ADEBA and Exhaustive REBMIX&EM. The REBMIX stand-alone outperformed all the used strategies and the ADEBA algorithm in terms of computation efficacy.

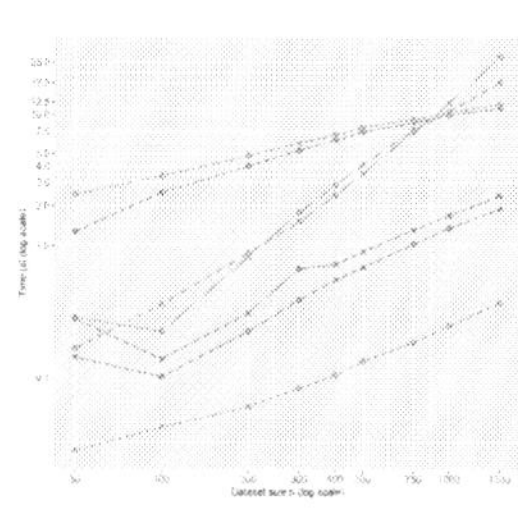

Figure 8. Mean computation times of Exhaustive REBMIX&EM strategy, Best REBMIX&EM strategy, Single REBMIX&EM strategy, Kmeans&EM, Random&EM, and ADEBA for each dataset size n.

Image-Segmentation Tasks

The image segmentations were carried out using the following strategies: the Single REBMIX&EM strategy, where the number of bins v was estimated using the Knuth rule; the Single REBMIX&EM strategy, where the number of bins v was taken to be v = 255; the Best REBMIX&EM strategy, with the range of the number of bins being 3–100; and the Kmeans&EM strategy and the Random&EM strategy. The number of bins v = 255 was taken as the color-channel value. The intensities of the RGB images only have integer values and, therefore, the frequency of each color channel is estimated. Furthermore, using v = 255, the marginal distributions of the color-channel value's intensity are represented. The maximum number of components for each strategy is kept the same, i.e., cmax = 20. The threshold value for the EM algorithm was 0.0001 and the maximum allowed number of EM iterations was 1000. The numerical results in terms of the number of components c and the ARI value are given in Table 3. From the numerical results, one of the REBMIX&EM strategies always had the best ARI values. Additionally, the Single REBMIX&EM with v = 255 yielded very good results in terms of ARI value on the flower image (353013), herd image (38092), and woman image (216053); therefore, the intuitive guess of v = 255 was shown to be beneficial. Finally, the selected GMMs with all strategies had a large number of components c, which can be related to the BIC model selection criteria. To establish some qualitative comparison, the segmentations of the flower and herd images are given in Figure 9. First, the issue of the black or white colored background. Namely, the black or white colored background caused a high density on the edges of the feature space, creating a heavy-tailed distribution, for which GMMs are not suitable. The GMMs estimated with each strategy needed at least two components to address the heavy tails of distribution; therefore, a background like on the flower image (label 353013) was mostly over-segmented and caused a lower ARI value. Next, the strategies mostly had problems with the objects that occupied a large part of the image scene and had little variation in color (for example, the flower object on the flower image or soil on the herd image). Namely, the object that occupied larger parts of the scene caused a high-density peak in the distribution. This high-density peak caused a certain asymmetry in the distribution and the strategies tried to fit multiple components of the GMM for that part. This caused a lot of noise in the segmentation, which can be seen in Figure 9. This problem was slightly reduced by the Single REBMIX&EM strategy with v = 255, which mostly did tie the peaks with one component of the GMM (flower object in the flower image or the soil-ground part of the herd image in Figure 9). This also led to a much higher ARI value for those segmentations. The sky gradients mostly caused over-segmentations, which were not properly addressed by any strategy.

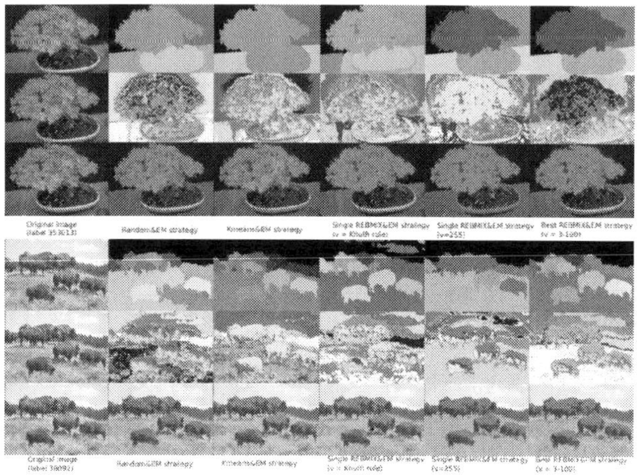

Figure 9. Showcase segmentation of flower image (label 353013) and herd image (label 38092. The first row is ground-truth segmentations; Second row: Segmentation using random colors for each component in estimated GMM; Third row: Segmentations using mean values of the components in the estimated GMM

Table 3. Numerical results of image segmentation tasks

Image Label/Strategy		353013	38092	216053	48055	22093
Random&EM	c	20	20	20	20	20
	ARI	0.29	0.28	0.47	0.37	0.34
Kmeans&EM	c	20	20	20	20	20
	ARI	0.24	0.28	0.24	0.36	0.35
Single REBMIX&EM	c	20	19	19	19	20
with Knuth rule	ARI	0.38	0.42	0.35	0.41	0.41
Single REBMIX&EM	c	18	20	20	20	20
with $v = 255$	ARI	0.67	0.67	0.63	0.45	0.34
Best	c	19	16	16	16	18
REBMIX&EM	ARI	0.44	0.54	0.25	0.47	0.38

Additionally, we also included results from purely clustering algorithms, such as the MeanShift (MS) algorithm or the newer Cutting edge spatial clustering (CutESC) [41], which exhibited good results in their study. We have used scikit-learn Python MS implementation [42] with default values. The segmentation was made on the original pixel-color values. For the CutESC algorithm, we followed their recommendation. First, we calculated superpixels of the images with scikit-image SLIC implementation [43] with defaults values and then constructed a shrunken dataset, containing the spatial and color values. Table 4 summarizes the ARI values obtained using their algorithms. In addition, to obtain consistently good segmentations for the CutESC algorithm, parameters α and β were set to 0.5. As expected, CutESC outperformed all the other strategies/algorithm. The GMM estimated with the REBMIX&EM strategies outperformed the MS clustering algorithm on images 353013, 38092 and 48055, and MS yielded slightly better results on images 216053 and 22093.

Table 4. ARI values obtained from MS and CutESC clustering algorithm.

Image/Algorithm	353013	38092	216053	48055	22093
MS	0.46	0.54	0.69	0.1	0.43
CutESC	0.86	0.71	0.92	0.79	0.76

To investigate the robustness of the different strategies in a clustering scenario of the image-segmentation tasks, we added different types of noise to the images used. Seven types of noise, salt, pepper, salt&pepper or shortly (s&p), Gaussian, Poisson, speckle, localvar implemented in the Scikit-image python module, and their effects can be seen in the herd image in Figure 10. The results of the ARI value are shown on the plots in Figure 11 for each strategy/method and the corresponding image and noise type.

Figure 10. Different types of noise added to the herd image (label 38092). First row: Original image, salt&pepper noise added, salt noise added, pepper noise added; Second row: Gaussian noise added, Poisson noise added, speckle noise added, localvar noise added.

As expected, the different types of noise contributed differently to the quality of the segmentation; however, surprisingly, not all the types of noise deteriorated the results in terms of reducing the ARI value. The Poisson had an improving effect on almost every image and every method/strategy used. Noises such as salt, pepper and salt&pepper generally had a deteriorating effect on the results of the segmentation. With certain methods/strategies, this type of noise deteriorated the quality of the segmentation by up to 20%. In order to better evaluate the results, we provided additional box-plots (Figure 12) of the different methods/strategies that show the percentage of difference between the results of the non-noisy images and the results of the noisy images. The box-plots are constructed as follows. The upper limit of the box is the maximum value of the difference (positive value means improvement), and the lower limit is the minimum value (negative value means deterioration) and the horizontal line within the box median value. Since the ARI measure is bounded between −1 and 1, the difference of 10% can be interpreted as 0.05 of the difference of the ARI value

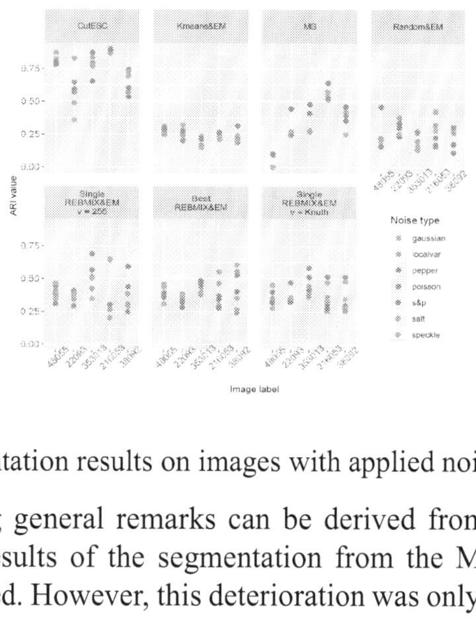

Figure 11. Segmentation results on images with applied noise.

The following general remarks can be derived from the box plots in Figure 12. The results of the segmentation from the MS algorithm have usually deteriorated. However, this deterioration was only minor and ranged from 5% up to 10%. On the other hand, the CutESC algorithm on the image 48055 mostly had improvements, while on the image 216053 it produced only a slight deterioration of the segmentation results. On the image 22093, there were slight improvements in the segmentation results for certain noise types, while there were large deteriorations for the others. On the images 353013 and 38092, the noise usually deteriorated the segmentation results from the CutESC algorithm. Forwardly, the noise had the following effects on the results of the segmentation using the Kmeans&EM and Random&EM strategies. Different noise types had little effect (either improving or deteriorating) on the segmentation results with the Kmeans&EM strategy, while the noise deteriorated the segmentation results with the Random&EM strategy in most cases, although the specific noise types improved the segmentation results for the images 48055 and 22093.

The REBMIX&EM strategies used were influenced by the noise and the different types of noise in different ways. We will start with the Single REBMIX&EM strategy with the number of bins v = 255. This strategy had the greatest deterioration for the images 353013, 216053, and 38092. Although it is still not as significant, since 20% represents the decrease of the ARI value of 0.1, it indicates that this method should be used with caution in the presence of noise. On the other hand, the Best REBMIX&EM and Single REBMIX&EM strategy with the number of bins v estimated with the Knuth rule had minor to mild (image 38092 with Best REBMIX&EM strategy) deterioration of the segmentation results, and, interestingly, the greatest improvements of the segmentation results in the presence of different types of noise. In fact, the Best REBMIX&EM only had improvements of the segmentation results for the image 216053 without any deterioration (the resulting rectangle in Figure 12 is above zero). The Single REBMIX&EM strategy with the number of bins v estimated with the Knuth rule had similar results for the 353013 image. This can be interpreted as follows. By adding noise to particular images, the density shape of the image pixel values became more similar to those that the GMM can generate, so the REBMIX&EM strategies were able to successfully estimate them.

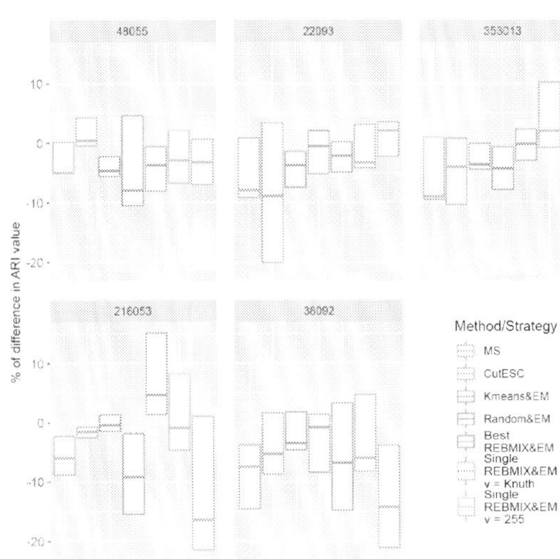

Figure 12. Difference in ARI value between noisy and non-noisy images for each strategy/algorithm.

The results of the different strategies/algorithms regarding image segmentation with noise applied to the images revealed some interesting facts, such as the fact that adding some types of noise could improve the results of the GMM-based image segmentation. The in-depth investigation of when and why this happened goes beyond the scope of this article and also goes beyond the topic, so we conclude with a final remark. In many cases, the noise will deteriorate the results of the segmentation; however, this is to be expected. For most of the strategies/algorithms used, the amount of deterioration was low. If the present amount of noise is large, the Random&EM strategy or the Single REBMIX&EM with the number of bins v = 255 is not recommended, while the others are safe to use. It should also be considered as to whether the addition of some artificial noise can improve the result of the segmentation.

Comparison of the Time Complexity of REBMIX and k-Means Algorithm

As the k-means algorithm is more or less the fastest used algorithm to initialize the parameters for the EM estimation of the GMM, we compared the time complexity of the k-means algorithm and the REBMIX algorithm. It is hard to estimate theoretically the time complexity for both algorithms. Judging by [31], the theoretical time complexity of the k-means is O(tcnd), where t is the number of iterations needed, c is the number of components, n is the number of observations, and d is the dimension of the dataset. It is hard to know how many iterations t will the k-means algorithm need. On the other hand, the time complexity of the REBMIX algorithm depends on multiple factors. First, the most important is definitely the number of non-empty bins v * in the histogram preprocessing. Forward, it depends on the chosen maximum value of the components cmax and the dimension of the problem d. Finally, it depends on the different number of bins v for the histogram preprocessing, e.g., the 3–100 range has 97 values. Therefore, we have chosen to give an empirical comparison of both algorithms in terms of the number of observations n, the number of dimensions d, and the maximum number of components cmax. The results are presented in Figure 13. For the REBMIX algorithm, we used the REBMIX function provided by the rebmix R package. For the k-means algorithm, we used the default R kmeans implementation provided in the base R package. Although the k-means algorithm needs a different number of iterations, here we considered 100 iterations of the k-means algorithm. This seems reasonable due to the fact that repetitions of the k-means are often preferable [31]. In addition, due to the fact that, for the GMM initialization, the k - means algorithm needs to be repeated $c_{max} - c_{min}$ times, we have taken this into consideration by repeating the k - means algorithm for each $c \in \{c_{min}, c_{min} + 1, \ldots, c_{max}\}$. For the REBMIX algorithm, we have used the setting in which the number of bins v is not known and the range used was 3–100. In addition, when the time complexity with respect to the one parameter is evaluated, the other two were kept constant. In other words, when comparing the time complexity with respect to the number of dimensions d, the number of observations n = 1000 and the maximum number of components c_{max} = 10 was kept constant. For the comparison with respect to the number of observations n, the number of dimensions was d = 5 and the maximum number of components c_{max} = 10. Finally, for the comparison with respect to the maximum number of components cmax, the number of dimensions was d = 5 and the number of observations was n = 1000.

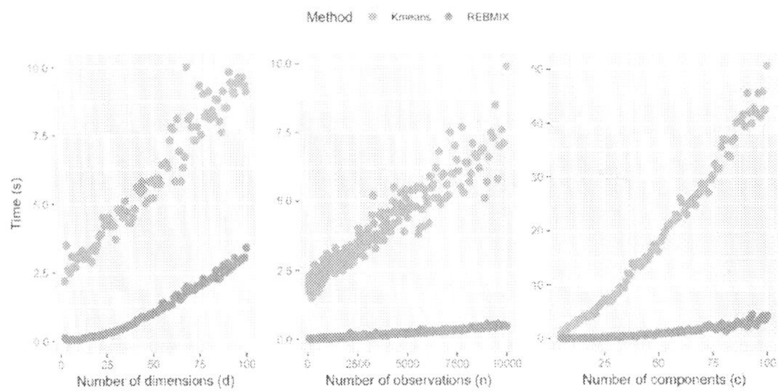

Figure 13. Time complexity of REBMIX algorithm vs. k-means algorithm

First, we will start with a comparison with respect to the number of dimensions d, which is given in the first plot of Figure 13. We can see that the time complexity of k-means is linear with respect to this parameter. Therefore, it agrees with the theoretical assumption. On other hand, the REBMIX seems to have a nonlinear complexity with respect to the parameter d. The nonlinearity with respect to this parameter comes from the definition of the multivariate Gaussian distribution in Equation (2). To calculate the PDF of the GMM, the inverse of the covariance matrix Σ for each component is required. The matrix inversion is a nonlinear operation; therefore, nonlinear dependence in time is justified. The second parameter for which the comparison was made was the number of observations n in the dataset. The results are shown on second plot of Figure 13. Both the REBMIX and k-means algorithms exhibited a linear dependency in respect to the number of observations n, and, finally, k-means had a linear dependency with respect to the c_{max} parameter, while the REBMIX has nonlinear. The nonlinear complexity in terms of the parameter c_{max} comes from its empirical stopping criteria, which does not guarantee that each iteration adds exactly one more component to the GMM (details of the stopping criteria of the REBMIX algorithm are explained in the last two paragraphs of Section 3.2). The nonlinear dependencies of the REBMIX algorithm with respect to the parameters d and cmax are troublesome, given that k-means has a linear dependence. This gives the k-means algorithm an advantage over the REBMIX algorithm. However, if the time complexity with respect to the number of observations n is observed, they have proven to be equal competitors. In the end, with the obtained values of time spent for the REBMIX algorithm and the k-means algorithm, the REBMIX algorithm was generally faster.

Note on Selection of Hyperparameters for the REBMIX&EM Strategies and Their Impact on Performance

In this section, we will discuss the hyperparameters of the REBMIX&EM strategies used for the presented experiments. There were five different hyperparameters used for all the experiments, in particular the number of bins v, the minimum number of components c_{min} in the GMM, the maximum number of components c_{max} in the GMM, the maximum number of iterations of the EM algorithm, and the threshold for convergence of the EM algorithm.

First, we will discuss the choice of the number of bins v for the histogram preprocessing. There is no general rule for estimating the number of bins v in the histogram [44]. Some common estimators are Sturges' rule $\log_2$ (n) + 1 originating in [45] and the RootN rule $\sqrt{}$ n from [46]. In [16], it was suggested to create the lower and upper bounds for the bin range K based on these rules. Strurges' rule always yields a smaller number of bins v and is suitable for the lower bound, while the RootN rule gives a larger number of bins v and thus can be used for the upper bound. The ranges K based on those two rules are narrow for datasets with a smaller number of observations n, e.g., for n = 100, the range is 7–10 bins. For the dataset with a large number of observations, e.g., n = 10,000 leads to the range of 14–100, which is very wide. The choice of the bin range 3–100 as a constant for each experiment was chosen to test whether these additional ranges can be used to better estimate the GMM parameters using the REBMIX algorithm and thus allow a better initialization of the EM algorithm, or only cause an unnecessary computational effort. For the estimation of the number of bins v for the Single REBMIX&EM strategy, we have used the Knuth rule. The Knuth rule uses a Bayesian approach to estimate the number of bins v in a histogram, thus making this estimation dependent on the observations from the dataset. Ref. [40] gives a straightforward extension to multivariate histograms. Although, in the original paper, the multivariate histograms can have a different number of bins for each dimension, i.e., vi , ∀i ∈ {1, 2, . . . , d}, it is worth noting that we only used the same number of bins v in each dimension. The number of bins v estimated with the Knuth rule for datasets from Section 4.1 is shown in Figure 14. It is clear from Figure 14 that the Knuth rule depends on both the dimension d of the dataset and the number of observations n, unlike the above-mentioned estimators. In addition, it had a different number of bins v for the datasets with the same dimension d and number of observations n (boxes span over multiple values of v), which indicates that n and d are not the only influences. In addition, it is clear that the number of bins v is greatly reduced as the dimensions of the dataset increased. This can be tied to what is called the curse of dimensionality [44]. There are many bins in a high-dimensional histogram grid. Just for the v = 3 and d = 10 settings, there are 9,765,625 bins. Such an extremely large number of bins leads to many empty bins, i.e., for which the frequency was $k_j = 0$, which were heavily penalized by the Knuth rule (it can be seen that, for d = 10, the estimated number of bins was mostly v = 2 or v = 3).

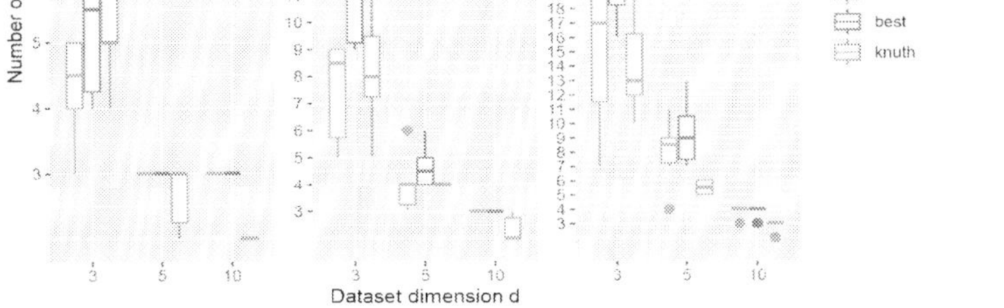

Figure 14. Optimal number of bins v selected with the different strategies and Knuth rule on the artificial datasets from Section 4.1.

Additionally, in Figure 14, the number of bins v that yielded the optimal GMM parameters using the Exhaustive and Best REBMIX&EM strategies given for the datasets. Although we have given a large range, judging by the values in Figure 14, it can be seen that both the Exhaustive and the Best REBMIX&EM strategies estimated the GMM parameters for a different number of bins v than what is proposed by the Knuth estimator, which leads in some experiments to benefits, while in another leads to some numerical instabilities. For example, the number of optimal GMM parameters estimated with the Exhaustive REBMIX&EM strategy was the greatest if we recall Figure 4. On the other hand, regarding what was found by comparing the density-estimation performance in Section 5.1.2 for a small number of observations n = 100, this caused some of the degeneracies in the estimated GMM parameters. The Best REBMIX&EM strategy favored a larger number of bins v in the histogram preprocessing in all the presented cases. This is also a useful insight because in the image-segmentation experiment (Table 3) the results were better than what was acquired with a Single REBMIX&EM strategy with the number of bins v estimated with the Knuth rule. For the density-estimation tasks in Section 5.2, the results seem to have not been very impacted by the chosen REBMIX&EM strategy, at least in terms of the MISE value (Figure 7). Let us recall that we also used the intuitive v = 255 for the image-segmentation tasks. As we already said, this intuition was made based on the fact that digital images have 255 different intensity values for each color channel. This assumption was shown to be good, judging by the results we obtained with it; however, it was probably most sensitive to the noise along with the Random&EM strategy.

To conclude, the Single REBMIX&EM strategy with the Knuth rule can be safely used when a short computation time is preferred, and the dimension of the dataset is small. The datasets in the density-estimation tasks in Section 5.2 had d = 2 or d = 3. For the Exhaustive and Best REBMIX&EM strategies, we recommend the RootN rule for the upper bound and the Knuth rule for the lower bound. The RootN rule will mostly overestimate the number of bins for larger n, judging by the obtained values in Figure 14. However, due to the fact that a smaller number of EM iterations was needed with the REBMIX, as an initialization technique, this remains acceptable. If the time efficacy is a major concern, to obtain the best-possible results for the datasets with a larger number of observations n, the Best REBMIX&EM strategy could be used to obtain the upper bound for the histogram preprocessing number of bins v range. Then, the Exhaustive REBMIX&EM strategy could be used with this value as the upper bound, while keeping the Knuth rule for the lower bound, for the histogram preprocessing number of bins v range

We would like to mention some points about the selection of the c_{min} and c_{max} values for the GMM or any other MM parameter estimation. The number of MM parameters increases with the increasing dimension, especially the GMM parameters. It is, therefore, always advisable to check to see whether the number of observations n in the dataset is larger than the number of parameters M, i.e., $n \geq M$. For a dataset with a smaller number of observations n, this can be used as a rule-of-thumb selector of c_{max}. Choosing a smaller value for c_{max} when the number of observations n is also small can additionally reduce the probability of estimating the degenerated solution. Otherwise, the maximum number of components c_{max} can be as large as required, depending on the application of the GMM. For example, in the image-segmentation tasks, the number of observations was large. As can be seen in Table 3, the estimated number of components c was always high. The images used had a resolution of 480×320 or 320×480, which resulted in a number of observations equal to n =153,600. Based on our above rule of thumb, the value of c_{max} can go as high as $c_{max} = 15,000$. However, this would be numerically unstable due to the floating arithmetic. For this kind of example, where the number of observations n is high and we can expect a large number of components in the GMM, this hyperparameter can also be fine-tuned. Some lower value can be used as the starting point to obtain some information about how the IC (in our case the BIC) is changing with respect to an increase in the number of components c in the GMM. If there are some large fluctuations in the value of BIC, or the slope of the BIC-c curve seems too steep, we can try to increase the value of c_{max} until the curve stabilizes and the values of BIC slowly start to increase. If some other IC is used instead of the BIC, then the same analogy applies. As for the c_{min}, it is advisable to use $c_{min} = 1$ for the minimum value to check whether the dataset fits well into the non-mixture distribution.

For the values of the maximum number of iterations of the EM algorithm and the threshold of the EM algorithm, we have chosen 1000 and 0.0001, respectively. These two values are chosen in this paper based on some empirical evidence from other software that implemented the EM algorithm (for example, Sklearn's EM implementation for the GMM estimation has a default value of 100 maximum iterations and 0.001 for the average log-likelihood threshold). Choosing these values should always be based on pragmatic grounds, i.e., we expect that this threshold can be reached in this amount of iterations. For example, the authors of [25] argue that even perhaps the most efficient way of managing the convergence of the EM algorithm, for the GMM parameter estimation, is to use only the specified number of iterations as stopping criteria because setting small threshold values can be hazardous in terms of the computation times due to the unboundedness of the log-likelihood function. However, in our experiments, we have observed that the REBMIX algorithm tends to estimate the initial parameters close to the final ones and we usually did not even use half of the expected iterations (see the second plot in Figure 4). To conclude, we can safely recommend these values as starting points; however, if necessary, fine-tuning can also be applied.

CONCLUSIONS

We have proposed the use of the REBMIX algorithm as an initialization technique for the EM algorithm, as applied to the MM parameter-estimation problem. Additionally, three different strategies are proposed when the number of components in the MM is unknown. The proposed strategies were thoroughly tested on artificially created datasets (Section 5.1), density-estimation tasks (Section 5.2), and image-segmentation tasks (Section 5.3). Our proposals showed benefits in terms of accuracy as well as with the computational efficacy. While preserving good or even better results from various applications of the GMM, fewer of the more expensive EM iterations were needed for the EM to converge. Moving forward, the proposed initialization and strategies when applied to real-world tasks, such as density estimation and image segmentation, were again competitive and even better with the state-of-the-art initialization strategies and, additionally, with the state-of-the-art algorithms for density estimation and image-segmentation (clustering), respectively. Moreover, the proposed initialization and strategies provided comparable or better results compared to the ADEBA algorithm in terms of computational effectiveness, as was shown in Figure 8. In addition, we provided some insights into the time complexity of the REBMIX algorithm in Section 5.4. Even though we did not provide a theoretical proof of the time complexity, which was due to multiple stochastic factors impacting on its performance, from the empirical evaluation, it can be seen that REBMIX has linear time with respect to the parameter n and a nonlinear time with respect to the parameters c and d. On other hand, the k-means algorithm was linear with respect to all three parameters. However, the actual value of the computation times in Figure 13 showed that REBMIX was faster than k-means for this specific case. On the third plot in Figures 4 and 8, we have provided some empirical insights into the time efficacy of the proposed REBMIX&EM strategies. Here, the theoretical estimation of the time complexity classes, due to the use of the EM algorithm, which does not have clear termination criteria, was not provided. That being said, we have given our best effort to provide enough critical empirical insights with which we can state that our proposal is more time efficient with regard to the state-of-the-art approaches

From the paragraph above, the main advantages of our proposal with regard to the state of the art can be extracted. Our proposal was time efficient and had good results. The current state-of-the-art approach, nevertheless, of the initialization technique used, is to try the EM algorithm on as many initializations as possible in a selected/given amount of time. Therefore, algorithms/initialization that can provide multiple repeats are preferred. The REBMIX heuristic is deterministic and cannot be improved with multiple restarts. We see this as the main disadvantage with regard to the state-of-the-art approaches. For a given number of bins v, it will always estimate the same parameters of the GMM. In addition, this parameter, i.e., the number of bins v, is usually not known in advance. Therefore, we proposed the use of the Exhaustive and Best REBMIX&EM strategies. Both strategies are more computational demanding when compared to the Single REBMIX&EM strategy. Furthermore, the choices we made for the number of bins v are thoroughly described in Section 5.5. We see this task of choosing the appropriate range of the number of bins, or some specific value or set of specific values, as the largest overhead introduced with our proposal. Other state-of-the-art strategies do not need this kind of parameter, only the rest of the hyperparameters (c_{max}, c_{min}, maximum number of iterations, and threshold). Some smaller limitations were also encountered. Namely, datasets with a large dimension d and small number of observations n led to some degeneracies in the estimates. Therefore, at this point, our proposal cannot be recommended for such problems. We did not include any safeguards for the EM algorithm and thus this happened. The empirically evaluated time complexity of the REBMIX algorithm showed some problems. The nonlinearity in terms of the dimension of the dataset d was to be expected, while the nonlinearity in terms of the maximum number of components c_{max} was troublesome. This overhead introduced with our proposal can be burdensome, especially with larger values of the parameter c_{max}.

To conclude this article, let us say that we have implemented all our proposals in the rebmix R package (https://cran.r-project.org/web/packages/rebmix/index.html), and they are ready to use. In addition, let us explain how our work can be extended. First, the Bayesian regularization for the EM algorithm could be used to additionally avoid degeneracies and constrict the likelihood [14]. Second, the use of the GMM for certain image-segmentation tasks and the density-estimation tasks was questionable, like for the flower image or the Bend distribution. Although we have used the GMM as a proof of concept, our work can be extended to other types of MMs. For the extension to the other parametric family of the MM, the equations for the REBMIX and EM algorithms need to be derived. In [17], it was shown how the REBMIX algorithm can be extended to other types of MM, particularly to the von-Mises MM. For the EM algorithm, the M-step equations need to be derived. For example, in [18], the derivation of the Weibull-Normal MM parameters is given. In addition, there is a plethora of literature about how the EM can be extended to other parametric families [9,10] to name just a few. We will strive to provide some more theoretical insights into the REBMIX algorithm. Specifically, the empirical equation for decreasing the parameter D_{min} should be improved to avoid nonlinear time complexity in terms of the parameter c_{max}. The hyperparameter, the number of bins v for the histogram preprocessing, needs special care. For the histogram preprocessing, we only used the same number of bins in each dimension, although the number of bins can be different in each dimension, or, finally, some adaptive bandwidth for the histogram preprocessing can be applied, like the ones used for a kernel density estimation in [11]. Some other REBMIX preprocessing capabilities like kernel density or k-nearest neighbor could be used. In the end, the application of the MM for the image segmentation should be more thoroughly researched. Throughout the testing of the proposals for the image-segmentation tasks, we encountered some interesting facts outlined in Section 5.3. As this was not the primary problem of this article, we left this topic open to be revisited.

REFERENCES

1. Dempster, A.P.; Laird, N.M.; Rubin, D.B. Maximum likelihood from Incomplete Data via the EM Algorithm. J. R. Stat. Soc. 1977, 39, 1–38.
2. Yu, J.; Chaomurilige, C.; Yang, M.S. On convergence and parameter selection of the EM and DA-EM algorithms for Gaussian mixtures. Pattern Recognit. 2018, 77, 188–203.
3. Ma, J.; Jiang, X.; Jiang, J.; Gao, Y. Feature-guided Gaussian mixture model for image matching. Pattern Recognit. 2019, 92, 231–245.
4. Liu, C.; Li, H.C.; Fu, K.; Zhang, F.; Datcu, M.; Emery, W.J. Bayesian estimation of generalized Gamma mixture model based on variational EM algorithm. Pattern Recognit. 2019, 87, 269–284.
5. Du, Y.; Gui, W. Goodness of Fit Tests for the Log-Logistic Distribution Based on Cumulative Entropy under Progressive Type II Censoring. Mathematics 2019, 7, 361.
6. Pagès-Zamora, A.; Cabrera-Bean, M.; Díaz-Vilor, C. Unsupervised online clustering and detection algorithms using crowdsourced data for malaria diagnosis. Pattern Recognit. 2019, 86, 209–223.
7. Yu, L.; Yang, T.; Chan, A.B. Density-Preserving Hierarchical EM Algorithm: Simplifying Gaussian Mixture Models for Approximate Inference. IEEE Trans. Pattern Anal. Mach. Intell 2019, 41, 1323–1337.
8. Gebru, I.D.; Alameda-Pineda, X.; Forbes, F.; Horaud, R. EM Algorithms for Weighted-Data Clustering with Application to Audio-Visual Scene Analysis. IEEE Trans. Pattern Anal. Mach. Intell 2016, 38, 2402–2415.
9. McLachlan, G.; Peel, D. Finite Mixture Models, 1st ed.; John Wiley & Sons: Hoboken, NJ, USA, 2000.
10. McLachlan, G.; Krishnan, T. The EM Algorithm and Extensions, 2nd ed.; John Wiley & Sons: Hoboken, NJ, USA, 2007.
11. Bäcklin, C.L.; Andersson, C.; Gustafsson, M.G. Self-tuning density estimation based on Bayesian averaging of adaptive kernel density estimations yields state-of-the-art performance. Pattern Recognit. 2018, 78, 133–143.
12. Yang, M.S.; Lai, C.Y.; Lin, C.Y. A robust EM clustering algorithm for Gaussian mixture models. Pattern Recognit. 2012, 45, 3950–3961.
13. Celeux, G.; Govaert, G. Gaussian parsimonious clustering models. Pattern Recognit. 1995, 28, 781–793.
14. Baudry, J.P.; Celeux, G. EM for mixtures. Stat. Comput. 2015, 25,

713–726.

15. Ng, S.K.; McLachlan, G.J. Speeding up the EM algorithm for mixture model-based segmentation of magnetic resonance images. Pattern Recognit. 2004, 37, 1573–1589.
16. Nagode, M. Finite Mixture Modeling via REBMIX. J. Algorithms Optim. 2015, 3, 14–28.
17. Ye, X.; Xi, P.; Nagode, M. Extension of REBMIX algorithm to von Mises parametric family for modeling joint distribution of wind speed and direction. Eng. Struct. 2019, 183, 1134–1145.

CHAPTER

2

A Fast Incremental Gaussian Mixture Model

Rafael Coimbra Pinto ,Paulo Martins Engel

Instituto de Informática, Universidade Federal do Rio Grande do Sul, Porto Alegre, RS, Brazil

ABSTRACT

This work builds upon previous efforts in online incremental learning, namely the Incremental Gaussian Mixture Network (IGMN). The IGMN is capable of learning from data streams in a single-pass by improving its model after analyzing each data point and discarding it thereafter. Nevertheless, it suffers from the scalability point-of-view, due to its asymptotic time complexity of $O(NKD^3)$ for N data points, K Gaussian components and D dimensions, rendering it inadequate for high-dimensional data. In this work, we manage

Citation: Pinto, R. C., & Engel, P. M. (2015). A Fast Incremental Gaussian Mixture Model. PloS one, 10(10), e0139931-e0139931. DOI: https://doi.org/10.1371/journal.pone.0139931

to reduce this complexity to $O(NKD^2)$ by deriving formulas for working directly with precision matrices instead of covariance matrices. The final result is a much faster and scalable algorithm which can be applied to high dimensional tasks. This is confirmed by applying the modified algorithm to high-dimensional classification datasets.

INTRODUCTION

The Incremental Gaussian Mixture Network (IGMN) [1, 2] is a supervised algorithm which approximates the EM algorithm for Gaussian mixture models [3], as shown in [4]. It creates and continually adjusts a probabilistic model of the joint input-output space consistent to all sequentially presented data, after each data point presentation, and without the need to store any past data points. Its learning process is aggressive, meaning that only a single scan through the data is necessary to obtain a consistent model.

IGMN adopts a Gaussian mixture model of distribution components that can be expanded to accommodate new information from an input data point, or reduced if spurious components are identified along the learning process. Each data point assimilated by the model contributes to the sequential update of the model parameters based on the maximization of the likelihood of the data. The parameters are updated through the accumulation of relevant information extracted from each data point. New points are added directly to existing Gaussian components or new components are created when necessary, avoiding merge and split operations, much like what is seen in the Adaptive Resonance Theory (ART) algorithms [5]. It has been previously shown in [6] that the algorithm is robust even when data is presented in random order, having similar performance and producing similar number of clusters in any order. Also, [4] has shown that the resulting models are very similar to the ones produced by the batch EM algorithm.

The IGMN is capable of supervised learning, simply by assigning any of its input vector elements as outputs. In other words, any element can be used to predict any other element, like auto-associative neural networks [7] or missing data imputation [8]. This feature is useful for simultaneous learning of forward and inverse kinematics [9], as well as for simultaneous learning of a value function and a policy in reinforcement learning [10].

Previous successful applications of the IGMN algorithm include time-series prediction [11–13], reinforcement learning [2, 14], mobile robot control and mapping [1, 15, 16] and outlier detection in data streams [17].

However, the IGMN suffers from cubic time complexity due to matrix inversion operations and determinant computations. Its time complexity is of $O(NKD^3)$, where N is the number of data points, K is the number of Gaussian components and D is the problem dimension. It makes the algorithm prohibitive for high-dimensional tasks (like visual tasks) and thus of limited use. One solution would be to use diagonal covariance matrices, but this decreases the quality of the results, as already reported in previous works [6, 11]. In this work, we propose the use of rank-one updates for both inverse matrices and determinants applied to full covariance matrices, thus reducing the time complexity to $O(NKD^2)$ for learning while keeping the quality of a full covariance matrix solution.

For the specific case of the IGMN algorithm, to the best of our knowledge, this has not been tried before, although we can find similar efforts for related algorithms. In [18], rank-one updates were applied to an iterated linear discriminant analysis algorithm in order to decrease the complexity of the algorithm. Rank-one updates were also used in [19], where Gaussian models are employed for feature selection. Finally, in [20], the same kind of optimization was applied to Maximum Likelihood Linear Transforms (MLLT).

The next Section describes the algorithm in more detail with the latest improvements to date. Section 3 describes our improvements to the algorithm. Section 4 shows the experiments and results obtained from both versions of the IGMN for comparison, and Section 5 finishes this work with concluding remarks.

INCREMENTAL GAUSSIAN MIXTURE NETWORK

In the next subsections we describe the current version of the IGMN algorithm, a slightly improved version of the one described in [15].

Learning

The algorithm starts with no components, which are created as necessary (see subsection 2.2). Given input $\mathbf{x}$ (a single instantaneous data point), the IGMN algorithm processing step is as follows. First, the squared Mahalanobis distance $d^2(\mathbf{x}, j)$ for each component j is computed:

$$d_M^2(\mathbf{x}, j) = (\mathbf{x} - \boldsymbol{\mu}_j)^T \mathbf{C}_j^{-1} (\mathbf{x} - \boldsymbol{\mu}_j) \tag{1}$$

where $\boldsymbol{\mu}_j$ is the jth component mean, $\mathbf{C}_j$ its full covariance matrix. If any $d^2(\mathbf{x}, j)$ is smaller than than $_{\chi 2D,1-\beta}$ (the $1-\beta$ percentile of a chi-squared distribution with D degrees-of-freedom, where D is the input dimensionality and β is a user defined meta-parameter, e.g., 0.1), an update will occur, and posterior probabilities are calculated for each component as follows:

$$p(\mathbf{x}|j) = \frac{1}{(2\pi)^{D/2}\sqrt{|\mathbf{C}_j|}} \exp\left(-\frac{1}{2} d_M^2(\mathbf{x}, j)\right) \quad (2)$$

$$p(j|\mathbf{x}) = \frac{p(\mathbf{x}|j)p(j)}{\sum_{k=1}^{K} p(\mathbf{x}|k)p(k)} \quad \forall j \quad (3)$$

where K is the number of components. Now, parameters of the algorithm must be updated according to the following equations:

$$v_j(t) = v_j(t-1) + 1 \quad (4)$$

$$sp_j(t) = sp_j(t-1) + p(j|\mathbf{x}) \quad (5)$$

$$\mathbf{e}_j = \mathbf{x} - \boldsymbol{\mu}_j(t-1) \quad (6)$$

$$\omega_j = \frac{p(j|\mathbf{x})}{sp_j} \quad (7)$$

$$\Delta\boldsymbol{\mu}_j = \omega_j \mathbf{e}_j \quad (8)$$

$$\boldsymbol{\mu}_j(t) = \boldsymbol{\mu}_j(t-1) + \Delta\boldsymbol{\mu}_j \quad (9)$$

$$\mathbf{e}_j^* = \mathbf{x} - \boldsymbol{\mu}_j(t) \quad (10)$$

$$\mathbf{C}_j(t) = (1-\omega_j)\mathbf{C}_j(t-1) + \omega_j \mathbf{e}_j^* \mathbf{e}_j^{*T} - \Delta\boldsymbol{\mu}_j \Delta\boldsymbol{\mu}_j^T \quad (11)$$

$$j) = \frac{sp_j}{\sum_{q=1}^{M} sp_q} \quad (12)$$

where sp_j and v_j are the accumulator and the age of component j, respectively, and $p(j)$ is its prior probability. The equations are derived using the Robbins-Monro stochastic approximation [21] for maximizing the likelihood of the model. This derivation can be found in [4, 22].

Creating New Components

If the update condition in the previous subsection is not met, then a new component j is created and initialized as follows:

$$\boldsymbol{\mu}_j = \mathbf{x};\ sp_j = 1;\ v_j = 1;\ p(j) = \frac{1}{\sum_{i=1}^{K} sp_i};\ \mathbf{C}_j = \boldsymbol{\sigma}_{ini}^2 \mathbf{I}$$

where K already includes the new component and $\boldsymbol{\sigma}_{ini}$ can be obtained by:

$$\boldsymbol{\sigma}_{ini} = \delta std(\mathbf{x}) \tag{13}$$

where δ is a manually chosen scaling factor (e.g., 0.01) and *std* is the standard deviation of the dataset. Note that the IGMN is an online and incremental algorithm and therefore it may be the case that we do not have the entire dataset to extract descriptive statistics. In this case the standard deviation can be just an estimation (e.g., based on sensor limits from a robotic platform), without impacting the algorithm.

Removing Spurious Components

Optionally, a component j is removed whenever $v_j > v_{min}$ and $sp_j < sp_{min}$, where v_{min} and sp_{min} are manually chosen (e.g., 5.0 and 3.0, respectively). In that case, also, $p(k)$ must be adjusted for all $k \in K$, $k \neq j$, using Eq (12). In other words, each component is given some time v_{min} to show its importance to the model in the form of an accumulation of its posterior probabilities sp_j. Those components are entirely removed from the model instead of merged with other components, because we assume they represent outliers. Since the removed components have small accumulated activations, it also implies that their removal has almost no negative impact on the model quality, often producing positive impact on generalization performance due to model simplification (a more throughout analysis of parameter sensibility for the IGMN algorithm can be found in [6]).

Inference

In the IGMN, any element can be predicted by any other element. In other words, inputs and targets are presented together as inputs during training. Thus, inference is done by reconstructing data from the target elements ($\mathbf{x}_t$, a slice of the entire input vector $\mathbf{x}$) by estimating the posterior probabilities using only the given elements ($\mathbf{x}_i$, also a slice of the entire input vector $\mathbf{x}$), as follows:

$$p(j|\mathbf{x}_i) = \frac{p(\mathbf{x}_i|j)p(j)}{\sum_{q=1}^{M} p(\mathbf{x}_i|q)p(q)} \quad \forall j \tag{14}$$

It is similar to Eq (3), except that it uses a modified input vector $\mathbf{x}_i$ with the target elements $\mathbf{x}_t$ removed from calculations. After that, $\mathbf{x}_t$ can be reconstructed using the conditional mean equation:

$$\hat{\mathbf{x}}_t = \sum_{j=1}^{M} p(j|\mathbf{x}_i)(\boldsymbol{\mu}_{j,t} + \mathbf{C}_{j,ti}\mathbf{C}_{j,i}^{-1}(\mathbf{x}_i - \boldsymbol{\mu}_{j,i})) \tag{15}$$

where $\mathbf{C}_{j,ti}$ is the sub-matrix of the *j*th component covariance matrix associating the unknown and known parts of the data, $\mathbf{C}_{j,i}$ is the sub-matrix corresponding to the known part only and $\boldsymbol{\mu}_{j,i}$ is the *j*th's component mean without the element corresponding to the target element. This division can be seen below:

$$\mathbf{C}_j = \left(\begin{array}{c|c} \mathbf{C}_{j,i} & \mathbf{C}_{j,it} \\ \hline \mathbf{C}_{j,ti} & \mathbf{C}_{j,t} \end{array}\right)$$

FAST IGMN

One of the contributions of this work lies in the fact that Eq (1) (the squared Mahalanobis distance) requires a matrix inversion, which has a asymptotic time complexity of $O(D^3)$, for D dimensions ($O(D^{\log_2 7+O1})$ for the Strassen algorithm or at best $O(D^{2.3728639})$ with the most recent algorithms to date [23]). This renders the entire IGMN algorithm as impractical for high-dimension tasks. Here we show how to work directly with the inverse of covariance matrix (also called the precision or concentration matrix) for the entire procedure, therefore avoiding costly inversions.

Firstly, let us denote $\mathbf{C}^{-1} = \mathbf{\Lambda}$, the precision matrix. Our task is to adapt all equations involving $\mathbf{C}$ to instead use $\mathbf{\Lambda}$.

We now proceed to adapt Eq (11) (covariance matrix update). This equation can be seen as a sequence of two rank-one updates to the $\mathbf{C}$ matrix, as follows:

$$\bar{\mathbf{C}}_j(t) = (1-\omega_j)\mathbf{C}_j(t-1) + \omega_j\mathbf{e}_j^*\mathbf{e}_j^{*T} \tag{16}$$

$$\mathbf{C}_j(t) = \bar{\mathbf{C}}_j(t) - \Delta\boldsymbol{\mu}_j \Delta\boldsymbol{\mu}_j^T \tag{17}$$

This allows us to apply the Sherman-Morrison formula [24]:

$$(\mathbf{A} + \mathbf{u}\mathbf{v}^T)^{-1} = \mathbf{A}^{-1} - \frac{\mathbf{A}^{-1}\mathbf{u}\mathbf{v}^T\mathbf{A}^{-1}}{1 + \mathbf{v}^T\mathbf{A}^{-1}\mathbf{u}} \tag{18}$$

This formula shows how to update the inverse of a matrix plus a rank-one update. For the second update, which subtracts, the formula becomes:

$$(\mathbf{A} - \mathbf{u}\mathbf{v}^T)^{-1} = \mathbf{A}^{-1} + \frac{\mathbf{A}^{-1}\mathbf{u}\mathbf{v}^T\mathbf{A}^{-1}}{1 - \mathbf{v}^T\mathbf{A}^{-1}\mathbf{u}} \tag{19}$$

In the context of IGMN, we have $\mathbf{A} = (1-\omega)\mathbf{C}_j(t-1) = (1-\omega)\boldsymbol{\Lambda}_j^{-1}(t-1)$ and u = $\mathbf{v} = \sqrt{\omega}\mathbf{e}^*$ for the first update, while for the second one we have $\mathbf{A} = \bar{\mathbf{C}}_j(t)$ and $\mathbf{u} = \mathbf{v} = \Delta\boldsymbol{\mu}_j$. Rewriting Eqs (18) and (19) we get (for the sake of compactness, assume all subscripts for $\boldsymbol{\Lambda}$ and $\Delta\boldsymbol{\mu}$ to be j):

$$\bar{\boldsymbol{\Lambda}}(t) = \frac{\boldsymbol{\Lambda}(t-1)}{1-\omega} - \frac{\frac{\omega}{(1-\omega)^2}\boldsymbol{\Lambda}(t-1)\mathbf{e}^*\mathbf{e}^{*T}\boldsymbol{\Lambda}(t-1)}{1 + \frac{\omega}{1-\omega}\mathbf{e}^{*T}\boldsymbol{\Lambda}(t-1)\mathbf{e}^*} \tag{20}$$

$$\boldsymbol{\Lambda}(t) = \bar{\boldsymbol{\Lambda}}(t) + \frac{\bar{\boldsymbol{\Lambda}}(t)\Delta\boldsymbol{\mu}\Delta\boldsymbol{\mu}^T\bar{\boldsymbol{\Lambda}}(t)}{1 - \Delta\boldsymbol{\mu}^T\bar{\boldsymbol{\Lambda}}(t)\Delta\boldsymbol{\mu}} \tag{21}$$

These two equations allow us to update the precision matrix directly, eliminating the need for the covariance matrix **C**. They have $O(N^2)$ complexity due to matrix-vector products.

Following on the adaptation of the IGMN equations, Eq (1) (the squared Mahalanobis distance) allows for a direct substituion, yielding the following new equation:

$$d_M^2(\mathbf{x}, j) = (\mathbf{x} - \boldsymbol{\mu}_j)^T \boldsymbol{\Lambda}_j (\mathbf{x} - \boldsymbol{\mu}_j) \tag{22}$$

which now has a $O(N^2)$ complexity, since there is no matrix inversion as the original equation. Note that the Sherman-Morrison identity is exact, thus the Mahalanobis computation yields exactly the same result, as will be shown in the experiments. After removing the cubic complexity from this step, the determinant computation will be dealt with next.

Since the determinant of the inverse of a matrix is simply the inverse of the determinant, it is sufficient to invert the result. But computing the determinant itself is also a $O(D^3)$ operation, so we will instead perform rank-one updates using the Matrix Determinant Lemma [25], which states the following:

$$|\mathbf{A} + \mathbf{u}\mathbf{v}^T| = |\mathbf{A}|(1 + \mathbf{v}^T\mathbf{A}^{-1}\mathbf{u}) \tag{23}$$

$$|\mathbf{A} - \mathbf{u}\mathbf{v}^T| = |\mathbf{A}|(1 - \mathbf{v}^T\mathbf{A}^{-1}\mathbf{u}) \tag{24}$$

Since the IGMN covariance matrix update involves a rank-two update, adding a term and then subtracting one, both rules must be applied in sequence, similar to what has been done with the $\mathbf{\Lambda}$ equations. Eqs (16) and (17) may be reused here, together with the same substitutions previously showed, leaving us with the following new equations for updating the determinant (again, j subscripts were dropped):

$$|\bar{\mathbf{C}}(t)| = (1-\omega)^D |\mathbf{C}(t-1)| \left(1 + \frac{\omega}{1-\omega} \mathbf{e}^{*T} \mathbf{\Lambda}(t-1)\mathbf{e}^*\right) \tag{25}$$

$$|\mathbf{C}(t)| = |\bar{\mathbf{C}}(t)|(1 - \Delta\boldsymbol{\mu}^T \bar{\mathbf{\Lambda}}(t)\Delta\boldsymbol{\mu}) \tag{26}$$

This was the last source of cubic complexity, which is now quadratic.

Finishing the adaptation in the learning part of the algorithm, we just need to define the initialization for $\mathbf{\Lambda}$ for each component. What previously was $\mathbf{C}_j = \sigma_{ini}^2\mathbf{I}$ now becomes $\mathbf{\Lambda}_j = \boldsymbol{\sigma}_{ini}^{-2}\mathbf{I}$, the inverse of the variances of the dataset. Since this matrix is diagonal, there are no costly inversions involved. And for initializing the determinant $|\mathbf{C}|$, just set it to $\prod \sigma_{ini}^2$, which again takes advantage of the initial diagonal matrix to avoid costly operations. Note that we keep the precision matrix $\mathbf{\Lambda}$, but the determinant of the covariance matrix $\mathbf{C}$ instead. See algorithms 1 to 3 for a summary of the new learning algorithm (excluding pruning, for brevity).

```
Algorithm 1 Fast IGMN Learning

Input: δ, β, X
  K > 0, σ_ini^{-1} = (δ std(X))^{-1}, M = ∅
  for all input data vector x ∈ X do
    if K = 0 or ∃j, d_M^2(x, j) < χ^2_{D,1-β} then
      update(x)
    else
      M ← M ∪ create(x)
    end if
  end for
```

Finally, the inference Eq (15) must also be updated in order to allow the IGMN to work in supervised mode. This can be accomplished by the use of a block matrix decomposition (note that here C is just another sub-matrix, not the covariance matrix as used before):

$$\Lambda = \begin{bmatrix} A & B \\ C & D \end{bmatrix}^{-1} = \begin{bmatrix} X & Y \\ Z & W \end{bmatrix} = \begin{bmatrix} (A - BD^{-1}C)^{-1} & -A^{-1}B(D - CA^{-1}B)^{-1} \\ -D^{-1}C(A - BD^{-1}C)^{-1} & (D - CA^{-1}B)^{-1} \end{bmatrix}$$

Here, according to Eq (15), we need **C** and **A** $^{-1}$. But since the terms that constitute these sub-matrices are relative to the original covariance matrix (which we do not have), they must be extracted from the precision matrix directly. Looking at the decomposition, it is clear that $\mathbf{YW}^{-1} = -\mathbf{A}^{-1}\mathbf{B} = -\mathbf{C}\mathbf{A}^{-1}$ (the terms between parenthesis in **Y** and **W** cancel each other, while $\mathbf{B} = \mathbf{C}^T$ due to symmetry). So Eq (15) can be rewritten as:

$$\hat{\mathbf{x}}_t = \sum_{j=1}^{M} p(j|\mathbf{x}_i)(\boldsymbol{\mu}_{j,t} - \mathbf{YW}^{-1}(\mathbf{x}_i - \boldsymbol{\mu}_{j,i})) \tag{27}$$

where **Y** and **W** can be extracted directly from **Λ**. However, we still need to compute the inverse of **W**. So we can say that this particular implementation has O(NKD^2) complexity for learning and O(NKD^3) for inference. The reason for us to not worry about that is that $d = i + o$, where i is the number of inputs and o is the number of outputs. The inverse computation acts only upon the output portion of the matrix. Since, in general, $o \ll i$ (in many cases even o = 1), the impact is minimal, and the same applies to the $\mathbf{YW}^{-1}$ product. In fact, Weka (the data mining platform used in this work [26]) allows for only 1 output, leaving us with just scalar operations.

EXPERIMENTS

The first experiment was meant to verify that both IGMN implementations produce exactly the same results. They were both applied to 7 standard datasets distributed with the Weka software (Table 1). Parameters were set to $\delta = 0.5$ (chosen by 2-fold cross-validation) and $\beta = 4.9E - 324$, the smallest possible double precision number available for the Java Virtual Machine (and also the default value for this implementation of the algorithm), such that Gaussian components are created only when strictly necessary. The same parameters were used for all datasets. Results were obtained from 10-fold cross-validation (resulting in training sets with 90% of the data and test sets with the remaining 10%) and statistical significances came from paired t-tests with $p = 0.05$. As can be seen in Table 2, both IGMN and FIGMN algorithms produced exactly the same results, confirming our expectations. The number of clusters created by them was also the same, and the exact quantity for each dataset is shown in Table 3. The Weka packages for both variations of the IGMN algorithm, as well as the datasets used in the experiments can be found at [27]. The MNIST dataset can be found at http://yann.lecun.com/exdb/mnist/, while the CIFAR10 dataset is available at http://www.cs.toronto.edu/~kriz/cifar.html.

Table 1: Datasets.

Dataset	Instances (N)	Attributes (D)	Classes
breast-cancer	286	9	2
pima-diabetes	768	8	2
Glass	214	9	7
ionosphere	351	34	2
iris	150	4	3
labor-neg-data	57	16	2
soybean	683	35	19
MNIST [28]	70000	784	10
CIFAR-10 [29]	60000	3072	10

Table 2: Accuracy of different algorithms on standard datasets.

Dataset	RF	NN	Lin. SVM	RBF SVM	IGMN	FIGMN
breast-cancer	69.6 ± 9.1	75.2 ± 6.5	69.3 ± 7.5	70.6 ± 1.5	71.4 ± 7.4	71.4 ± 7.4

pima-diabetes	75.8 ± 3.5	74.2 ± 4.9	77.5 ± 4.4	65.1 ± 0.4•	73.0 ± 4.5	73.0 ± 4.5
Glass	79.9 ± 5.0	53.8 ± 7.4•	62.7 ± 7.8•	68.8 ± 8.7•	65.4 ± 4.9•	65.4 ± 4.9•
iono-sphere	92.9 ± 3.6	92.6 ± 2.4	88.0 ± 3.5	93.5 ± 3.0	92.6 ± 3.8	92.6 ± 3.8
iris	95.3 ± 4.5	95.3 ± 5.5	96.7 ± 4.7	96.7 ± 3.5	97.3 ± 3.4	97.3 ± 3.4
labor-neg-data	89.7 ± 14.3	89.7 ± 14.3	93.3 ± 11.7	93.3 ± 8.6	94.7 ± 8.6	94.7 ± 8.6
soybean	93.0 ± 3.1	93.0 ± 2.4	94.0 ± 2.2	88.7 ± 3.0•	91.5 ± 5.4	91.5 ± 5.4
Average	85.2	82.0	83.1	82.4	83.7	83.7

• statistically significant degradation

Table 3: Number of Gaussian components created.

Dataset	# of Components
breast-cancer	14.2 ± 1.9
pima-diabetes	19.4 ± 1.3
Glass	15.9 ± 1.1
ionosphere	74.4 ± 1.4
iris	2.7 ± 0.7
labor-neg-data	12.0 ± 1.2
soybean	42.6 ± 2.2

Besides the confirmation we wanted, we could also compare the IGMN/FIGMN classification accuracy for the referred datasets against other 4 algorithms: Random Forest (RF), Neural Network (NN), Linear SVM and RBF SVM. The neural network is a parallel implementation of a state-of-the-art Dropout Neural Network [30] with 100 hidden neurons, 50% dropout for the hidden layer and 20% dropout for the input layer (this specific implementation can be found at https://github.com/amten/NeuralNetwork). The 4 algorithms were kept with their default parameters. The IGMN algorithms produced competitive results, with just one of them (Glass) being statistically significant below the accuracy produced by the Random Forest algorithm. This value was significantly inferior for all other algorithms too. On average, the IGMN algorithms were the second best from the set, losing only to the Random Forest. Note, however, that the Random Forest is a batch algorithm, while the IGMN learns incrementally from each data point. Also, the resulting Random Forest model used 6 times more memory than the IGMN model. We also tested the FIGMN accuracy on the MNIST dataset, but even after parameter tuning, the results where not on par with the state-of-the-art (above 99%), reaching a maximum of around 93% accuracy.

A second experiment was performed in order to evaluate the speed performance of the proposed algorithm, both the original and improved IGMN algorithms, using the parameters $\delta = 1$ and $\beta = 0$, such that a single component was created and we could focus on speedups due only to dimensionality (this also made the algorithm highly insensitive to the δ parameter). They were applied to the 2 highest dimensional datasets in Table 1, namely, the MNIST and CIFAR-10 datasets. The MNIST dataset was split into a training set with 60000 data points and a testing set containing 10000 data points, the standard procedure in the machine learning community [28]. Similarlly, the CIFAR-10 dataset was split into 50000 training data points and 10000 testing data points, also a standard procedure for this dataset [29].

Results can be seen in Table 4. Training time for the MNIST dataset was 20 times smaller for the fast version while the testing time was 16 times smaller. It makes sense that the testing time has shown a bit less improvement, since inference only takes advantage from the incremental determinant computation but not from the incremental inverse computation. For the CIFAR-10 dataset, it was impractical to run the original IGMN algorithm on the entire dataset, requiring us to estimate the total time, linearly projecting it from 100 data points (note that, since the model always uses only 1 Gaussian component during the entire training, the computation time per data point does not increase over time). It resulted in 32 days of CPU time estimated for the original algorithm against 15545s (~ 4*h*) for the improved algorithm, a speedup above 2 orders of magnitude. Testing time is not available for the original algorithm on this dataset, since the training could not be concluded. Additionally, we compared a pure clustering version of the FIGMN algorithm on the MNIST training set against batch EM (the implementation found in the Weka software). While the FIGMN algorithm took ~ 7.5*h* hours to finish, using 208 Gaussian components, the batch EM algorithm took ~ 1.3*h* to complete a *single iteration* (we set the fixed number of components to 208 too) using 4 CPU cores. Besides generally requiring more than one iteration to achieve best results, the batch algorithm required the entire dataset in RAM. The FIGMN memory requirements were much lower.

Table 4: Training and testing running times (in seconds).

Dataset	IGMN Training	FIGMN Training	IGMN Testing	FIGMN Testing
MNIST	32,544.69	1,629.81	3,836.06	230.92
CIFAR-10	2,758,252*	15,545.05	-	795.98

* estimated time projected from 100 data points

Finally, both versions of the IGMN algorithm with $\delta = 1$ and $\beta = 0$ were compared on 11 synthetic datasets generated by Weka. All datasets have 1000 data points drawn from a single Gaussian distribution (90% training, 10% testing) and an exponentially growing number of dimensions: 1, 2, 4, 8, 16, 32, 64, 128, 256, 512 and 1024. This experiment was performed in order to compare the scalability of both algorithms. Results for training and testing can be seen in Fig 1:

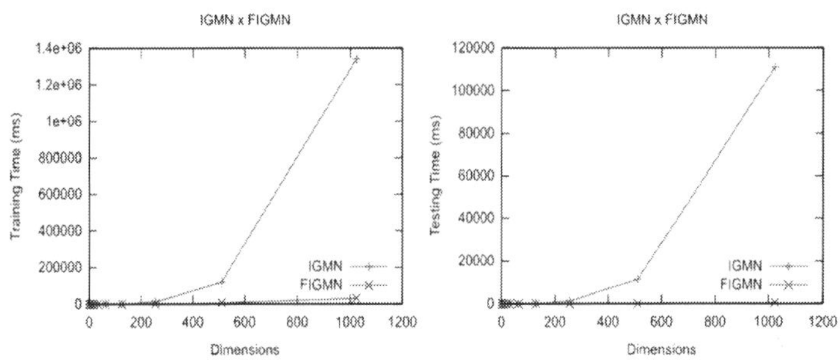

Figure 1: Training and testing times for both versions of the IGMN algorithm with growing number of dimensions.

CONCLUSION

We have shown how to work directly with precision matrices in the IGMN algorithm, avoiding costly matrix inversions by performing rank-one updates. The determinant computations were also avoided using a similar method, effectively eliminating any source of cubic complexity for the learning algorithm. This resulted in substantial speedups for high-dimensional datasets, turning the IGMN into a good option for this kind of tasks. The inference operation still has cubic complexity, but we argue that it has a much smaller impact on the total runtime of the algorithm, since the number of outputs is usually much smaller than the number of inputs. This was confirmed in the experiments.

In general, we could see that the fast IGMN is a good option for supervised learning, with low runtimes and good accuracy. It should be noted that this is achieved with a single-pass through the data, making it also a valid option for data streams.

Funding Statement

The authors have no support or funding to report.

Data Availability

Data are available at Figshare (http://figshare.com/articles/A_Fast_Incremental_Gaussian_Mixture_Model/1552030). The MNIST data set is available at (http://yann.lecun.com/exdb/mnist/) and the CIFAR10 data set is available at (http://www.cs.toronto.edu/~kriz/cifar.html). The software binaries and source code are deposited to GitHub (https://github.com/amten/NeuralNetwork).

REFERENCES

1. Heinen MR, Engel PM, Pinto RC. Using a Gaussian mixture neural network for incremental learning and robotics. In: Neural Networks (IJCNN), The 2012 International Joint Conference on. IEEE; 2012. p. 1–8.
2. Heinen MR, Engel PM, Pinto RC. IGMN: An incremental gaussian mixture network that learns instantaneously from data flows. Proc VIII Encontro Nacional de Inteligência Artificial (ENIA2011). 2011;.
3. Dempster AP, Laird NM, Rubin DB, et al. Maximum likelihood from incomplete data via the EM algorithm. Journal of the Royal Statistical Society Series B (Methodological). 1977;39(1):1–38.
4. Engel P, Heinen M. Incremental learning of multivariate gaussian mixture models. Advances in Artificial Intelligence–SBIA 2010. 2011;p. 82–91.
5. Grossberg S. Competitive learning: From interactive activation to adaptive resonance. Cognitive science. 1987;11(1):23–63. 10.1111/j.1551-6708.1987.tb00862.x
6. Heinen MR. A connectionist approach for incremental function approximation and on-line tasks. Universidade Federal do Rio Grande do Sul; 2011.
7. Rumelhart DE, McClelland JL. Parallel distributed processing. MIT Pr; 1986.
8. Ghahramani Z, Jordan MI. Supervised learning from incomplete data via an EM approach. In: Advances in neural information processing systems 6. Citeseer; 1994.
9. Damas B, Santos-Victor J. An online algorithm for simultaneously learning forward and inverse kinematics. In: Intelligent Robots and Systems (IROS), 2012 IEEE/RSJ International Conference on. IEEE; 2012. p. 1499–1506.
10. Heinen MR, Engel PM. IGMN: An incremental connectionist approach for concept formation, reinforcement learning and robotics. Journal of Applied Computing Research. 2011;1(1):2–19. 10.4013/jacr.2011.11.01.
11. Pinto RC, Engel PM, Heinen MR. Echo State Incremental Gaussian Mixture Network for Spatio-Temporal Pattern Processing. In: Proceedings of the IX ENIA-Brazilian Meeting on Artificial Intelligence, Natal (RN); 2011.

12. Pinto RC, Engel PM, Heinen MR. Recursive incremental gaussian mixture network for spatio-temporal pattern processing. Proc 10th Brazilian Congr Computational Intelligence (CBIC) Fortaleza, CE, Brazil (Nov 2011);.
13. Flores JHF, Engel PM, Pinto RC. Autocorrelation and partial autocorrelation functions to improve neural networks models on univariate time series forecasting. In: Neural Networks (IJCNN), The 2012 International Joint Conference on. IEEE; 2012. p. 1–8.
14. Heinen MR, Bazzan AL, Engel PM. Dealing with continuous-state reinforcement learning for intelligent control of traffic signals. In: Intelligent Transportation Systems (ITSC), 2011 14th International IEEE Conference on. IEEE; 2011. p. 890–895.
15. Pinto RC, Engel PM, Heinen MR. One-shot learning in the road sign problem. In: Neural Networks (IJCNN), The 2012 International Joint Conference on. IEEE; 2012. p. 1–6.
16. de Pontes Pereira R, Engel PM, Pinto RC. Learning Abstract Behaviors with the Hierarchical Incremental Gaussian Mixture Network. In: Neural Networks (SBRN), 2012 Brazilian Symposium on. IEEE; 2012. p. 131–135.
17. Santos ADPd, Wives LK, Alvares LO. Location-Based Events Detection on Micro-Blogs. arXiv preprint arXiv:12104008. 2012;.
18. Salmen J, Schlipsing M, Igel C. Efficient update of the covariance matrix inverse in iterated linear discriminant analysis. Pattern Recognition Letters. 2010;31(13):1903–1907. Meta-heuristic Intelligence Based Image Processing 10.1016/j.patrec.2010.03.001
19. Lefakis L, Fleuret F. Jointly Informative Feature Selection. In: Proceedings of the Seventeenth International Conference on Artificial Intelligence and Statistics; 2014. p. 567–575.
20. Olsen PA, Gopinath RA. Extended mllt for gaussian mixture models. Transactions in Speech and Audio Processing. 2001;.
21. Robbins H, Monro S. A stochastic approximation method. The annals of mathematical statistics. 1951;p. 400–407. 10.1214/aoms/1177729586
22. Engel PM. INBC: An incremental algorithm for dataflow segmentation based on a probabilistic approach. INBC: an incremental algorithm for dataflow segmantation based on a probabilistic approach. 2009;.
23. Gall FL. Powers of tensors and fast matrix multiplication. arXiv preprint arXiv:14017714. 2014;.

24. Sherman J, Morrison W J. Adjustment of an Inverse Matrix Corresponding to a Change in One Element of a Given Matrix. The Annals of Mathematical Statistics. 1950. March;21(1):124–127. Available from: 10.1214/aoms/1177729893 10.1214/aoms/1177729893
25. Harville DA. Matrix algebra from a statistician's perspective.
26. Hall M, Frank E, Holmes G, Pfahringer B, Reutemann P, Witten IH. The WEKA data mining software: an update. ACM SIGKDD explorations newsletter. 2009;11(1):10–18. 10.1145/1656274.1656278
27. Pinto, RC. Experiment Data for "A Fast Incremental Gaussian Mixture Model". 10.6084/m9.figshare.1552030
28. LeCun Y, Bottou L, Bengio Y, Haffner P. Gradient-based learning applied to document recognition. Proceedings of the IEEE. 1998;86(11):2278–2324. 10.1109/5.726791
29. Krizhevsky A, Hinton G. Learning multiple layers of features from tiny images. Computer Science Department, University of Toronto, Tech Rep. 2009;.
30. Hinton GE, Srivastava N, Krizhevsky A, Sutskever I, Salakhutdinov RR. Improving neural networks by preventing co-adaptation of feature detectors. arXiv preprint arXiv:12070580. 2012;.

CHAPTER

3

Detection of Emerging Faults on Industrial Gas Turbines Using Extended Gaussian Mixture Models

Yu Zhang,[1] Chris Bingham,[1] Miguel Martínez-García,[1] and Darren Cox[2]

[1]School of Engineering, University of Lincoln, Lincoln LN6 7TS, UK

[2]Siemens Industrial Turbomachinery Ltd., Lincoln LN5 7FD, UK

This paper extends traditional Gaussian mixture model (GMM) techniques to provide recognition of operational states and detection of emerging faults for industrial systems. A variational Bayesian method allows a GMM to cluster with its mixture components to facilitate the extraction of steady-state operational behaviour; this is recognised as being a primary factor in reducing the susceptibility of alternative prognostic/diagnostic techniques, which would initiate false-alarms resulting from control set-point and load changes. Furthermore, a GMM with an outlier component is discussed and

Citation: Zhang, Y., Bingham, C., Martínez-García, M., & Cox, D. (2017). Detection of emerging faults on industrial gas turbines using extended Gaussian mixture models. International Journal of Rotating Machinery, 2017. DOI: https://doi.org/10.1155/2017/5435794

applied for direct novelty/fault detection. An advantage of the variational Bayesian method over traditional predefined thresholds is the extraction of steady-state data during both full- and part-load cases, and a primary advantage of the GMM with an outlier component is its applicability for novelty detection when there is a lack of prior knowledge of fault patterns. Results obtained from the real-time measurements on the operational industrial gas turbines have shown that the proposed technique provides integrated preprocessing, benchmarking, and novelty/fault detection methodology.

INTRODUCTION

Industrial gas turbines (IGTs) are utilised globally with units generally acting as prime-movers for either pumps, generators, or compressors, for both on- and off-shore systems. Various root-cause factors responsible for failures on such systems include vibration, shock, noise, heat, cold, dust, corrosion, humidity, rain, oil debris, flow, pressure, and excessive operating speed [1]. A key challenge of condition monitoring in order to provide an "early warning" of faults is to distinguish between sensor-based failure, component failure, and the normal operational transient behaviour of the system, for example, due to control or load changes. With advances in instrumentation, communications hardware, and computational capability [2], there has been an increased ability to realize such prognostic and diagnostic methods and provide remedial action and informed flexible maintenance scheduling prior to encountering unplanned downtime. This is especially pertinent as a result of the increasing operational speeds of IGTs [3]. Two popular categories of monitoring techniques have emerged over recent years, namely, model-based and signal processing-based approaches [4]. Model-based approaches construct models (or virtual sensors) to estimate physical variables from which residuals are calculated and used as indicators of emerging failure modes [5]. However, to build an accurate dynamic model that can accommodate the full operating envelope of IGTs is, in general, computationally demanding. In such circumstances, direct signal processing and data fusion methods often provide for more practical and effective monitoring solutions [6]. It is this latter category of techniques that is considered here. Traditional signal processing-based methodologies use techniques such as principal component analysis (PCA) [7, 8], artificial neural networks (ANNs) [9], and data filter methods [10]. However, monitoring systems based on these algorithms are generally only applicable during steadystate operating conditions, since measurement transients caused by changes of loading or control action can generate "false alarms." Many techniques are therefore only effective under very constrained operating regimes. For instance, [11] employed ANNs for fault detection on gas turbines during the engine start-up phase, whilst [12] only considers solutions during steady-state operation. Typically, such techniques do not attempt to address the issue by incorporating implicit methods that discriminate between steady-state and transient behaviour as part of the fault detection system, and it is this aspect that is initially considered here [13]. By recognizing that steady-state measurement data is often superimposed by noise having a characteristic Gaussian distribution [14], that is, containing practical "measurement transients" that are not due to operational variations, the signals can be modelled through use of Gaussian Mixture Models (GMMs) [15]. This characterization of signals has previously been reported [16, 17] and used in a related family of clustering methods based on GMMs, regarded as "soft clustering" techniques, with the benefits for condition monitoring and fault detection explored in [16–18]. Here then, it is initially shown that GMMs provide a convenient mechanism to effectively discriminate between data relating to steady-state operation and that relating operation with transients and are specifically regarded in this instance as a preprocessing tool for subsequent operational pattern discrimination [13]. An essential part of developing the GMM methodology is parameter fitting, which is often carried out using an expectation-maximization (EM) method [19]. However, this requires the number of the mixture components (MCs) in the model to be fixed a priori [18]. Consequently, Bayesian-based frameworks are commonly used to provide a probabilistic inference on the data [20], and Variational Bayesian GMM techniques (VBGMMs) have been proposed [21] to provide improved performance [22, 23] by automatically selecting the number of MCs in the GMMs. VBGMMs are able to classify steady-state operation that can occur under full- or part-load conditions (e.g., 50% load). Additionally therefore, as well as identifying steadystate operation, the remaining data, including that associated with start-ups, shutdowns, and load changing conditions, is also naturally separated and it can therefore be used as a preanalysis tool for alternative dynamic scenarios, as reported in [24, 25], for instance. An additional property of GMMs that facilitates novelty/fault detection is the ability to filter novel class samples when the machine learning system has no a priori knowledge [26]. Moreover, GMMs have been previously reported for the detection of "novel" vibration signatures with experimental results showing good fault detection properties with known classifications [27]. However, the technique was sensitive to vibration characteristics that are not associated with the detection of wear or damage and so remained susceptible to initiating false alarms. As a consequence of these features, the outlier components are now included as background "noise" for the GMM [28, 29], and the resulting technique is considered as a GMM with an outlier component (GMMOC). When using GMMOC, measurements remain characterized by the GMM, but novel characteristics, that is, measurements that have a very low probability of being clustered into existing distributions, are considered as outliers. It is the identification of such outliers that provides a robust mechanism for IGT fault detection considered here to provide an "early warning" fault detection tool. Key contributions of this paper are summarized as follows:

(1) An extension to the use of GMMs, including VBGMM and GMMOC, is proposed for novelty/fault detection on industrial systems.
(2) Steady-state operation is discriminated from transient operation using VBGMM.
(3) GMMOC is used to indicate the presence of outliers and hence the emergence of faults.
(4) The efficacy of the proposed techniques is demonstrated from case studies (CSs) on IGTs, where CS1 is considered as a feasibility study of VBGMM and GMMOC for novelty detection and CS2 demonstrates a real bearing fault case study.

METHODOLOGY

The stages in the proposed methodology are depicted in Table 1. To provide an application focus for the development of the algorithms, measurements taken from bearing vibration probes on IGTs during commissioning are used as an illustrative example. Operational pattern separation is achieved through the use of VBGMM, where the steadystate data are distinguished for further analysis in this paper. Datasets from the identified transient operation can also be used for fault detection through start-up analysis and shutdown analysis and during load changes [6, 24, 25], which is not included in the current paper. The most relevant features are then extracted from the steady-state data and a statistical "fingerprint" for the extracted features is obtained through the application of GMMOC.

Underlying Principles of GMM.

The empirical probability distribution of sampled data can be estimated by a GMM using a linear combination of Gaussian distributions $N(x \mid \mu, \sigma)$ [15], for example, as a sum of K Gaussian distributions with mean μk and standard deviation σk. The GMM with $K \times$ MCs is expressed as

$$p(x \mid \theta, \mu, \sigma) = \sum_{k=1}^{K} \theta_k N(x \mid \mu_k, \sigma_k), \tag{1}$$

where x is a multidimensional variable and θ_k are the mixing coefficients that need to be chosen.

Let $X \equiv \{x_m\}$ represent M data samples, $m = 1, \ldots, M$, where each sample consists of a multidimensional variable x_m. Provided that xm are statistically independent, the probability function of X can be expressed as

$$p(X \mid \theta, \mu, \sigma) = \prod_{m=1}^{M} \left(\sum_{k=1}^{K} \theta_k N(x_m \mid \mu_k, \sigma_k) \right), \tag{2}$$

Table 1: Stages of the proposed methodology

Step	Process	Purpose	Output
1	Sensor measurement collection (sampled or batch)	----	Collate data for analysis
2	VBGMM	Preprocess data to remove transient measurements	Measurements only of steady state operation
3	Statistical analysis	Extraction of important features within the data	Features for identifying emerging faults
4	GMMOC	Novelty detection of emerging faults	Early warning of emerging or imminent fault conditions

which constitutes the likelihood function and by taking the natural logarithm gives

$$L(X \mid \theta) = \ln p(X \mid \theta, \mu, \sigma) = \sum_{m=1}^{M} \ln \left(\sum_{k=1}^{K} \theta_k N(x_m \mid \mu_k, \sigma_k) \right). \tag{3}$$

The mixture density is then

$$\rho_{mk} = \theta_k N(x_m \mid \mu_k, \sigma_k), \tag{4}$$

where k indicates the MC index and m indicates the data sample index. The conditional probability is calculated as

$$p_{mk} = \frac{\rho_{mk}}{\sum_{j=1}^{K} \rho_{mj}}, \qquad (5)$$

for a selected component k on the given data sample x_m. Evaluation of (3) typically necessitates an EM optimization procedure to maximize the log-likelihood function [18] from the maximization step (termed the M-step):

$$\frac{\partial L}{\partial \mu_k} = 0,$$

$$\frac{\partial L}{\partial \sigma_k} = 0,$$

$$\frac{\partial L}{\partial \theta_k} = 0. \qquad (4)$$

The unknowns in (3) are solved in the expectation step (E- step), from (6), as follows:

(1) Choose an initialization of $\theta_k^{(0)}, \mu_k^{(0)}$, and $\sigma_k^{(0)}$.

(2) Iteratively update pmk, θ_k, μ_k, and σ_k until convergence to a desired tolerance, using

$$p_{mk}^{(i)} = \frac{\theta_k^{(i)} N\left(x_m \mid \mu_k^{(i)}, \sigma_k^{(i)}\right)}{\sum_{k=1}^{K} \theta_k^{(i)} N\left(x_m \mid \mu_k^{(i)}, \sigma_k^{(i)}\right)},$$

$$\mu_k^{(i+1)} = \frac{\sum_{m=1}^{M} p_{mk}^{(i)} x_m}{\sum_{m=1}^{M} p_{mk}^{(i)}},$$

$$\sigma_k^{(i+1)} = \sqrt{\frac{1}{M} \frac{\sum_{m=1}^{M} p_{mk}^{(i)} \left\| x_m - \mu_k^{(i+1)} \right\|^2}{\sum_{m=1}^{M} p_{mk}^{(i)}}},$$

$$\theta_k^{(i+1)} = \frac{1}{M} \sum_{m=1}^{M} p_{mk}^{(i)}, \qquad (7)$$

where M is the dimensionality of the data. For the special case k=1, the whole dataset belongs to only 1 cluster, and the problem reduces to that of a Gaussian distribution fit.

Extension to VBGMM

A Variational Bayesian (VB) method can be used to determine the required number of MCs. Specifically, $M \times K$ binary latent variables $z_{mk} \in \{0, 1\}$ are used to indicate which MCs the data sample clusters into. When forming a GMM using classical methods, K is selected a priori. However, when using a VB method, K is resolved from the solutions of $\{z_{mk}\}$, termed Z. The joint probability distribution function of all variables is therefore

$$p(X, Z, \theta, \mu, \sigma) = p(X \mid Z, \mu, \sigma)\, p(Z \mid \theta)\, p(\theta)\, p(\mu \mid \sigma)\, p(\sigma). \quad (8)$$

A lower bound on $(X \mid \theta) = p(X \mid Z, \mu, \sigma)p(Z \mid \theta)$ can be determined using the VB method reported in [21]. Let $\Psi \equiv \{Z, \mu, \sigma\}$, and the marginal likelihood is expressed as

$$p(X \mid \theta) = \int p(X, \Psi \mid \theta)\, d\Psi. \quad (9)$$

Next, a variational distribution is introduced, that is, $(\Psi) \approx p(\Psi \mid \theta)$. Through the use of Kullback-Leibler (KL) divergence [22], that is, $D_{KL}(q \parallel p)$, (9) becomes

$$\ln p(X \mid \theta) = D_{\mathrm{KL}}(q \parallel p) + L(q), \quad (10)$$

where $L(q)$ indicates the lower bound. Minimization of the KL divergence can be achieved by maximizing (q) by selecting appropriate q distributions. (Ψ) can be rewritten as the product of $q_i(\Psi_i)$, in terms of the subsets $\{\Psi_i\} = \{Z, \mu, \sigma\}$, giving

$$q(Z, \mu, \sigma) = q_Z(Z)\, q_\mu(\mu)\, q_\sigma(\sigma). \quad (11)$$

The best distribution for each term $q_i^*(\Psi_i)$, can be solved using

$$q_i^*(\Psi_i) = \frac{\exp\left(E_{j \neq i}[\ln p(X, \Psi)]\right)}{\int \exp\left(E_{j \neq i}[\ln p(X, \Psi)]\right) d\Psi_i}, \quad (12)$$

where $E_{j \neq i}[\cdot]$ denotes the expectation of $q_j(\Psi_j)$ for all $j/=i$.

(1) Initialize the parameters, which are normally set to be small, real, and positive [30].

(2) Calculate (Ψ) in (11) through use of (12).

(3) Iteratively update until the predefined tolerance is met.

After calculating (Ψ), $L(q)$ can be obtained. Maximization of (q), which minimizes $\ln p(X \mid \theta)$, is again achieved by using the EM procedure. The M-step for calculating θ can be derived from (7), and the optimized distributions qi can be obtained through the E-step as mentioned above. For brevity, the reader is directed to [30] for a more in-depth discussion of the VB framework on GMMs.

Principles of GMMOC.

GMMOC extends the original GMM [28, 29] approach by adding an outlier component that is modelled by a uniform distribution. The hybrid GMMOC model is then written as

$$p(x) = \sum_{k=1}^{K} \theta_k N(x \mid \mu_k, \sigma_k) + \left(1 - \sum_{k=1}^{K} \theta_k\right) U(x). \tag{13}$$

Since (x) is uniformly distributed, the EM procedure described previously can still be employed to solve for the required parameters. The outlier component is normally assumed to be small initially (e.g., 0.01, and therefore the initialization of the mixing coefficients in GMM satisfies $\sum_{k=1}^{K} \theta_k = 0.99$).

In this case, there are (K+1) clusters, including K Gaussian distributions and one uniform distribution, the outlier component.

Parameters μ_k, σ_k, and θ_k can be estimated from (7), and if the probability of a data sample, belonging to any of the K Gaussian MCs, is smaller than a predefined threshold, it is clustered to the outlier component and therefore indicates a warning of an emerging fault, or facilitate novelty detection.

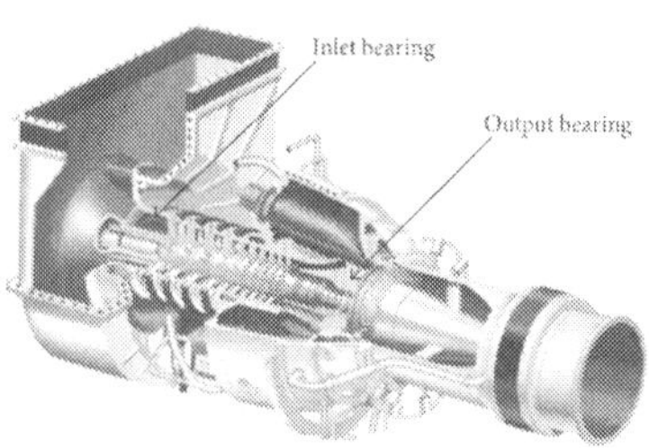

Figure 1: Locations of IGT bearing vibration sensors.

Feature Extraction

Feature extraction provides an essential tool for reducing the dimensionality of raw data whilst keeping informative features [25]. Many feature extraction techniques have been reported and successfully applied, including the use of the Fast Fourier Transform (FFT) and Discrete Wavelet Transform (DWT), all involving elaborate time-frequency transforms [31, 32]. However, the data used in the following studies are taken from IGT units in the field with sampling rates in the order of minutes, which excludes the use of frequency domain based methods as they do not satisfy traditional sampling rate criteria for the measured variables. In this case, statistical features of the data in the time domain are used, for example, the peak value, root mean square, crest factor, kurtosis, clearance factor, impulse factor, shape factor, and skewness, and the most informative features can be identified through optimization methods [33]. For this study, in order not to divert the focus from the use of extended GMMs, only the most basic statistical attributes are employed; the mean (which carries information about measurement equilibrium) and standard deviation (which carries information about signal power) are used here as a proof of concept and also for practical reasons since it has been observe.

APPLICATION CASE STUDIES

The proposed methods are applied to monitor the vibration characteristics of fluid-film inlet- and output-bearings which typically support the compressor rotors of sub-15MW IGTs (Figure 1). The thrust bearings and journal bearings have operating speeds in excess of ~10,000 rpm. Radial and axial positions are monitored using noncontact probes. Two experimental case studies are now considered, both of which adopt the procedure in Table 1. CS1 uses measurements taken over a relatively short time period (1-month) and demonstrates the effectiveness of VBGMM for operational state discrimination (steady-state/transient behaviour), feature extraction, and the initial setup of a benchmarking ellipse using GMMOC, whilst CS2 considers the analysis of longer periods of measurement data (12-month) and aims to show the efficacy of GMMs for identifying longer-term emerging faults.

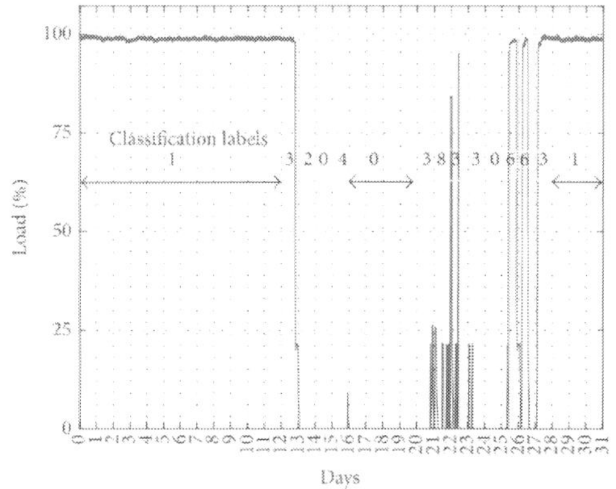

Figure 2: [CS1] Plot of load (%) versus time (days): identification of clusters in power/load measurements (month 1—normal operation).

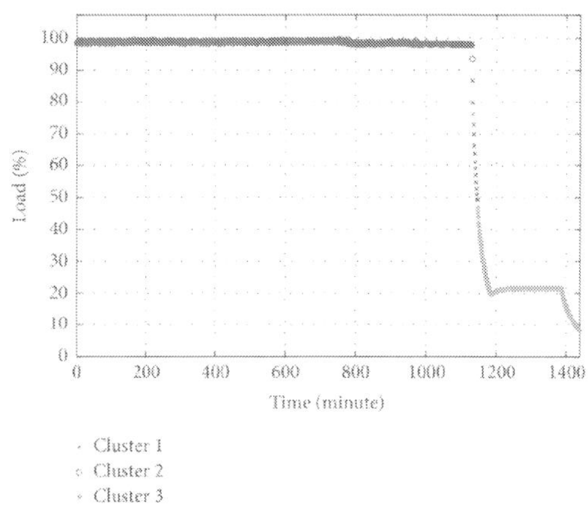

Figure 3: [CS1] VBGMM clustering result: day 13, 3 clusters identified, that is, non-steady-state data incorporating load changes

Case Study 1

Operation State Discrimination, Feature Extraction, and Novelty Detection. VBGMM is used to cluster measurements of output power (in terms of loading percentage) from a sub-15MW IGT to classify the unit's operational behaviour (Figure 2). One month (31 days) of daily data, each containing 1440 sample measurements, is used (i.e., sampling period = 1 minute). The resulting clusters from the power/load measurements are also shown in Figure 2 after applying VBGMM to each individual set of daily data.

In line with the algorithm description, it can be seen that when the unit is considered to be operating normally, in steady-state, a classification label of 1 is assigned and that classification labels > 1 are assigned for all other detected cases (cluster result given in Figure 3). A classification label of 0 indicates constant null readings and is precluded from further pattern analysis as the unit is considered to be shut down.

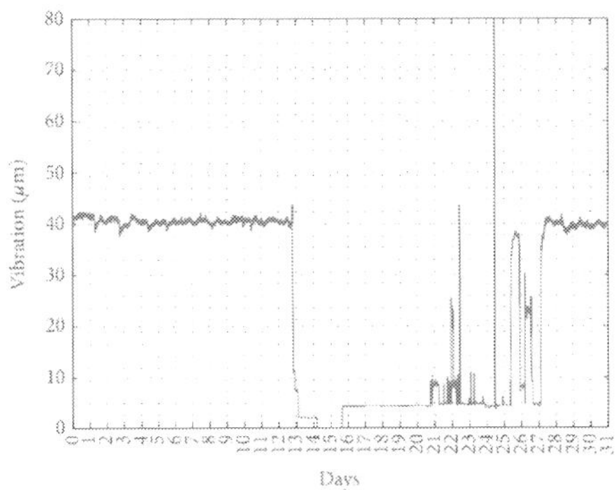

Figure 4: [CS1] Plot of vibration measurements versus time (days) (month 1—normal operation).

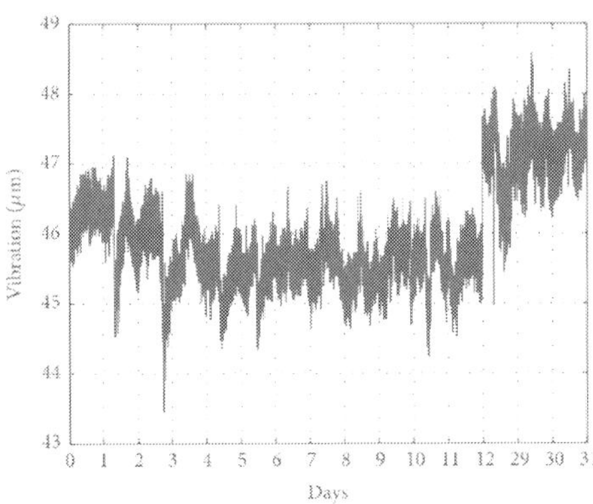

Figure 5: [CS1] Vibration data identified as steady-state (month 1—normal operation).

Corresponding inlet bearing vibration measurements taken for the same 31-day period are shown in Figure 4. Having discriminated between transient and steady-state operation using VBGMM, the steady-state data (days 1–12 and days 29–31) shown in Figure 5 can be used for subsequent novelty detection. Having identified appropriate datasets, feature extraction is used to capture important characteristics present in the data. The mean and the standard deviation of each day's data for months 1–3 are calculated and given in Figure 6, to present a benchmarking envelope representing normal operation. Having effectively obtained an operational fingerprint of behaviour, measurements taken over subsequent months are then used and compared to the fingerprint. Considering only the magnitude of vibration amplitudes, levels up to ~50 μm are typically considered normal, with warnings at ~70 μm and unit shutdown occurring at ~90 μm.

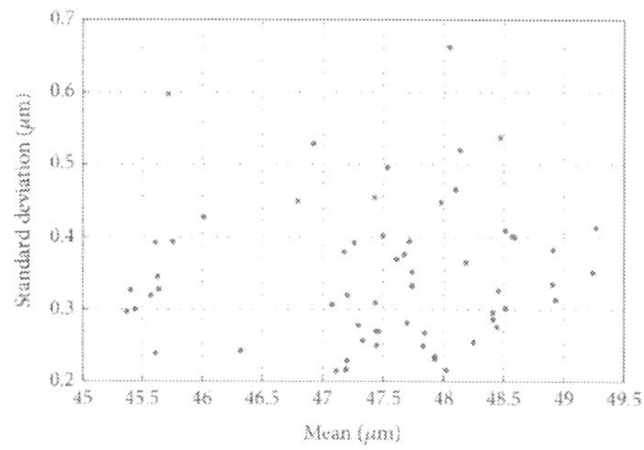

Figure 6: [CS1] Extracted features for 3 months' steady-state data (months 1–3—normal operation).

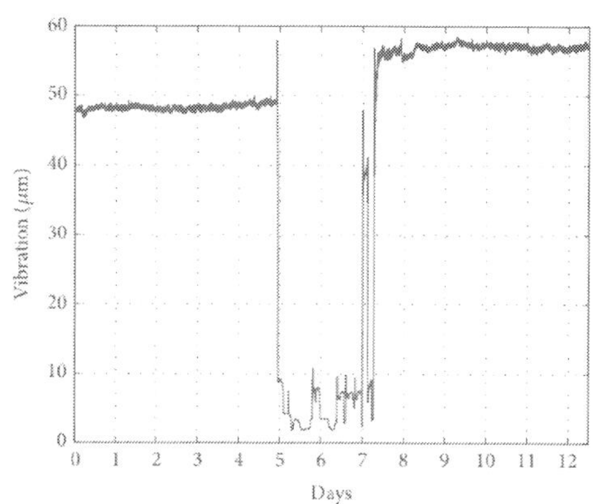

Figure 7: [CS1] Test data for novelty detection (in month 4).

Having obtained a fingerprint representing normal operation, GMMOC is applied to subsequent periods of data on a daily basis (in this case in month 4, Figure 7). It is well known that gradual bearing wear leading to failure is often preceded by gradual changes in vibration characteristics. Figure 8 provides an ellipse boundary drawn according to the 1-cluster GMM model (see (13)). In general, the confidence level to identify outliers will be set according to application. For instance, in this case, a 99.99% confidence level does not discriminate the outliers (here, it indicates outliers in normal operation which may be caused by sensor malfunctioning); however, a 95% confidence level is clearly seen to be appropriate in this instance. The results from GMMOC are compared with the envelopes drawn from the original GMM, as shown in Figure 8, and the advantage of GMMOC over the original GMM is evident, since the outliers, days 9 and 10, are clearly identified even with a 99.99% confidence level of GMMOC in this case. By considering the following month of measurement data, it is notable that the measurements have correctly been identified as outliers (day 9 and day 10 in this case), whereas measurements from days 1–4 are correctly considered to correspond to normal behaviour. Although it is not known at that stage if the increase in vibration in days 9 and 10 is related to a component fault, the measurements are considered as anomalies.

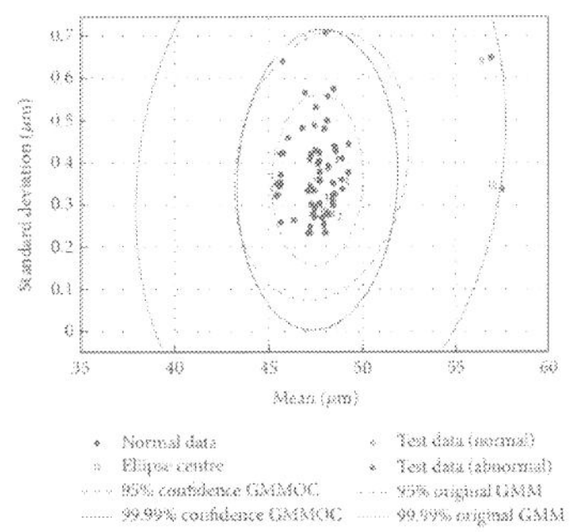

Figure 8: [CS1] GMMOC novelty detection results compared to those of the original GMM.

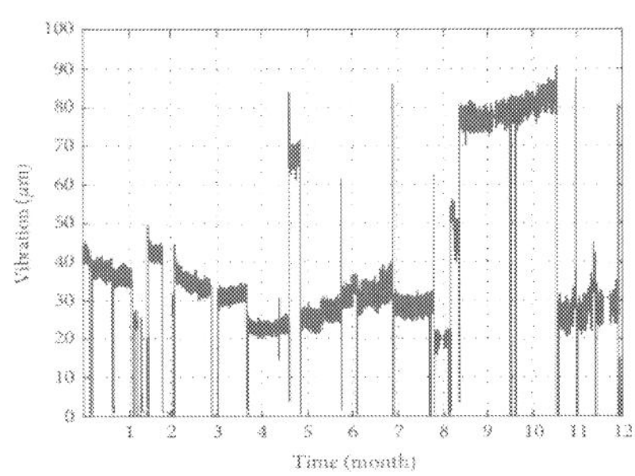

Figure 9: [CS2] IGT inlet bearing vibration information (months 1–12—all operations including transients).

Although CS1 is used as a proof of concept application of VBGMM and GMMOC for novelty detection, the bearing considered in the study did fail around 3 months after initially being identified using the extended GMM.

Case Study 2

CS2 uses measurements taken over a longer period of 12 months using a lower sampling rate; specifically 1 data sample taken every 9 minutes, as shown in Figure 9. Again the procedure depicted in Table 1 is applied.

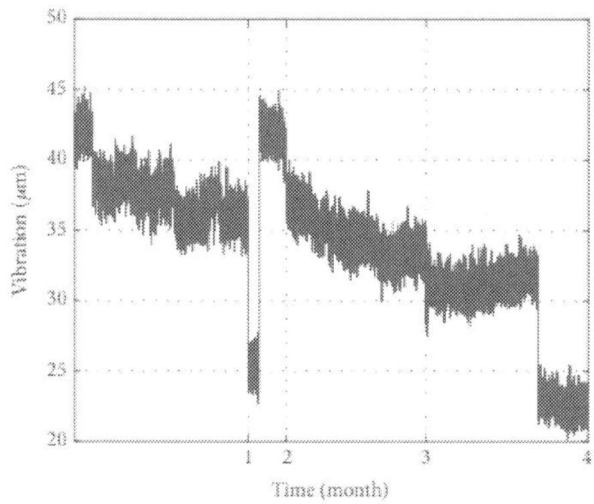

Figure 10: [CS2] Extracted steady-state vibration data for normal operation (months 1–4).

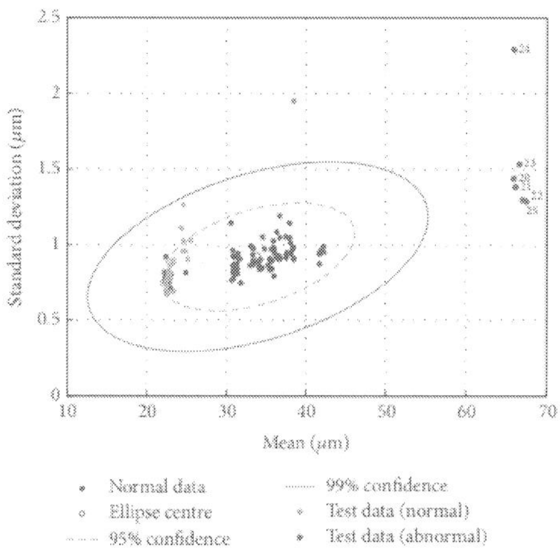

Figure 11: [CS2] Fingerprint envelope and novelty detection through GMMOC (months 1–4 and month 5).

Measurements from months 1–4 are deemed to describe normal operational characteristics. VBGMM is then applied to identify what is considered to be steady-state operational data for further analysis (Figure 10). Important characteristics are then determined; once again it is known from empirical studies that for this application case the mean and standard deviation are effective measures of underlying behaviour. Through the application of GMMOC and by clustering the extracted features into a single cluster, a fingerprint of normal operation is obtained, as shown in Figure 11, where the 95% confidence envelope is considered as an early warning boundary in this instance and the 99% confidence envelope as a fault detection boundary. Using measurements from month 5 as testing data, it is shown in Figure 11 that, between days 20 and 25 of the 5th month, operation falls outside the "normal fingerprint" ellipsoid and is therefore identified as an outlier, thereby providing an early warning of expected failure, which was evidenced during the field service.

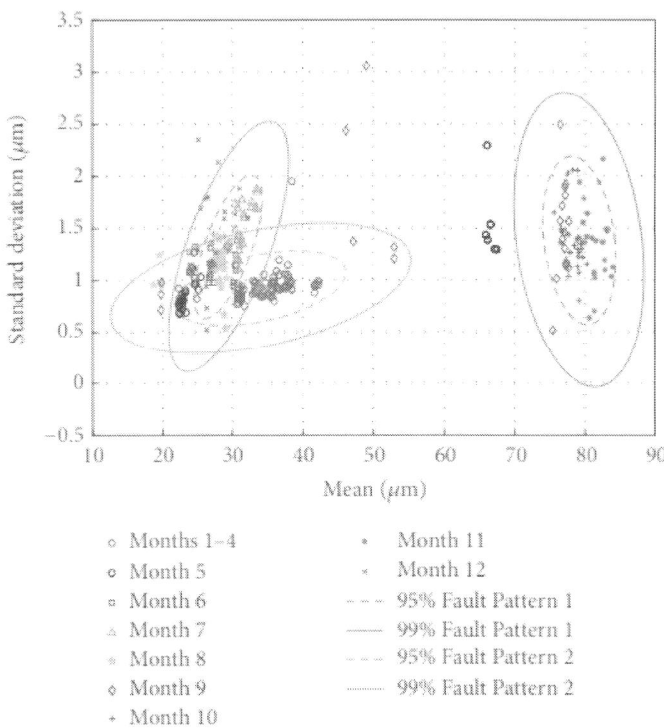

Figure 12: [CS2] Novelty detection and fault pattern location based on GM-MOC (months 1–12).

Referring again to measurements shown in Figure 9, on day 140 of operation (day 20 of month 5) a transient in the vibration level is evident, and although the mean level ultimately returns to normal after month 5, the variance continued to indicate evidence of emerging failure during and after month 6 (Figure 12). Considering months 6–12 as a period of emerging fault, the number of fault patterns can be discriminated using VBGMM (2 clusters for faults in this case) and the fault pattern locations identified using GMMOC, as shown in Figure 12. It can be seen that Fault Pattern 2 overlaps with the original fingerprint from months 1–4, indicating in this case that ideally other feature extraction indices (e.g., those involved with loading conditions) could be used to further isolate this type of fault characteristic from that of normal operation. Between the 2 sets of fault patterns (Figure 12), some anomalies are apparent which are due to the increasing levels of vibration as the bearing deteriorates, such as the 20th to 25th days of the 5th month's operation. In this instance the unit was shut down for maintenance in order to prevent a catastrophic failure. The bearing was known to be undamaged in month 1 as it was assessed in the previous service check (see Figure 13). Through subsequent decommissioning of the unit and investigation, vibration damage is evident on the inlet bearing, with excessive wear on the tilt pad shown in Figure 14. It can be clearly seen that there has been abnormal wear from the markings shown on both the shaft (Figure 15) and bearing pads (Figure 14). Through root-cause analysis, the damage was attributed to an incorrectly specified lubricant oil cooler causing high temperatures in the lubricant oil.

Figure 13: Normal IGT bearing with undamaged tilt pads.

Figure 14: IGT inlet bearing showing vibration damage.

Figure 15: Compressor rotor inlet bearing shaft showing vibration damage.

CONCLUSION

The paper has developed and demonstrated extensions of GMMs to provide a highly practical preprocessing and novelty/fault detection tool. The main contributions of the paper are (1) an automatic clustering method for VBGMM which identifies steady-state operational behaviour from transient operation, allowing the extraction of steady-state measurement segments for subsequent condition monitoring and (2) a GMMOC method that has been proposed and shown to provide a valuable tool for use as an early warning system of emerging failure through novelty detection. The presented techniques are currently being utilised in an industrial environment to monitor the operational status of a global fleet of IGTs. Although the experimental trials have focused on IGTs and bearing vibration measurements in this instance, the proposed methods are much more widely applicable to other industrial components and systems for pattern analysis, benchmarking, and novelty/fault detection.

Nomenclature

IGT: Industrial gas turbine

GMM: Gaussian Mixture Model

VBGMM: Variational Bayesian Gaussian Mixture Model

GMMOC: Gaussian Mixture Model with an outlier component

PCA: Principal component analysis

ANN: Artificial neural network

MC: Mixture component

EM: Expectation-maximization

M-step: Maximization step

E-step: Expectation step

KL: Kullback-Leibler

CS: Case study.

REFERENCES

1. R. Kurz, K. Brun, and M. Wollie, "Degradation effects on industrial gas turbines," Journal of Engineering for Gas Turbines and Power, vol. 131, no. 6, Article ID 062401, 2009.
2. M. A. Zaidan, A. R. Mills, R. F. Harrison, and P. J. Fleming, "Gas turbine engine prognostics using Bayesian hierarchical models: A variational approach," Mechanical Systems and Signal Processing, vol. 70-71, pp. 120–140, 2016.
3. M. Maalouf, "Gas turbine vibration monitoring," Orbit Article, vol. 25, no. 1, pp. 1–7, 2005.
4. F. Lu, J. Huang, and Y. Xing, "Fault diagnostics for turbo-shaft engine sensors based on a simplified on-board model," Sensors (Switzerland), vol. 12, no. 8, pp. 11061–11076, 2012.
5. H. Hanachi, J. Liu, A. Banerjee, and Y. Chen, "Sequential state estimation of nonlinear/non-Gaussian systems with stochastic input for turbine degradation estimation," Mechanical Systems and Signal Processing, vol. 72-73, pp. 32–45, 2016.
6. Y. Zhang, C. Bingham, Z. Yang, B. W.-K. Ling, and M. Gallimore, "Machine fault detection by signal denoising - With application to industrial gas turbines," Measurement: Journal of the International Measurement Confederation, vol. 58, pp. 230– 240, 2014.
7. S. Yin, S. X. Ding, A. Haghani, H. Hao, and P. Zhang, "A comparison study of basic data-driven fault diagnosis and process monitoring methods on the benchmark Tennessee Eastman process," Journal of Process Control, vol. 22, no. 9, pp. 1567–1581, 2012.
8. Y. Zhang, C. M. Bingham, and M. Gallimore, "Fault detection and diagnosis based on extensions of PCA," Advances in Military Technology, vol. 8, no. 2, pp. 27–41, 2013.
9. G. Romesis and K. Mathioudakis, "Setting up of a probabilistic neural network for sensor fault detection including operation with component faults," Journal of Engineering for Gas Turbines and Power, vol. 125, no. 3, pp. 634–641, 2003.
10. D. L. Simon and J. S. Litt, "A data filter for identifying steadystate operating points in engine flight data for condition monitoring applications," Journal of Engineering for Gas Turbines and Power, vol. 133, no. 7, Article ID 071603, 2011.
11. O. Uluyol, Kyusung K., and E. O. Nwadiogbu, "Synergistic use of soft

computing technologies for fault detection in gas turbine engines," IEEE Transactions on Systems, Man and Cybernetics, Part C (Applications and Reviews), vol. 36, no. 4, pp. 476–484, 2006

12. Y. Zhang, C. M. Bingham, and M. Gallimore, "Applied sensor fault detection,identification and data reconstruction, identification and data reconstruction," Advances in Military Technology, vol. 8, no. 2, pp. 13–26, 2013.
13. Y. Zhang, C. Bingham, M. Gallimore, and J. Chen, "Steady-state and transient operation discrimination by Variational Bayesian Gaussian Mixture Models," in 2013 16th IEEE International Workshop on Machine Learning for Signal Processing, MLSP 2013, gbr, September 2013.
14. A. Jablonski, T. Barszcz, M. Bielecka, and P. Breuhaus, "Modeling of probability distribution functions for automatic threshold calculation in condition monitoring systems," Measurement, vol. 46, no. 1, pp. 727–738, 2013
15. D. A. Reynolds, Encyclopedia of Biometrics Recognition, Springer, Heidelberg, Germany, 2008
16. G. Yu, C. Li, and J. Sun, "Machine fault diagnosis based on Gaussian mixture model and its application," International Journal of Advanced Manufacturing Technology, vol. 48, no. 1-4, pp. 205–212, 2010
17. J. Yu, "Bearing performance degradation assessment using locality preserving projections and Gaussian mixture models," Mechanical Systems and Signal Processing, vol. 25, no. 7, pp. 2573–2588, 2011
18. C. M. Bishop, Pattern Recognition and Machine Learning, Information Science and Statistics, Springer, Heidelberg, Germany, 2006.
19. G. J. McLachlan and T. Krishnan, The EM Algorithm and Extensions, Wiley-Interscience, New York, NY, UK, Second edition, 2008.
20. A. R. Groves, Bayesian Learning Methods for Modelling Functional MRI, Department of Clinical Neurology, University of Oxford, Oxford, UK, 2010.
21. A. Corduneanu and C. M. Bishop, "Variational Bayesian model selection for mixture distributions," Artificial Intelligent Statistics, pp. 27–34, 2001.
22. K. Watanabe and S. Watanabe, "Stochastic complexities of Gaussian mixtures in variational Bayesian approximation," Journal of Machine Learning Research, vol. 7, pp. 625–644, 2006.

23. Y. Zhang, C. Bingham, M. Gallimore, and D. Cox, "Novelty detection based on extensions of GMMs for industrial gas turbines," in 2015 IEEE International Conference on Computational Intelligence and Virtual Environments for Measurement Systems and Applications, CIVEMSA 2015, chn, June 2015
24. Y. Zhang, S. Cruz-Manzo, and A. Latimer, "Start-up vibration analysis for novelty detection on industrial gas turbines," in Proceedings of 2016 International Symposium on Industrial Electronics, Banja Luka, Bosnia and Herzegovina, November 2016.
25. Y. Zhang, M. Gallimore, C. Bingham, J. Chen, and Y. Xu, "Hybrid hierarchical clustering — piecewise aggregate approximation, with applications," International Journal of Computational Intelligence and Applications, vol. 15, no. 4, Article ID 1650019, 26 pages, 2016
26. M. Markou and S. Singh, "A neural network-based novelty detector for image sequence analysis," IEEE Transactions on Pattern Analysis and Machine Intelligence, vol. 28, no. 10, pp. 1664–1677, 2006
27. T. Heyns, P. S. Heyns, and J. P. de Villiers, "Combining synchronous averaging with a Gaussian mixture model novelty detection scheme for vibration-based condition monitoring of a gearbox," Mechanical Systems and Signal Processing, vol. 32, pp. 200–215, 2012.
28. M. K. Titsias, Unsupervised Learning of Multiple Objects in Images, School of Informatics, University of Edinburgh, UK, 2005
29. P. Coretto, The Noise Component in Model-Based Clustering, Department of Statistical Science, University of London, London, UK, 2008
30. N. Nasios and A. G. Bors, "Variational learning for Gaussian mixture models," IEEE Transactions on Systems, Man, and Cybernetics, Part B: Cybernetics, vol. 36, no. 4, pp. 849–862, 2006.
31. P. Jayaswal, A. K. Wadhwani, and K. B. Mulchandani, "Machine fault signature analysis," International Journal of Rotating Machinery, vol. 2008, Article ID 583982, 10 pages, 2008.
32. S. F. Ding, H. Zhu, W. K. Jia, and C. Y. Su, "A survey on feature extraction for pattern recognition," Artificial Intelligence Review, vol. 37, no. 3, pp. 169–180, 2012.
33. Z. Huo, Y. Zhang, P. Francq, L. Shu, and J. Huang, "Incipient fault diagnosis of roller bearing using optimized wavelet transform based multi-speed vibration signatures," IEEE Access, 2017

CHAPTER 4

Estimating Mixture Entropy with Pairwise Distances

Artemy Kolchinsky [1], **and Brendan D. Tracey** [1, 2]

Santa Fe Institute, Santa Fe, NM 87501, USA

Department of Aeronautics and Astronautics, Massachusetts Institute of Technology, Cambridge, MA 02139, USA

ABSTRACT

Mixture distributions arise in many parametric and non-parametric settings—for example, in Gaussian mixture models and in non-parametric estimation. It is often necessary to compute the entropy of a mixture, but, in most cases, this quantity has no closed-form expression, making some form of approximation necessary. We propose a family of estimators based on a pairwise distance function between mixture components, and show that this estimator class has

Citation: (APA): Kolchinsky, A., & Tracey, B. D. (2017). Estimating mixture entropy with pairwise distances. Entropy, 19(7), 361. (17 pages) DOI: https://doi.org/10.3390/e19070361 URL: https://www.mdpi.com/1099-4300/19/7/361/htm

many attractive properties. For many distributions of interest, the proposed estimators are efficient to compute, differentiable in the mixture parameters, and become exact when the mixture components are clustered. We prove this family includes lower and upper bounds on the mixture entropy. The Chernoff α -divergence gives a lower bound when chosen as the distance function, with the Bhattacharyaa distance providing the tightest lower bound for components that are symmetric and members of a location family. The Kullback–Leibler divergence gives an upper bound when used as the distance function. We provide closed-form expressions of these bounds for mixtures of Gaussians, and discuss their applications to the estimation of mutual information. We then demonstrate that our bounds are significantly tighter than well-known existing bounds using numeric simulations. This estimator class is very useful in optimization problems involving maximization/minimization of entropy and mutual information, such as MaxEnt and rate distortion problems.

INTRODUCTION

A mixture distribution is a probability distribution whose density function is a weighted sum of individual densities. Mixture distributions are a common choice for modeling probability distributions, in both parametric settings, for example, learning a mixture of Gaussians statistical model [**1**], and non-parametric settings, such as kernel density estimation.

It is often necessary to compute the differential entropy [2] of a random variable with a mixture distribution, which is a measure of the inherent uncertainty in the outcome of the random variable. Entropy estimation arises in image retrieval tasks [3], image alignment and error correction [4], speech recognition [5,6], analysis of debris spread in rocket launch failures [7], and many other settings. Entropy also arises in optimization contexts [4,8,9,10], where it is minimized or maximized under some constraints (e.g., MaxEnt problems). Finally, entropy also plays a central role in minimization or maximization of mutual information, such as in problems related to rate distortion [11].

Unfortunately, in most cases, the entropy of a mixture distribution has no known closed-form expression [12]. This is true even when the entropy of each component distribution does have a known closed-form expression. For instance, the entropy of a Gaussian has a well-known form, while the entropy of a mixture of Gaussians does not [13]. As a result, the problem of finding a tractable and accurate estimate for mixture entropy has been described as "a problem of considerable current interest and practical significance" [14].

One way to approximate mixture entropy is with Monte Carlo (MC) sampling. MC sampling provides an unbiased estimate of the entropy, and this estimate can become arbitrarily accurate by increasing the number of MC samples. Unfortunately, MC sampling is very computationally intensive, as, for each sample, the (log) probability of the sample location must be computed under every component in the mixture. MC sampling typically requires a large number of samples to estimate entropy, especially in high-dimensions. Sampling is thus typically impractical, especially for optimization problems where, for every parameter change, a new entropy estimate is required. Alternatively, it is possible to approximate entropy using numerical integration, but this is also computationally expensive and limited to low-dimensional applications [15,16].

Instead of Monte Carlo sampling or numerical integration, one may use an analytic estimator of mixture entropy. Analytic estimators have estimation bias but are much more computationally efficient. There are several existing analytic estimators of entropy, discussed in-depth below. To summarize, however, commonly-used estimators have significant drawbacks: they have large bias relative to the true entropy, and/or they are invariant to the amount of "overlap" between mixture components. For example, many estimators do not depend on the locations of the means in a Gaussian mixture model.

In this paper, we introduce a novel family of estimators for the mixture entropy. Each member of this family is defined via a pairwise-distance function between component densities. The estimators in this family have several attractive properties. They are computationally efficient, as long as the pairwise-distance function and the entropy of each component distribution are easy to compute. The estimation bias of any member of this family is bounded by a constant. The estimator is continuous and smooth and is therefore useful for optimization problems. In addition, we show that when the Chernoff α-divergence (i.e., a scaled Rényi divergence) is used as a pairwise-distance function, the corresponding estimator is a lower-bound on the mixture entropy. Furthermore, among all the Chernoff α-divergences, the Bhattacharyya distance (α=0.5) provides the tightest lower bound when the mixture components are symmetric and belong to a location family (such as a mixture of Gaussians with equal covariances). We also show that when the Kullback–Leibler [KL] divergence is used as a pairwise-distance function, the corresponding estimator is an upper-bound on the mixture entropy. Finally, our family of estimators can compute the exact mixture entropy when the component distributions are grouped into well-separated clusters, a property not shared by other analytic estimators of entropy. In particular, the bounds mentioned above converge to the same value for well-separated clusters.

The paper is laid out as follows. We first review mixture distributions and entropy estimation in **Section 2**. We then present the class of pairwise distance estimators in **Section 3**, prove bounds on the error of any estimator in this class, and show distance functions that bound the entropy as discussed above. In **Section 4**, we consider the special case of mixtures of Gaussians, and give explicit expressions for lower and upper bounds on the mixture entropy. When all the Gaussian components have the same covariance matrix, we show that these bounds have particularly simple expressions. In **Section 5**, we consider the closely related problem of estimating the mutual information between two random variables, and show that our estimators can be directly used to estimate and bound the mutual information. For the Gaussian case, these can be used to bound the mutual information across a type of Additive White Noise Gaussian channel. Finally, in **Section 6**, we run numerical experiments and compare the performance of our lower and upper bounds relative to existing estimators. We consider both mixtures of Gaussians and mixtures of uniform distributions.

Background and Definitions

We consider the differential entropy of a continuous random variable *X*, defined as

$$H(X) := -\int p_X(x) \ln p_X(x) \, dx,$$

where pX is a mixture distribution,

$$p_X(x) = \sum_{i=1}^{N} c_i p_i(x),$$

and where c_i indicates the weight of component $i \, (c_i \geq 0, \sum_i c_i = 1)$ and p_i the probability density of component i.

We can treat the set of component weights as the probabilities of outcomes 1 . . . N of a discrete random variable C, where $Pr(C = i) = c_i$. Consider the mixed joint distribution of the discrete random variable C and the continuous random variable X,

$$p_{X,C}(x,i) = p_i(x)c_i\,,$$

and note the following identities for conditional and joint entropy [17],

$$H(X,C) = H(X|C) + H(C) = H(C|X) + H(X)\,,$$

where we use H for discrete and differential entropy interchangeably. Here, the conditional entropies are defined as

$$H(X|C) = \sum_i c_i H(p_i) \qquad \text{and} \qquad H(C|X) = \int p_{X,C}(x,i)\log\frac{p_{X,C}(x,i)}{p_X(x)}\,dx\,.$$

Using elementary results from information theory [2], H(X) can be bounded from below by

$$H(X) \geq H(X|C)\,, \tag{1}$$

since conditioning can only decrease entropy. Similarly, H(X) can be bounded from above by

$$H(X) \leq H(X,C) = H(X|C) + H(C)\,, \tag{2}$$

following from H(X) = H(X, C) − H(C|X) and the non-negativity of the conditional discrete entropy H(C|X). This upper bound on the mixture entropy was previously proposed by Huber et al. [18]. It is easy to see that the bound in Equation (1) is tight when all the components have the same distribution, since then H(pX) = H(pi) for all i. The bound in Equation (2) becomes tight when H(C|X) = 0, i.e., when any sample from pX uniquely determines the component identity C. This occurs when the different mixture components have non-overlapping supports, pi(x) > 0 =⇒ $p_j(x) = 0$ for all x and $i \neq j$. More generally, the bound of Equation (2) becomes increasingly tight as the mixture distributions move farther apart from one another.

In the case where the entropy of each component density, $H(p_i)$ for i = 1 . . . N, has a simple closed form expression, the bounds in Equations (1) and (2) can be easily computed. However, neither bound depends on the "overlap" between components. For instance, in a Gaussian mixture model, these bounds are invariant to changes in the component means. The bounds are thus unsuitable for many problems; for instance, in optimization, one typically tunes parameters to adjust component means, but the above entropy bounds remain the same regardless of mean location.

There are two other estimators of the mixture entropy that should be mentioned. The first estimator is based on kernel density estimation [16,19]. It estimates the entropy using the mixture probability of the component means, μ_i,

$$\hat{H}_{\mathrm{KDE}}(X) := -\sum_i c_i \ln \sum_j c_j p_j(\mu_i)\,. \tag{3}$$

The second estimator is a lower bound that is derived using Jensen's inequality [2], $-\mathrm{E}[\ln f(X)] \geq -\ln[\mathrm{E}(f(X))]$, , giving

$$H(X) := -\int \sum_i c_i p_i(x) \ln \sum_j c_j p_j(x)\, dx \geq -\sum_i c_i \ln \sum_j c_j \left(\int p_i(x) p_j(x)\, dx \right) =: \hat{H}_{\mathrm{ELK}}. \tag{4}$$

In the literature, the term R pi(x)pj(x) dx has been referred to as the "Cross Information Potential" [20,21] and the "Expected Likelihood Kernel" [22,23] (ELK, we use this second acronym to label this estimator). When the component distributions are Gaussian, pi := N (μi , Σi), the ELK has a simple closed-form expression,

$$\hat{H}_{\mathrm{ELK}}(X) = -\sum_i c_i \ln \sum_j c_j q_{j,i}(\mu_i), \tag{5}$$

where each qj,i is a Gaussian defined as $q_{j,i} := \mathcal{N}(\mu_j, \Sigma_i + \Sigma_j)$. This lower bound was previously proposed for Gaussian mixtures in [18] and in a more general context in [12]. Both H^ KDE, Equation (3), and H^ ELK, Equation (5), are computationally efficient, continuous and differentiable, and depend on component overlap, making them suitable for optimization. However, as will be shown via numerical experiments (Section 6), they exhibit significant underestimation bias. At the same time, we will show that for Gaussian mixtures with equal covariance, $\hat{H}_{\mathrm{KDE}}$ is only an additive constant away from an estimator in our proposed class.

Estimators Based on Pairwise-Distances

OVERVIEW

Let $D(p_i \| p_j)$ be some (generalized) distance function between probability densities p_i and p_j. Formally, we assume that D is a premetric, meaning that it is non-negative and D(pikpj) = 0 if pi = pj . We do not assume that D is symmetric, nor that it obeys the triangle inequality, nor that it is strictly greater than 0 when pi 6= pj . For any allowable distance function D, we propose the following entropy estimator:

$$\hat{H}_D(X) := H(X|C) - \sum_i c_i \ln \sum_j c_j \exp(-D(p_i \| p_j)). \tag{6}$$

This estimator can be efficiently computed if the entropy of each component and $D(p_i \| p_j)$ for all i, j have simple closed-form expressions. There are many distribution-distance function pairs that satisfy these conditions (e.g., Kullback–Leibler divergence, Renyi divergences, Bregman divergences, f-divergences, etc., for Gaussian, uniform, exponential, etc.) [24–28]. It is straightforward to show that for any D, H^ D falls between the bounds of Equations (1) and (2)

$$H(X|C) \leq \hat{H}_D(X) \leq H(X,C). \tag{7}$$

To do so, consider the "smallest" and "largest" allowable distance functions,

$$D_{\min}(p_i\|p_j) = 0 \quad \text{and} \quad D_{\max}(p_i\|p_j) = \begin{cases} 0, & \text{if } p_i = p_j, \\ \infty, & \text{otherwise.} \end{cases} \tag{8}$$

For any D and p_i , p_j , $D_{\min}(p_i\|p_j) \leq D(p_i\|p_j) \leq D_{\max}(p_i\|p_j)$, thus

$$\hat{H}_{D_{\min}}(X) \leq \hat{H}_D(X) \leq \hat{H}_{D_{\max}}(X).$$

Plugging D_{min} into Equation (6) (noting that $\sum_j c_j = 1$) gives $\hat{H}_{D_{\min}}(X) = H(X|C)$, while plugging D_{max} into Equation (6) gives

$$\hat{H}_{D_{\max}} = H(X|C) - \sum_i c_i \ln\left(c_i + \sum_{j \neq i} c_j e^{-D_{\max}(p_i\|p_j)}\right) \leq H(X|C) - \sum_i c_i \ln c_i = H(X,C).$$

These two inequalities yield Equation (7). The true entropy, as shown in Section 2, also obeys $H(X|C) \leq H(X) \leq H(X,C)$ The magnitude of the bias of H^ D is thus bounded by

$$\left|\hat{H}_D(X) - H(X)\right| \leq H(X,C) - H(X|C) = H(C).$$

In the next two subsections, we improve upon the bounds suggested in Equations (1) and (2), by examining bounds induced by particular distance functions.

Lower Bound

The "Chernoff α-divergence" [28,29] for some real-valued α is defined as

$$C_\alpha(p\|q) := -\ln \int p^\alpha(x) q^{1-\alpha}(x)\, dx. \tag{9}$$

Note that $C_\alpha(p\|q) = (1-\alpha)R_\alpha(p\|q)$, where R_α is Rényi divergence of order α [30].

We show that for any $\alpha \in [0,1]$, $\hat{H}_{C_\alpha}(X)$ is a lower bound on the entropy (for $\alpha \notin [0,1]$, C_α is not a valid distance function). To do so, we make use of a derivation from [31] and write,

$$\begin{aligned} H(X) &= H(X|C) - \int \sum_i c_i p_i(x) \ln \frac{p_X(x)}{p_i(x)} dx \\ &= H(X|C) - \int \sum_i c_i p_i(x) \ln \frac{p_X(x)}{p_i(x)^\alpha \sum_j c_j p_j(x)^{1-\alpha}} dx - \int \sum_i c_i p_i(x) \ln \left(\frac{\sum_j c_j p_j(x)^{1-\alpha}}{p_i(x)^{1-\alpha}} \right) dx \\ &\overset{(a)}{\geq} H(X|C) - \int \sum_i c_i p_i(x) \ln \left(\frac{\sum_j c_j p_j(x)^{1-\alpha}}{p_i(x)^{1-\alpha}} \right) dx \\ &\overset{(b)}{\geq} H(X|C) - \sum_i c_i \ln \int p_i(x)^\alpha \sum_j c_j p_j(x)^{1-\alpha} dx \end{aligned}$$

$$= H(X|C) - \sum_i c_i \ln \sum_j c_j e^{-C_\alpha(p_i \| p_j)} := \hat{H}_{C_\alpha}(X). \tag{10}$$

The inequalities (a) and (b) follow from Jensen's inequality. This inequality is used directly in (b), while in (a) it follows from

$$\begin{aligned} -\int \sum_i c_i p_i(x) \ln \frac{p_X(x)}{p_i(x)^\alpha \sum_j c_j p_j(x)^{1-\alpha}} dx &\geq -\ln \int \sum_i c_i p_i(x) \frac{p_X(x)}{p_i(x)^\alpha \sum_j c_j p_j(x)^{1-\alpha}} dx \\ &= -\ln \int \sum_i c_i p_i(x)^{1-\alpha} \frac{p_X(x)}{\sum_j c_j p_j(x)^{1-\alpha}} dx \\ &= -\ln \sum_i c_i \int p_X(x) dx = 0. \end{aligned}$$

Note that Jensen's inequality is used in the derivations of both this lower bound as well as the lower bound $\hat{H}_{\text{ELK}}$ in Equation (4). However, the inequality is applied differently in the two cases, and, as will be demonstrated in Section 6, the estimators have different performance.

We have shown that using C_α as a distance function gives a lower bound on the mixture entropy for any $\alpha \in [0, 1]$. For a general mixture distribution, one could optimize over the value of α to find the tightest lower bound. However, we can show that the tightest bound is achieved for α = 0.5 in the special case when all of the mixture components pi are symmetric and come from a location family,

$$p_i(x) = b(x - \mu_i) = b(\mu_i - x).$$

Examples of this situation include mixtures of Gaussians with the same covariance ("homoscedastic" mixtures), multivariate t-distributions with the same covariance, location-shifted bounded uniform distributions, most kernels used in kernel density estimation, etc. It does not apply to skewed distributions, such as as the skew-normal distribution [12].

To show that α = 0.5 is optimal, first define the Chernoff α-coefficient as

$$c_\alpha(p_i \| p_j) := \int p_i(x)^\alpha p_j(x)^{1-\alpha} dx.$$

We show that for any pair pi , pj of symmetric distributions from a location family, $c_\alpha(p_j \| p_j)$ is minimized by α = 0.5. This means that all pairwise distances $C_\alpha(p_j \| p_i) = -\ln c_\alpha(p_j \| p_i)$ are maximized by α = 0.5, and, therefore, the entropy estimator HCα (Equation (6)) is maximized by α = 0.5. First, define a change of variables

$$y := \mu_i + \mu_j - x,$$

which gives $x - \mu_i = \mu_j - y$ and $x - \mu_j = \mu_i - y$. This allows us to write the Chernoff α-coefficient as

$$\begin{aligned} c_\alpha(p_i \| p_j) &= \int p_i(x)^\alpha p_j(x)^{1-\alpha} dx \\ &= \int b(x-\mu_i)^\alpha b(x-\mu_j)^{1-\alpha} dx \\ &\overset{(a)}{=} \int b(\mu_j - y)^\alpha b(\mu_i - y)^{1-\alpha} dy \\ &\overset{(b)}{=} \int b(y-\mu_j)^\alpha b(y-\mu_i)^{1-\alpha} dy \\ &= c_{1-\alpha}(p_i \| p_j), \end{aligned}$$

where, in (a), we have substituted variables, and in (b) we used the assumption that b(x) = b(−x). Since we have shown that $c_\alpha(p_i \| p_j) = c_{1-\alpha}(p_i \| p_j)$, c_α is symmetric in α about α = 0.5, we show that $c_\alpha(p \| q)$ is everywhere convex in α. Together, this means that $c_\alpha(p_i \| p_j)$ must achieve a minimum value at α = 0.5.

The Chernoff α-coefficient for α = 0.5 is known as the Bhattacharyya coefficient, with the corresponding Bhattacharyya distance [32] defined as

$$\mathrm{BD}(p \| q) := -\ln \int \sqrt{p(x) q(x)} dx = C_{0.5}(p \| q).$$

Since any Chernoff α-divergence is a lower bound for the entropy, we write the particular case of Bhattacharyya-distance lower bound as

$$H(X) \geq \hat{H}_{\mathrm{BD}}(X) := H(X|C) - \sum_i c_i \ln \sum_j c_j e^{-\mathrm{BD}(p_i \| p_j)}. \tag{11}$$

Upper Bound

The Kullback–Leibler [KL] divergence [2] is defined as

$$\mathrm{KL}(p \| q) := \int p(x) \ln \frac{p(x)}{q(x)} dx.$$

Using KL divergence as the pairwise distance provides an upper bound on the mixture entropy. We show this as follows:

$$\begin{aligned} H(X) &= -\sum_i c_i \mathrm{E}_{p_i}\left[\ln \sum_j c_j p_j(X)\right] \\ &\overset{(a)}{\leq} -\sum_i c_i \ln \sum_j c_j e^{\mathrm{E}_{p_i}[\ln p_j(X)]} \end{aligned}$$

$$\begin{aligned} &= -\sum_i c_i \ln \sum_j c_j e^{-H(p_i \| p_j)} \\ &= \sum_i c_i H(p_i) - \sum_i c_i \ln \sum_j c_j e^{-\mathrm{KL}(p_i \| p_j)}, \end{aligned}$$

where E_{pi} indicates expectation when X is distributed according to p_i, $H(\cdot \| \cdot)$ indicates the cross-entropy function, and we employ the identity $H(p_i \| p_j) = H(p_i) + \mathrm{KL}(p_i \| p_j)$ The inequality in step (a) uses a variational lower bound on the expectation of a log-sum [5,33],

$$\mathrm{E}\left[\ln \sum_j Z_j\right] \geq \ln \sum_j \exp\left(\mathrm{E}\left[\ln Z_j\right]\right).$$

Combining yields the upper boun

$$H(X) \leq \hat{H}_{\mathrm{KL}} := H(X|C) - \sum_i c_i \ln \sum_j c_j e^{-\mathrm{KL}(p_i \| p_j)}. \tag{12}$$

Exact Estimation in the "Clustered" Case

In the previous sections, we derived lower and upper bounds on the mixture entropy, using estimators based on Chernoff α-divergence and KL divergence, respectively.

There are situations in which the lower and upper bounds become similar. Consider a pair of component distributions, pi and pj. By applying Jensen's inequality to Equation (9), we can derive the inequality $C_\alpha(p_i \| p_j) \leq \alpha \mathrm{KL}(p_i \| q_j)$ There are two cases in which a pair of components contributes similarly to the lower and upper bounds. The first case is when $C_\alpha(p_i \| p_j)$ is very large, meaning that the KL is also very large. By Equation (6), distances enter into our estimators as exp $(-D(p_i \| p_j))$, and, in this case, exp $(-\mathrm{KL}(p_i \| p_j)) \approx \exp(-C_\alpha(p_i \| p_j)) \approx 0$. In the second case, $\mathrm{KL}(p_i \| p_j) \approx 0$, meaning that $C_\alpha(p_i \| p_j)$ must also be near zero, and, in this case, exp $(-\mathrm{KL}(p_i \| p_j)) \approx \exp(-C_\alpha(p_i \| p_j)) \approx 1$. Thus, the lower and upper bounds become similar when all pairs of components are either very close together or very far apart. In this section, we analyze this special case. Specifically, we consider the situation when mixture components are "clustered", meaning that there is a grouping of component distributions such that distributions in the same group are approximately the same and distributions assigned to different groups are very different from one another. We show that in this case our lower and upper bounds become equal and our pairwise-distance estimate of the entropy is tight. Though this situation may seem like an edge case, clustered distributions do arise in mixture estimation, e.g., when there are repeated data points, or as solutions to information-theoretic optimization problems [11]. Note that the number of groups is arbitrary, and therefore this situation includes the extreme cases of a single group (all component distributions are nearly the same) as well as N different groups (all component distributions are very different).

Formally, let the function g(i) indicate the group of component i. We define that the components are "clustered" with respect to grouping g if $\mathrm{KL}(p_i \| p_j) \leq \kappa$, κ whenever g(i) = g(j) for some small κ, and B $(p_i \| p_j) \geq \beta$ whenever $g(i) \neq g(j)$ some large β. We use the notation $p_G(k) = \sum_i \delta_{g(i),k} c_i$ to indicate the sum of the weights of the components in group k, where δ_{ij} indicates the Kronecker delta function. For technical reasons, below we only consider C_α where α is strictly greater than 0. We show that when κ is small and β is large, both $\hat{H}_{C_\alpha}$ for α ∈ (0, 1] and $\hat{H}_{KL}$ approach

$$H(X|C) - \sum_k p_G(k) \ln p_G(k) .$$

Since one is a lower bound and one is an upper bound on the true entropy, the estimators become exact as they converge in value

Recall that $\mathrm{BD}(p\|q) = C_{0.5}(p\|q)$. For $\alpha \in (0,1]$, $\alpha^{-1} C_\alpha(p\|q)$ is a monotonically decreasing function for α ∈ (0, 1 [34], meaning that $C_\alpha \geq 2\alpha \mathrm{BD}(p\|q)$ for $\alpha \in (0, 0.5]$ $(1-\alpha)^{-1} C_\alpha(p\|q)$ is a monotonically increasing function for $\alpha > 0$ [34], thus $C_\alpha \geq 2(1-\alpha)\mathrm{BD}(p\|q)$ for α ∈ [0.5, 1]. Using the assumption that BD $(p_i \| p_j) \geq \beta$ and combining gives the bound

$$C_\alpha(p\|q) \geq (1 - |1 - 2\alpha|)\beta$$

for α ∈ (0, 1], leading to

$$\begin{aligned} \hat{H}_{C_\alpha}(X) &:= H(X|C) - \sum_i c_i \ln \sum_j c_j \exp(-C_\alpha(p_i \| p_j)) \\ &\geq H(X|C) - \sum_i c_i \ln \left[\sum_j \delta_{g(i),g(j)} c_j + \sum_j \left(1 - \delta_{g(i),g(j)}\right) c_j \exp(-C_\alpha(p_i \| p_j)) \right] \\ &\geq H(X|C) - \sum_k p_G(k) \ln \left[p_G(k) + (1 - p_G(k)) \exp(-(1 - |1 - 2\alpha|)\beta) \right] . \end{aligned}$$

In the second line, for the summation over i, j in the same group, we've used the non-negativity of $C_\alpha(p_i \| p_j) \geq 0$

For the upper bound H^ KL, we use that KL $(p_i \| p_j) \leq \kappa$ for i and j in the same group, and otherwise exp $(-\mathrm{KL}(p_i \| p_j)) \geq 0$. This gives the bound

$$
\begin{aligned}
\hat{H}_{\mathrm{KL}}(X) &:= H(X|C) - \sum_i c_i \ln \sum_j c_j \exp(-\mathrm{KL}(p_i \| p_j)) \\
&\leq H(X|C) - \sum_i c_i \ln \left[\sum_j \delta_{g(i),g(j)} c_j e^{-\kappa} + \sum_j \left(1 - \delta_{g(i),g(j)}\right) c_j \exp(-\mathrm{KL}(p_i \| p_j)) \right] \\
&\leq H(X|C) - \sum_k p_G(k) \ln p_G(k) e^{-\kappa}.
\end{aligned}
$$

The difference between the bounds is bounded by

$$
\begin{aligned}
\hat{H}_{\mathrm{KL}}(X) - \hat{H}_{C_\alpha}(X) &\leq \kappa + \sum_k p_G(k) \ln \left[1 + \frac{(1 - p_G(k)) \exp(-(1 - |1 - 2\alpha|)\beta)}{p_G(k)} \right] \\
&\leq \kappa + \sum_k p_G(k) \frac{(1 - p_G(k)) \exp(-(1 - |1 - 2\alpha|)\beta)}{p_G(k)} \\
&= \kappa + (|G| - 1) \exp(-(1 - |1 - 2\alpha|)\beta),
\end{aligned}
$$

where |G| is the number of groups. Thus, the difference decreases at least linearly in κ and exponentially in β. This shows that, in the clustered case, when κ ≈ 0 and β is very large, our lower and upper bounds become exact.

It also shows that any distance measure bounded between BD and KL also gives an exact estimate of entropy in the clustered case. Furthermore, the idea behind this proof can be extended to estimators induced by other bounding distances, beyond BD and KL, so as to show that a particular estimator converges to an exact entropy estimate in the clustered case. Note, however, that, for some distribution-distance pairs, the components will never be considered as "clustered"; e.g., the α-Chernoff distance for α = 0 between any two Gaussians is 0, and so a Gaussian mixture distribution will never be considered clustered according to this distance.

Finally, in the perfectly clustered case, we can show that our lower bound, $\hat{H}_{BD}$, is at least as good as the Expected Likelihood Kernel lower bound, $\hat{H}_{ELK}$, as defined in Equation (4). See Appendix B for details

Gaussian Mixtures

Gaussians are very frequently used as components in mixture distributions. Our family of estimators is well-suited to estimating the entropies of Gaussian mixtures, since the entropy of a d-dimensional Gaussian pi = N (μ_i, Σ_i) has a simple closed-form expression,

$$
H(p_i) = \frac{1}{2} \left[\ln |\boldsymbol{\Sigma}_i| + d \ln 2\pi + d \right],
$$

and because there are many distance functions between Gaussians with closed-form expressions (KL divergence, the Chernoff α-divergences [35], 2-Wasserstein distance [36,37], etc.). In this section, we consider Gaussian mixtures and state explicit expression for the lower and upper bounds on the mixture entropy derived in the previous section. We also consider these bounds in the special case where all Gaussian components have the same covariance matrix (homoscedastic mixtures)

We first consider the lower bound, $\hat{H}_{C\alpha}$, based on the Chernoff α-divergence distance function. For two multivariate Gaussians $p_1 \sim \mathcal{N}(\mu_1, \Sigma_1)$ and $p_2 \sim \mathcal{N}(\mu_2, \Sigma_2)$, , this distance is defined as [35]:

$$C_\alpha(p_1 \| p_2) = \frac{(1-\alpha)\alpha}{2}(\mu_1-\mu_2)^T\left((1-\alpha)\Sigma_1+\alpha\Sigma_2\right)^{-1}(\mu_1-\mu_2)+\frac{1}{2}\ln\left(\frac{|(1-\alpha)\Sigma_1+\alpha\Sigma_2|}{|\Sigma_1|^{1-\alpha}|\Sigma_2|^{\alpha}}\right). \tag{14}$$

(As a warning, note that most sources show erroneous expressions for the Chernoff and/or Rényi α-divergence between two multivariate Gaussians, including [27,29,38–40], and even a late draft of this manuscript.)

For the upper bound $\hat{H}_{KL}$, the KL divergence between two multivariate Gaussians $p_1 \sim \mathcal{N}(\mu_1, \Sigma_1)$ and $p_2 \sim \mathcal{N}(\mu_2, \Sigma_2)$ is

$$\mathrm{KL}(p_1 \| p_2) = \frac{1}{2}\left[\ln|\Sigma_2| - \ln|\Sigma_1| + (\mu_1-\mu_2)^T\Sigma_j^{-1}(\mu_1-\mu_2) + \mathrm{tr}\left(\Sigma_2^{-1}\Sigma_1\right) - d\right].$$

The appropriate lower and upper bounds are found by plugging in Equations (14) and (15) into Equation (6).

These bounds have simple forms when all of the mixture components have equal covariance matrices; i.e., $\Sigma_i = \Sigma$ for all i. First, define a transformation in which each Gaussian component pj is mapped to a different Gaussian $\tilde{p}_{j,\alpha}$, which has the same mean but where the covariance matrix is rescaled by $\frac{1}{\alpha(1-\alpha)}$,

$$p_j := \mathcal{N}(\mu_j, \Sigma) \quad \mapsto \quad \tilde{p}_{j,\alpha} := \mathcal{N}\left(\mu_j, \frac{1}{\alpha(1-\alpha)}\Sigma\right).$$

Then, the lower bound of Equation (10) can be written a

$$\hat{H}_{C_\alpha} = \frac{d}{2} + \frac{d}{2}\ln(\alpha(1-\alpha)) - \sum_i c_i \ln \sum_j c_j \tilde{p}_{j,\alpha}(\mu_i).$$

This is derived by combining the expressions for C_α, Equation (14), the entropy of a Gaussian, Equation (13), and the Gaussian density function. For a homoscedastic mixture, the tightest lower bound among the Chernoff α-divergences is given by $\alpha = 0.5$, corresponding to the Bhattacharyya distance

$$\hat{H}_{\mathrm{BD}} = \frac{d}{2} + \frac{d}{2}\ln\frac{1}{4} - \sum_i c_i \ln \sum_j c_j \tilde{p}_{j,0.5}(\mu_i).$$

(This is derived above in Section 3.2.)

For the upper bound, when all Gaussians have the same covariance matrix, we again combine the expressions for KL, Equation (15), the entropy of a Gaussian, Equation (13), and the Gaussian density function to give

$$\hat{H}_{\mathrm{KL}}(X) = \frac{d}{2} - \sum_i c_i \ln \sum_j c_j p_j(\mu_i).$$

Note that this is exactly the expression for the kernel density estimator $\hat{H}_{\mathrm{KDE}}$ (Equation (3)), plus a dimensional correction. Thus, surprisingly $\hat{H}_{\mathrm{KDE}}$ is a reasonable entropy estimator for homoscedastic Gaussian mixtures, since it is only an additive constant away from KL-distance based estimator $\hat{H}_{\mathrm{KL}}$ (which has various beneficial properties, as described above). This may explain why $\hat{H}_{\mathrm{KDE}}$ has been used effectively in optimization contexts [4,8–10], where the additive constant is often irrelevant, despite lacking a principled justification in terms of being a a bound on entropy.

ESTIMATING MUTUAL INFORMATION

It is often of interest, for example in rate distortion and related problems [11], to calculate thc mutual information across a communication channel,

$$MI(X;U) = H(X) - H(X|U),$$

where U is the distribution of signals sent across the channel, and X is the distribution of messages received on the other end of the channel. As with mixture distributions, it is often easy to compute H(X|U), the entropy of the received signal given the sent signal (i.e., the distribution of noise on the channel). The marginal entropy of the received signals, H(X), on the other hand, is often difficult to compute.

In some cases, the distribution of U may be well approximated by a mixture model. In this case, we can estimate the entropy of the received signals, H(X), using our pairwise distance estimators, as discussed in Section 3. In particular, we have the lower bound

$$MI(X;U) \geq -\sum_i c_i \ln \sum_j c_j \exp(-C_\alpha(p_i \| p_j)),$$

and the upper bound

$$MI(X;U) \leq -\sum_i c_i \ln \sum_j c_j \exp(-\mathrm{KL}(p_i \| p_j)),$$

where p_i is the density of component i, and noting that the H(X|U) terms cancel in the expression. This also illuminates that the actual pairwise portion of the estimator, $-\sum_i c_i \ln \sum_j c_j \exp(-D(p_i \| p_j))$ is a measure of the mutual information between the random variable specifying the component identity and the random variable distributed as the mixture of the component densities. If the components are identical, this mutual information is zero, since knowing the component identity tells one nothing about the outcome of X. On the other hand, when all of the components are very different from one another, knowing the component that generated X is very informative, giving the maximum amount of information, H(C).

As a practical example, consider a scenario in which U is a random variable representing the outside temperature on any particular day. This temperature is measured with a thermometer with Gaussian measurement noise (the "Additive White Noise Gaussian channel"). This gives our measurement distribution

$$X = U + \mathcal{N}(0, \Sigma').$$

If the actual temperature U is (approximately or exactly) distributed as a mixture of M Gaussians, each one having mixture weight c_i, mean μ_i, and covariance matrix Σ_i, then X will also be distributed as a mixture of M Gaussians, each with weight c_i, mean μ_i, and covariance matri $\Sigma_i' := \Sigma_i + \Sigma'$. We can then use our estimators to estimate the mutual information between the actual temperature, U, and thermometer measurements, X.

NUMERICAL RESULTS

In this section, we run numerical experiments and compare estimators of mixture entropy under a variety of conditions. We consider two different types of mixtures, mixtures of Gaussians and mixtures of uniform distributions, for a variety of parameter values. We evaluate the following estimators:

1. The true entropy, H(X), as estimated by a Monte Carlo sampling of the mixture model. Two thousand samples were used for each MC estimate for the mixtures of Gaussians, and 5000 samples were used for the mixtures of uniform distributions.
2. Our proposed upper-bound, based on the KL divergence, $\hat{H}_{KL}$

(Equation (12))

3. Our proposed lower-bound, based on the Bhattacharyya distance, $\hat{H}_{BD}$ (Equation (11))
4. The kernel density estimate based on the component means, $\hat{H}_{KDE}$ (Equation (3))
5. The lower bound based on the "Expected Likelihood Kernel", $\hat{H}_{ELK}$ (Equation (4))
6. The lower bound based on the conditional entropy, $H(X|C)$ (Equation (1))
7. The upper bound based on the joint entropy, $H(X,C)$ (Equation (2)).

We show the values of the estimators 1–5 as line plots, while the region between the conditional (6) and joint entropy (7) is shown in shaded green. The code for these figures can be found at [41], and uses the Gonum numeric library [42].

Mixture of Gaussians

In the first experiment, we evaluate the estimators on a mixture of randomly placed Gaussians, and look at their behavior as the distance between the means of the Gaussians increases. The mixture is composed of 100 10-dimensional Gaussians, each Gaussian distributed as $p_i = \mathcal{N}\left(\mu_i, I_{(10)}\right)$, where $I_{(d)}$ indicates the d × d identity matrix. Means are sampled from $\mu_i \sim \mathcal{N}(0, \sigma I_{(10)})$ Figure 1A depicts the change in estimated entropy as the means grow farther apart, in particular a function of ln(σ). We see that our proposed bounds are closer to the true entropy than the other estimators over the whole range of σ values, and in the extremes, our bounds approach the exact value of the true entropy. This is as expected, since as $\sigma \to 0$ all of the Gaussian mixture components become identical, and as $\sigma \to \infty$ all of the Gaussian components grow very far apart, approaching the case where each Gaussian is in its own "cluster". The ELK lower bound is a strictly worse estimate than $\hat{H}_{BD}$, in this experiment. As expected, the KDE estimator differs by exactly d/2 from the KL estimator.

In the second experiment, we evaluate the entropy estimators as the covariance matrices change from less to more similar. We again generate 100 10-dimensional Gaussians. Each Gaussian is distributed as $p_i = \mathcal{N}(\mu_i, \Sigma_i)$, where now $\mu_i \sim \mathcal{N}(0, I_{(10)})$ and $\Sigma_i \sim \mathcal{W}(\frac{1}{10+n} I_{(10)}, n)$, where $\mathcal{W}(V, n)$ is a Wishart distribution with scale-matrix V and n degrees of freedom.

Figure 1B compares the the estimators with the true entropy as a function of ln(n). When n is small, the Wishart distribution is broad and the covariance matrices differ significantly from one another, while as $n \rightarrow \infty$, all the covariance matrices become close to the identity I(10) . Thus, for small n, we essentially recover a "clustered" case, in which every component is in its own cluster and our lower and upper bounds give highly accurate estimates. For large n, we converge to the $\sigma = 1$ case of the first experiment.

In the third experiment, we again generate a mixture of 100 10-dimensional Gaussians. Now, however, the Gaussians are grouped into five "clusters", with each Gaussian component randomly assigned to one of the clusters. We use $g(i) \in \{1 \ldots 5\}$ to indicate the group of each Gaussian's component $i \in \{1 \ldots 100\}$, and each of the 100 Gaussians is distributed as $p_i = \mathcal{N}(\tilde{\mu}_{g(i)}, I_{(10)})$ The cluster centers μ˜ k for $k \in \{1 \ldots 5\}$ are drawn from N (0, σI(10)).

The results are depicted in Figure 1C as a function of ln(σ). In the first experiment, we saw that the joint entropy H(X,C) became an increasingly better estimator as the Gaussians grew increasingly far apart. Here, however, we see that there is a significant difference between H(X,C) and the true entropy, even as the groups become increasingly separated. Our proposed bounds, on the other hand, provide accurate estimates of the entropy across the entire parameter sweep. As expected, they become exact in the limit when all clusters are at the same location, as well as when all clusters are very far apart from each other.

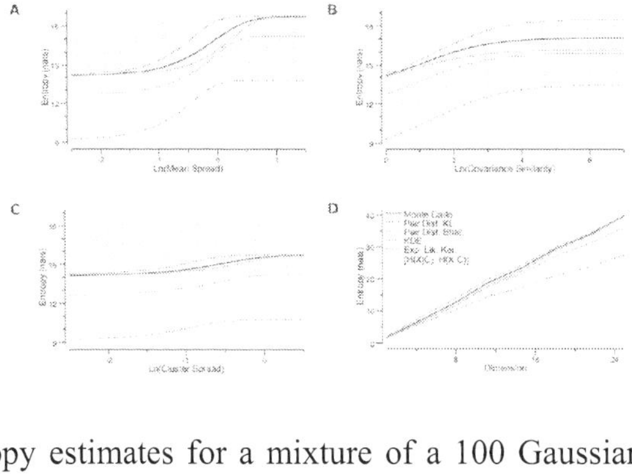

Figure 1. Entropy estimates for a mixture of a 100 Gaussians. In each plot, the vertical axis shows the entropy of the distribution, and the horizontal axis changes a feature of the components: (A) the distance between means is increased; (B) the component covariances become more similar (at the right side of the plot, all Gaussians have covariance matrices approximately equal to the

identity matrix); (C) the components are grouped into five "clusters", and the distance between the locations of the clusters is increased; (D) the dimension is increased.

Finally, we evaluate the entropy estimators while changing the dimension of the Gaussian components. We again generate 100 Gaussian components, each distributed as $p_i = \mathcal{N}(\mu_i, I_{(d)})$, with $\mu_i \sim \mathcal{N}(0, \sigma I_{(d)})$. We vary the dimensionality d from 1 to 60. The results are shown in Figure 1D. First, we see that when d = 1, the KDE estimator and the KL-divergence based estimator give a very similar prediction (differing only by 0.5), but as the dimension increases, the two estimates diverge at a rate of d/2. Similarly, $\hat{H}_{ELK}$ grows increasingly less accurate as the dimension increases. Our proposed lower and upper bounds provide good estimates of the mixture entropy across the whole sweep across dimensions.

As previously mentioned, our lower and upper bounds tend to perform best at the "extremes" and worse in the intermediate regimes. In particular, in Figure 1A,C,D, the distances between component means increase from left to right. On the left hand side of these figures, all of the component means are close and the component distributions overlap, as evidenced by the fact that the mixture entropy is ≈ H(X|C), i.e., I(X;C) ≈ 0. In this regime, when there is essentially a single "cluster", and our bounds become tight (see Section 3.4). On the right hand side of these figures, the components' means are all far apart from each other, and the mixture entropy ≈ H(X,C), i.e., I(X;C) ≈ H(C) (in Figure 1C, it is the five clusters that become far apart, and the mixture entropy ≈ H(X|C) + ln 5). In this regime where there are many well-separated clusters, our bounds again become tight. In between these two extremes, however, there is no clear clustering of the mixture components, and the entropy bounds are not as tight.

As noted in the previous paragraph, the extremes in three out of the four subfigures approach the perfectly clustered case. In this situation, we show in Appendix B that the BD-based estimator is a better bound on the true entropy than the Expected Likelihood Kernel estimator. We see confirmation of this in the experimental results, where $\hat{H}_{ELK}$ performs worse than the pairwise-distance based estimators.

Mixture of Uniforms

In the second set of experiments, we consider a mixture of uniform distributions. Unlike Gaussians, uniform distributions are bounded within a hyper-rectangle and do not have full support over the domain. In particular, a uniform distribution p = U(a, b) over d dimensions is defined as

$$p(x) \propto \begin{cases} 1, & \text{if } x_i \in [a_i, b_i] \, \forall i = 1 \ldots d, \\ 0, & \text{otherwise}, \end{cases} .$$

where x, a, and b are d-dimensional vectors, and the subscript xi refers to value of x on dimension i. Note that when pX is a mixture of uniforms, there can be significant regions where pX(x) > 0, but $p_i(x) = 0$ for some i

Here, we list the formulae for pairwise distance measure between uniform distributions. In the following, we use $V_i := \int_x 1\{p_i(x) > 0\}dx$ to indicate the "volume" of distribution p_i. Uniform components have a constant p(x) over their support, and so $p_i(x) = 1/V_i$ for all x where $p_i(x) > 0$.

Similarly, we use Vi∩j as the "volume of overlap" between pi and p_j, i.e., the volume of the intersection of the support of p_i and p_j, $V_{i\cap j} := \int_x 1\{p_i(x) > 0\} 1\{p_j(x) > 0\}dx$. The distance measures between uniforms are then

$$H(p_i) = \ln V_i,$$

$$\mathrm{KL}(p_i \| p_j) = \begin{cases} \ln(V_j / V_i), & \text{supp } p_i \subseteq \text{supp } p_j, \\ \infty, & \text{otherwise}, \end{cases}$$

$$\mathrm{BD}(p_i \| p_j) = 0.5 \ln V_i + 0.5 \ln V_j - \ln V_{i\cap j}, \tag{16}$$

$$\hat{H}_{\mathrm{ELK}} = -\sum_i c_i \ln \sum_j c_j \frac{V_{i\cap j}}{V_i V_j}. \tag{17}$$

Like the Gaussian case, we run four different computational experiments and compare the mixture entropy estimates to the true entropy, as determined by Monte Carlo sampling. In the first experiment, the mixture consists of 100 10-dimensional uniform components, with $p_i = \mathcal{U}(\mu_i - \mathbf{1}_{(10)}, \mu_i + \mathbf{1}_{(10)})$, and $\mu_i \sim \mathcal{N}(0, \sigma \boldsymbol{I}_{(10)})$, refers to a d-dimensional vector of 1s. Figure 2A depicts the change in entropy as a function of ln(σ). For very small σ, the distributions are almost entirely overlapping, while for large σ they tend very far apart. As expected, the entropy increases with σ. Here, we see that the prediction of $\hat{H}_{\mathrm{KL}}$ is identical to H(X,C), which arises because $\mathrm{KL}(p_i \| p_j)$ is infinite whenever the support of pi is not entirely contained in the support of pj . Uniform components with equal size and non-equal means must have some region of non-overlap, and so the KL is infinite between all pairs of components, thus KL is effectively Dmax (Equation (8)).

In contrast, we see that $\hat{H}_{\mathrm{BD}}$ estimates the true entropy quite well. This example demonstrates that getting an accurate estimate of mixture entropy may require selecting a distance function that works will with the component distributions. Finally, it turns out that, for uniform components of equal size, $\hat{H}_{\mathrm{ELK}} = \hat{H}_{\mathrm{BD}}$. This can be seen by combining Equations (6) and (16), and comparing to Equation (17) (note that Vi = Vj when the components have equal size).

In the second experiment, we adjust the variance in the size of the uniform components. We again use 100 10-dimensional components, $p_i = \mathcal{U}(\mu_i - \gamma_i \mathbf{1}_{(10)}, \mu_i + \gamma_i \mathbf{1}_{(10)})$, where $\mu_i \sim \mathcal{N}(0, \boldsymbol{I}_{(10)})$, and $\gamma_i \sim \Gamma(1+\sigma, 1+\sigma)$, where Γ(α, β) is the Gamma distribution with shape parameter α and rate parameter β. Figure 2B shows the change in entropy estimates as a function of ln(σ). When σ is small, the sizes have significant spread, while as σ grows the distributions become close to equally sized. We again see that $\hat{H}_{\mathrm{BD}}$ is a good estimator of entropy, outperforming all of the other estimators. Generally, not all supports will be non-overlapping, so $\hat{H}_{\mathrm{KL}}$ will not necessarily be equal to H(X,C), though we find the two to be numerically quite close. In this experiment, we find that the lower and upper bounds specified by $\hat{H}_{\mathrm{BD}}$ and $\hat{H}_{\mathrm{KL}}$ provide a tight estimate of the true entropy.

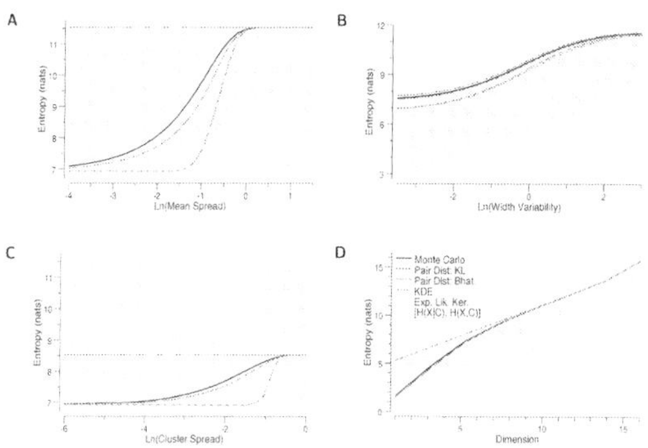

Figure 2. Entropy estimates for a mixture of a 100 uniform components. In each plot, the vertical axis shows the entropy of the distribution, and the horizontal axis changes a feature of the components: (A) the distance between means is increased; (B) the component sizes become more similar (at the right side of the

plot, all components have approximately the same size); (C) the components are grouped into five "clusters", and the distance between these clusters is increased; (D) the dimension is increased.

In the third experiment, we again consider a clustered mixture, and evaluate the entropy estimators as these clusters grow apart. Here, there are 100 components with $p_i = \mathcal{U}(\tilde{\mu}_{g(i)} - \mathbf{1}_{(10)}, \tilde{\mu}_{g(i)} + \mathbf{1}_{(10)})$, where g(i) ∈ {1 . . . 5} is the randomly assigned cluster identity of component i. The cluster centers μ˜ k for k ∈ {1 . . . 5} are generated according to N (0, σI(10)). Figure 2C shows the change in entropy as the clusters locations move apart. Note that, in this case, the upper bound $\hat{H}_{KL}$ significantly outperforms H(X,C), unlike in the first and second experiment, because in this experiment, components in the same cluster have perfect overlap. We again see that $\hat{H}_{BD}$ provides a relatively accurate lower bound for the true entropy.

In the final experiment, the dimension of the components is varied. There are again 100 components, with $p_i = \mathcal{U}(\mu_i - \mathbf{1}_{(d)}, \mu_i + \mathbf{1}_{(d)})$, and $\mu_i \sim \mathcal{N}(0, \sigma \mathbf{I}_{(d)})$. Figure 2D shows the change in entropy as the dimension increases from d = 1 to d = 16. Interestingly, in the low-dimensional case, H(X|C) is a very close estimate for the true entropy, while in the high-dimensional case, the entropy becomes very close to H(X,C). This is because in higher dimensions, there is more 'space' for the components to be far from each other. As in the first experiment, $\hat{H}_{KL}$ is equal to H(X,C). We again observe that H^BD provides a tight lower bound on the mixture entropy, regardless of dimension.

DISCUSSION

We have presented a new class of estimators for the entropy of a mixture distribution. We have shown that any estimator in this class has a bounded estimation bias, and that this class includes useful lower and upper bounds on the entropy of a mixture. Finally, we show that these bounds become exact when mixture components are grouped into well-separated clusters. Our derivation of the bounds make use of some existing results [5,31]. However, to our knowledge, these results have not been previously used to estimate mixture entropies. Furthermore, they have not been compared numerically or analytically to better-known bounds.

We evaluated these estimators using numerical simulations of mixtures of Gaussians as well as mixtures of bounded (hypercube) uniform distributions. Our results demonstrate that our estimators perform much better than existing well-known estimators. This estimator class can be especially useful for optimization problems that involve minimization of entropy or mutual information. If the distance function used in the pairwise estimator class is continuous and smooth in the parameters of the mixture components, then the entropy estimate is also continuous and smooth. This permits our estimators to be used within gradient-based optimization techniques, for example gradient descent, as often done in machine learning problems. In fact, we have used our upper bound to implement a non-parametric, nonlinear version of the "Information Bottleneck" [43]. Specifically, we minimized an upper bound on the mutual information between input and hidden layer in a neural networks [11]. We found that the optimal distributions were often clustered (Section 3.4). That work demonstrated practically the value of having an accurate, differentiable upper bound on mixture entropy that performs well in the clustered regime. Note that we have not proved that the bounds derived here are the best possible. Identifying better bounds, or proving that our results are optimal within some class of bounds, remains for future work.

REFERENCES

1. McLachlan, G.; Peel, D. Finite Mixture Models; John Wiley & Sons: Hoboken, NJ, USA, 2004.
2. Cover, T.M.; Thomas, J.A. Elements of Information Theory; John Wiley & Sons: Hoboken, NJ, USA, 2012.
3. Goldberger, J.; Gordon, S.; Greenspan, H. An Efficient Image Similarity Measure Based on Approximations of KL-Divergence Between Two Gaussian Mixtures. In Proceedings of the 9th International Conference on Computer Vision, Nice, France, 13–16 October 2003; Volume 3, pp. 487–493.
4. Viola, P.; Schraudolph, N.N.; Sejnowski, T.J. Empirical Entropy Manipulation for Real-World Problems. In Advances in Neural Information Processing Systems; The MIT Press: Cambridge, MA, USA, 1996; pp. 851–857.
5. Hershey, J.R.; Olsen, P.A. Approximating the Kullback Leibler divergence between Gaussian mixture models. In Proceedings of the 2007 IEEE International Conference on Acoustics, Speech and Signal Processing, Honolulu, HI, USA, 15–20 April 2007; Volume 4, pp. 317–320.
6. Chen, J.Y.; Hershey, J.R.; Olsen, P.A.; Yashchin, E. Accelerated monte carlo for kullback-leibler divergence between gaussian mixture models. In Proceedings of the IEEE International Conference on Acoustics, Speech and Signal Processing, Las Vegas, NV, USA, 31 March–4 April 2008; pp. 4553–4556.
7. Capristan, F.M.; Alonso, J.J. Range Safety Assessment Tool (RSAT): An analysis environment for safety assessment of launch and reentry vehicles. In Proceedings of the 52nd Aerospace Sciences Meeting, National Harbor, MD, USA, 13–17 January 2014; p. 304.
8. Schraudolph, N.N. Optimization of Entropy with Neural Networks. Ph.D. Thesis, University of California, San Diego, CA, USA, 1995.
9. Schraudolph, N.N. Gradient-based manipulation of nonparametric entropy estimates. IEEE Trans. Neural Netw. 2004, 15, 828–837.
10. Shwartz, S.; Zibulevsky, M.; Schechner, Y.Y. Fast kernel entropy estimation and optimization. Signal Process. 2005, 85, 1045–1058.
11. Kolchinsky, A.; Tracey, B.D.; Wolpert, D.H. Nonlinear Information Bottleneck. arXiv 2017, arXiv:1705.02436.

12. Contreras-Reyes, J.E.; Cortés, D.D. Bounds on Rényi and Shannon Entropies for Finite Mixtures of Multivariate Skew-Normal Distributions: Application to Swordfish (Xiphias gladius Linnaeus). Entropy
13. 2016, 18, 382.
14. Carreira-Perpinan, M.A. Mode-finding for mixtures of Gaussian distributions. IEEE Trans. Pattern Anal. Mach. Intell. 2000, 22, 1318–1323.
15. Zobay, O. Variational Bayesian inference with Gaussian-mixture approximations. Electron. J. Stat. 2014, 8, 355–389.
16. Beirlant, J.; Dudewicz, E.J.; Györfi, L.; van der Meulen, E.C. Nonparametric entropy estimation: An overview. Int. J. Math. Stat. Sci. 1997, 6, 17–39.
17. Joe, H. Estimation of entropy and other functionals of a multivariate density. Ann. Inst. Stat. Math. 1989, 41, 683–697.
18. Nair, C.; Prabhakar, B.; Shah, D. On entropy for mixtures of discrete and continuous variables. arXiv 2006, arXiv:cs/0607075.
19. Huber, M.F.; Bailey, T.; Durrant-Whyte, H.; Hanebeck, U.D. On entropy approximation for Gaussian mixture random vectors. In Proceedings of the IEEE International Conference on Multisensor Fusion and Integration for Intelligent Systems, Seoul, Korea, 20–22 August 2008; pp. 181–188.
20. Hall, P.; Morton, S.C. On the estimation of entropy. Ann. Inst. Stat. Math. 1993, 45, 69–88.
21. Principe, J.C.; Xu, D.; Fisher, J. Information theoretic learning. Unsuperv. Adapt. Filter 2000, 1, 265–319.
22. Xu, J.W.; Paiva, A.R.; Park, I.; Principe, J.C. A reproducing kernel Hilbert space framework for information-theoretic learning. IEEE Trans. Signal Process. 2008, 56, 5891–5902.
23. Jebara, T.; Kondor, R. Bhattacharyya and expected likelihood kernels. In Learning Theory and Kernel Machines; Springer: Berlin, Germany, 2003; pp. 57–71.
24. Jebara, T.; Kondor, R.; Howard, A. Probability product kernels. J. Mach. Learn. Res. 2004, 5, 819–844.
25. Banerjee, A.; Merugu, S.; Dhillon, I.S.; Ghosh, J. Clustering with Bregman divergences. J. Mach. Learn. Res. 2005, 6, 1705–1749.

26. Cichocki, A.; Zdunek, R.; Phan, A.H.; Amari, S.I. Similarity Measures and Generalized Divergences.
27. In Nonnegative Matrix and Tensor Factorizations: Applications to Exploratory Multi-Way Data Analysis and Blind Source Separation; JohnWiley & Sons: Hoboken, NJ, USA, 2009; pp. 81–129.
28. Cichocki, A.; Amari, S.I. Families of alpha-beta-and gamma-divergences: Flexible and robust measures of similarities. Entropy 2010, 12, 1532–1568.
29. Gil, M.; Alajaji, F.; Linder, T. Rényi divergence measures for commonly used univariate continuous distributions. Inf. Sci. 2013, 249, 124–131.
30. Crooks, G.E. On Measures of Entropy and Information. Available online: http://threeplusone.com/on_ information.pdf (accessed on 12 July 2017).
31. Nielsen, F. Chernoff information of exponential families. arXiv 2011, arXiv:1102.2684.
32. Van Erven, T.; Harremos, P. Rényi divergence and Kullback-Leibler divergence. IEEE Trans. Inf. Theory 2014, 60, 3797–3820.
33. Haussler, D.; Opper, M. Mutual information, metric entropy and cumulative relative entropy risk. Ann. Stat. 1997, 25, 2451–2492.
34. Fukunaga, K. Introduction to Statistical Pattern Recognition, 2nd ed.; Academic Press: Boston, MA, USA, 1990.
35. Paisley, J. Two Useful Bounds for Variational Inference; Princeton University: Princeton, NJ, USA, 2010.
36. Sason, I.; Verdú, S. f-Divergence Inequalities. IEEE Trans. Inf. Theory 2016, 62, 5973–6006.
37. Hero, A.O.; Ma, B.; Michel, O.; Gorman, J. Alpha-Divergence for Classification, Indexing and Retrieval. Available online: https://pdfs.semanticscholar.org/6d51/fbf90c59c2bb8cbf0cb609a224f53d1b68fb.pdf (accessed on 14 July 2017).
38. Dowson, D.; Landau, B. The Fréchet distance between multivariate normal distributions. J. Multivar. Anal. 1982, 12, 450–455.
39. Olkin, I.; Pukelsheim, F. The distance between two random vectors with given dispersion matrices. Linear Algebra Appl. 1982, 48, 257–263.
40. Pardo, L. Statistical Inference Based on Divergence Measures; CRC Press: Boca Raton, FL, USA, 2005.

41. Hobza, T.; Morales, D.; Pardo, L. Rényi statistics for testing equality of autocorrelation coefficients. Stat. Methodol. 2009, 6, 424–436.
42. Nielsen, F. Generalized Bhattacharyya and Chernoff upper bounds on Bayes error using quasi-arithmetic means. Pattern Recognit. Lett. 2014, 42, 25–34.
43. GitHub. Available online: https://www.github.com/btracey/mixent (accessed on 14 July 2017).
44. Gonum Numeric Library. Available online: https://www.gonum.org (accessed on 14 July 2017).
45. Tishby, N.; Pereira, F.; Bialek, W. The information bottleneck method. In Proceedings of the 37th Annual Allerton Conference on Communication, Control, and Computing, Monticello, IL, USA, 22–24 September 1999.

CHAPTER

5

Dependent Gaussian Mixture Models for Source Separation

Alicia Quiros´ [1] and Simon P Wilson [2]

[1] Department of Signal Theory and Communications, Universidad Rey Juan Carlos, C° Molino s/n, Fuenlabrada, Madrid, Spain

[2] Department of Statistics, Trinity College Dublin, Dublin 2, Ireland.

Citation (APA): Copyright:

ABSTRACT

Source separation is a common task in signal processing and is often analogous to factor analysis. In this study, we look at a factor analysis model for source separation of multi-spectral image data where prior information about the sources and their dependencies is quantified as a multivariate Gaussian mixture model with an unknown number of factors. Variational

Citation: (APA): Quirós, A., & Wilson, S. P. (2012). Dependent Gaussian mixture models for source separation. EURASIP Journal on Advances in Signal Processing, 2012(1), 239. (11 pages) DOI: https://doi.org/10.1186/1687-6180-2012-239

Bayes techniques for model parameter estimation are used. The development of this methodology is motivated by the need to bring an efficient solution to the separation of components in the microwave radiation maps that are being obtained by the satellite mission Planck which has the objective of uncovering cosmic microwave background radiation. The proposed algorithm successfully incorporates a rich variety of prior information available to us in this problem in contrast to many previous solutions that assume completely blind separation of the sources. Results on realistic simulations of Planck maps and on Wilkinson microwave anisotropy probe fifth year images are shown. The technique suggested is easily applicable to other source separation applications by modifying some of the priors.

INTRODUCTION

The discovery of the cosmic microwave background (CMB) is a strong evidence for the Big Bang theory of the formation and development of the universe. According to the theory, the early universe was smaller and hotter but cooled as it expanded. Once the temperature cooled to about 3000 K, photons were free to propagate without being scattered off ionized matter; the CMB is an image of this event and is visible across the entire sky. Three satellites have been launched to measure the CMB: the cosmic background explorer, Wilkinson microwave anisotropy probe (WMAP) and most recently the Planck surveyor. Planck is the highest resolution data to date, of the order of 10^7 pixels across the sky measured at nine channels.

Unfortunately, the signals measured by these satellites as shown in Figure 1 contain radiation not only from CMB, but also contributions from a number of other sources, namely foreground radiations and extragalactic sources in addition to antenna receiver noise. Foreground sources from our galaxy include synchrotron, dust, and free–free emission. Therefore, the separation of the CMB signal from other sources is an important stage in the production of CMB maps [1].

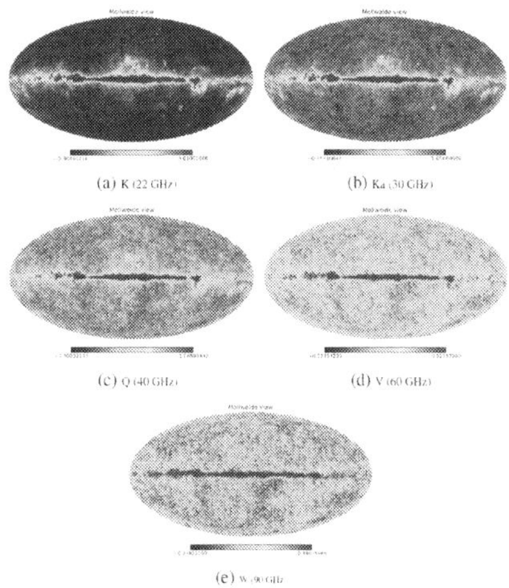

Figure 1 Observed WMAP 7 year-data. The data were taken from the NASA WMAP website [2]

To date, there have been several attempts to achieve it in a Bayesian framework using both (a) Gaussian mixture model (GMM) prior [3], and (b) Markov Random Field (MRF) prior [4, 5]. Full sky maps at low resolution through MCMC, using masks to reduce the effect of the signal in the galactic plane, were described in [6]. Some of these are fully Bayesian source separation methods which are developed to separate the underlying CMB from the mixed observed signals of extraterrestrial microwaves made at several frequencies.

A common assumption among works in the literature is the independence of the cosmological sources. Although it is well known that CMB is independent from the rest of the sources, the galactic sources demonstrate significant statistical dependence among themselves, as stated in [1]. Recently, a small number of researchers have started addressing this problem [7, 8]. Various dependent component analysis approaches are compared in [9], demonstrating their superior performance with respect to classical ICA.

In this study, we present a dependent components model for source separation of multi-spectral image data, where prior information about the sources and between-source dependencies is quantified as a multivariate GMM, using variational Bayes techniques for model parameter estimation. This article can thus be considered as an extension of [3], modeling dependencies between-sources through generalizing the prior to multivariate GMM.

The rest of the article is structured as follows. The next section gives the model for the mixing problem and describes the hierarchical Bayesian model that we use, including the prior we assume for the sources. Section "Implementing the source separation" describes the variational Bayes approach we use for the implementation of the separation. Section "Examples" provides results on both synthetic Planck and real WMAP images. Finally, we provide a discussion of the results in the last section.

MODEL

The model description is defined in terms of the microwave source separation problem, where there are n_f maps of the sky at frequencies $(\nu_1, \ldots, \nu_{n_f})$, each map consisting of J pixels. The data are denoted $\mathbf{d}_j \in \mathbb{R}^{n_f}, j =$ j=1,…,J. The source model consists of n_s sources and is represented by the vectors $\mathbf{s}_j \in \mathbb{R}^{n_s}$, with each component representing the amplitude of a physical source of microwaves. We assume that the $\mathbf{d}_j$ can be represented as a linear combination of the $\mathbf{s}_j$:

$$\mathbf{d}_j = \mathbf{A}\mathbf{s}_j + \mathbf{e}_j, \tag{1}$$

where $\mathbf{A}$ is an $n_f \times n_s$ "mixing" matrix and $\mathbf{e}_j$ is a vector of n_f independent Gaussian error terms with precisions (inverse variances) $\tau = (\tau_1, \ldots, \tau_{n_f})$ For convenience, define

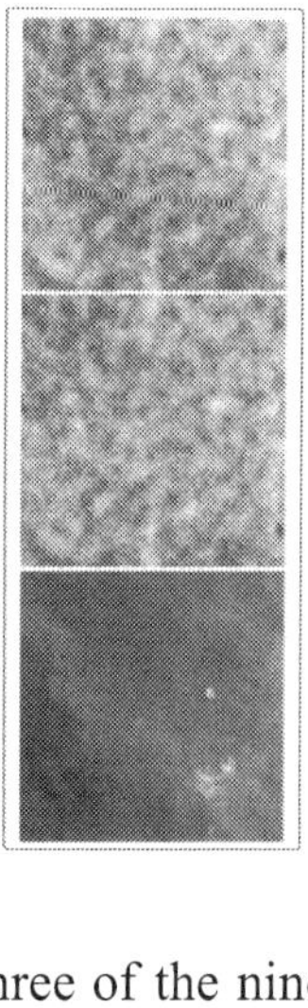

Figure 2 Observations on the three of the nine channels (lowest, middle, and highest frequencies are shown) of the data generated from the source separation model with realistic simulations of CMB, synchrotron, and galactic dust.

define

$$\mathbf{D} = \{d_{ij}|i = 1,\ldots,n_f, j = 1,\ldots,J\};$$
$$\mathbf{S} = \{s_{kj}|k = 1,\ldots,n_s, j = 1,\ldots,J\}$$

to represent all data and sources.

We assume dependence between the sources, defined by a prior distribution $p(\mathbf{S}|\psi)$ with parameters ψ. The goal is to estimate the **S** and the parameters ψ associated with the model for **S**, given observation of **D**. The noise variances $\boldsymbol{\tau}$ and the mixing matrix **A** are assumed known. GMM are used to represent the non-Gaussian sources, in which case it is an example of a model known as a mixture of factor analyzers [10]. As in [10], we adopt a Bayesian approach to the data fitting, implemented by a variational Bayes approach.

Bayesian inference will be based on the posterior distribution, which following the above description can be factorized as

$$p(\mathbf{S},\psi|\mathbf{A},\mathbf{D},\tau) \propto p(\mathbf{D}|\mathbf{S},\mathbf{A},\tau)p(\mathbf{S}|\psi)p(\psi). \tag{2}$$

Each element of this distribution is defined next in turn.

Noise structure

Gaussian error, $\mathbf{e}_j$, is assumed independent within and between pixels j and frequency, which gives

$$p(\mathbf{D}|\mathbf{S},\mathbf{A},\tau) = \prod_{j=1}^{J}\prod_{i=1}^{n_f}\sqrt{\frac{\tau_i}{2\pi}}\exp\left(-\frac{\tau_i}{2}(d_{ij}-\mathbf{A}_{i\cdot}\mathbf{s}_j)^2\right) \tag{3}$$

where $\mathbf{A}_{i\cdot}$ is the i th row of **A**.

Mixing matrix structure

In this application, **A** is parameterized and denoted $\mathbf{A}(\theta)$. Each column of $\mathbf{A}(\theta)$ is the contribution to the observation of a source at different frequencies, which is written as a function of the frequencies and θ. These parameterizations are approximations that come from the current state of knowledge about how the sources are generated. Here, we merely state the parameterization that we are going to use, and refer to [11] for a more detailed exposition on the background to them. Some restrictions are usually placed on $\mathbf{A}(\theta)$in order to force a unique solution; this is achieved here by setting the first row of $\mathbf{A}(\theta)$ to be ones.It is assumed that the CMB is the first source and therefore it corresponds to the first column of $\mathbf{A}(\theta)$. It is modeled as a black body at a temperature, and its contribution is a known constant at each frequency. The parametrization of the mixing matrix is given as

$$\mathbf{A}_{i1}(\theta) = \frac{g(\nu_i)}{g(\nu_1)}, \quad (4)$$

where

$$g(\nu_i) = \left(\frac{h\nu_i}{k_B T_0}\right)^2 \frac{\exp(h\nu_i/k_B T_0)}{(\exp(h\nu_i/k_B T_0) - 1)^2},$$

$T_0 = 2.725$K is the average CMB temperature, h is the Planck constant, and k_B is Boltzmann's constant. The ratio $g(\nu_i)/g(\nu_1)$ is designed to ensure that $\mathbf{A}_{11}(\theta) = 1$ as we constraint the first row of $\mathbf{A}(\theta)$ to be ones.

$$\mathbf{A}_{i2}(\theta) = \left(\frac{\nu_i}{\nu_1}\right)^{\kappa_s},$$

$$\mathbf{A}_{i3}(\theta) = \frac{\exp(h\nu_1/k_B T_1) - 1}{\exp(h\nu_i/k_B T_1) - 1}\left(\frac{\nu_i}{\nu_1}\right)^{1+\kappa_d}, \text{ and}$$

$$\mathbf{A}_{i4}(\theta) = \left(\frac{\nu_i}{\nu_1}\right)^{\kappa_f},$$

where $T_1 = 18.1$K is the assumed thermodynamical temperature of the dust grains, and column 2 corresponds to synchrotron, column 3 to galactic dust, and column 4 is free–free emission. There are three unknown model parameters for A, for synchrotron $\kappa s \in \{\kappa_s : -3.0 \leq \kappa_s \leq -2.3\}$, the spectral indices for dust $\kappa_d \in \{\kappa_d : 1 \leq \kappa_d \leq 2\}$, and for free–free emission $\kappa f \in \{\kappa f : -2.3 \leq \kappa f \leq -2.0\}$.

The sources

The distribution of $\mathbf{s}_j$ is modeled as a GMM with m factors. The model proposed allows for between-source dependence; the vector of sources at a pixel is a mixture of multivariate Gaussians

$$p(\mathbf{S}|\psi) = \prod_{j=1}^{J}\sum_{a=1}^{m} w_a p(\mathbf{s}_j|\mu_a, Q_a) \quad (5)$$

where

$$p(\mathbf{s}_j|\mu_a, Q_a) = \sqrt{\frac{|Q_a|}{(2\pi)^{n_s}}}\exp\left(-\frac{1}{2}(\mathbf{s}_j - \mu_a)^T Q_a(\mathbf{s}_j - \mu_a)\right)$$

for mixture component weights w_a, mean vectors μ_a, and precision matrices Q_a, so that ψ is all the w_a, μ_a and Q_a, with $a = 1,\ldots,m$. Note that, in the standard inflationary cosmological model the CMB is a single multivariate Gaussian ($m = 1$) while the Galactic foregrounds might require $m > 1$ to be correctly modeled. In order to fulfill this, we set the CMB for components

2,…,m of the multivariate mixture to be exactly zero in the implementation.

Priors

The remaining term in Equation (2) is $p(\psi)$. We use the conjugate prior distributions [12] that facilitate the computation of the posterior and yet flexible enough to incorporate good prior information: Gaussians for the component means, Dirichlet for the component weights, and Wishart for precision matrices. In the microwave source application, background knowledge about the magnitude of the sources can be incorporated through specifying values of the parameters of these prior distributions. This prior specification follows [13], who discuss how to specify these values in more detail.

IMPLEMENTING THE SOURCE SEPARATION

The posterior developed in the previous section does not lend itself to an analytical solution. MCMC techniques are one approach that let us evaluate complicated integrals by sampling rather than by analytical or numerical methods. The main criticism of Bayesian source separation with sampling methods, MCMC in particular, is their computational load and slow convergence. Regarding the speed, they cannot compete with methods such as FastICA [14, 15].

There are several approaches to speed up the algorithm, such as the strategies suggested in [16]. In the image source separation problem framework, the Langevin sampling scheme has been implemented [4], as a way to obtain a faster MC algorithm.

In this study, the source separation model presented in Section "Model" is implemented by a variational Bayesian approach [10, 17, 18], that allows for more efficient inference when dealing with large data when compared with MCMC techniques. In essence, given the data **D** and a model with parameters θ and latent variables **Z**, the variational Bayes method is based on approximating the posterior distribution $p(\mathbf{Z},\theta|\mathbf{D})$ with a factorial approximation $q(\mathbf{Z},\theta|\phi)=q(\mathbf{Z}|\phi_Z)q(\theta|\phi_\theta)$, where ϕ are the variational parameters. The approximation is fitted by minimizing the Kullback–Leibler divergence between q and p, or equivalently maximizing a lower bound on marginal log-likelihood of the data.

Attias [19] has recently developed a fully Bayesian approach to GMM with a variational approximation to the posterior that, when choosing conjugate priors, leads to the following components: Wishart densities for the precisions, Q_a; Normal densities for the means, μ_a; and a Dirichlet for the mixing coefficients, **p**; and a discrete distribution for the indicator posteriors, z_j, which indicates the component that explains information in pixel j. We further derived the variational approximation to the marginal posterior of sources, $\mathbf{s}_j$, which turns out to be a multivariate Gaussian distribution. In brief

$$\begin{aligned} q(\mathbf{s}_j) &\sim MVN(A_j^\star, B_j^\star) \\ q(\mathbf{p}) &\sim D(\lambda) \\ q(\mu_a|Q_a) &\sim N(\xi_a, \beta_a Q_a) \\ q(Q_a) &\sim W(\eta_a, V_a) \end{aligned}$$

and

$$\begin{aligned} q(z_j = a) \propto \exp &\left(\Psi(\lambda_a) - \Psi\left(\sum_{a'} \lambda_{a'}\right)\right) |V_a|^{\frac{1}{2}} 2^{\frac{n_s}{2}} \\ &\exp\left(\frac{1}{2}\sum_{i=1}^{n_s} \Psi\left(\frac{\eta_a + 1 - i}{2}\right)\right) \exp\left(-\frac{n_s}{2\beta_a}\right) \\ &\exp\left(-\frac{\eta_a}{2}\left((A_j^\star - \xi_a)^T V_a (A_j^\star - \xi_a) + tr(V_a (B_j^\star)^{-1})\right)\right) \end{aligned}$$

where *MVN* stands for multivariate normal distribution and Ψ denotes the digamma function. Note that $q(z_j = a)$ is the probability that component a is responsible for information in pixel j in sources, $\mathbf{s}_j$.

The quantities of interest, i.e., the hyper-parameters to be computed, $A_j^\star$, $B_j^\star$, λ, ξ_a, β_a, η_a, and V_a for $j = 1,\ldots,J$ and $a = 1,\ldots,m$, have the following values:

$$\begin{aligned} (B_j^\star)_{kl} &= \sum_{i=1}^{n_t} \tau_i A_{ik} A_{il} + \sum_{a=1}^{m} q(z_j = a)\eta_a (V_a)_{kl} \\ (A_j^\star)_k &= (B_j^\star)^{-1} \mathbf{v}(k), \text{ with} \\ \mathbf{v}(k) &= \sum_{i=1}^{n_t} \tau_i d_{ij} A_{ik} + \sum_{a=1}^{m} q(z_j = a)\eta_a \sum_{l=1}^{n_s} (V_a)_{kl} (\xi_a)_l \\ \lambda_a &= \sum_j q(z_j = a) + \lambda_a^{prior} \\ \xi_a &= \frac{\sum_j [q(z_j = a) A_j^\star] + \beta_a^{prior} \xi_a^{prior}}{\sum_j q(z_j = a) + \beta_a^{prior}} \\ \beta_a &= \sum_j q(z_j = a) + \beta_a^{prior} \\ \eta_a &= \sum_j q(z_j = a) + \eta_a^{prior} \end{aligned}$$

$$\begin{aligned} V_a = \sum_j \Bigg\{ q(z_j = a) \Bigg[&\left(1 - \frac{2q(z_j = a)}{\sum_j q(z_j = a)}\right) (B_j^\star)^{-1} + \Phi \\ &+ (A_j^\star - \bar{\mu}_a)(A_j^\star - \bar{\mu}_a)^T \Bigg] \Bigg\} + V_a^{prior} \\ &+ \frac{\beta_a^{prior} [\sum_j q(z_j = a)] [\Phi + (\bar{\mu}_a - \xi_a^{prior})(\bar{\mu}_a - \xi_a^{prior})^T}{\beta_a} \end{aligned}$$

with

$$\bar{\mu}_a = \frac{\sum_j [q(z_j = a) A_j^\star]}{\sum_j q(z_j = a)}$$

$$(\Phi)_{kl} = \frac{\sum_j \{[q(z_j = a)]^2 (B_j^\star)^{-1}_{kl}\}}{[\sum_j q(z_j = a)]^2}.$$

Computations were carried out using Matlab.

EXAMPLES

Analysis of simulated data

The synthetic Planck data were generated using the Planck Sky Model (PSM) package. The PSM is a set of IDL codes created by the Planck WG2 team in order to provide realistic simulations of the sky at the Planck frequencies [20]. Figure 2 shows data obtained from realistic simulations of CMB, synchrotron, and galactic dust on a 512×512 patch. The original sources are shown in Figure 3. The data were generated at the nine frequencies that are observed by Planck from 30 to 857 GHz. The mixing matrix used was as defined in Section "Mixing matrix structure" with $_{s}=-2.9$ and $_{d}=2.0$. Noise precisions were those published by the Planck research team [21]. After exploring several values for m, the number of components in the GMM source model was fixed to be $m=1$, as it provided the best fit, taking into account the compromise between fit and number of parameters in the model.

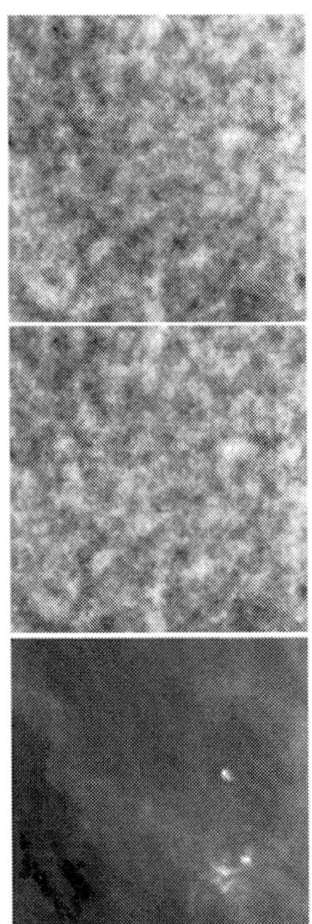

Figure 2: Observations on the three of the nine channels (lowest, middle, and highest frequencies are shown) of the data generated from the source separation model with realistic simulations of CMB, synchrotron, and galactic dust.

Figure 3: The simulated sources used to generate the simulated data in Figure 2 .

Figure 4 shows an estimate of CMB, along with a scatter plot of this estimate against the true value, as shown in Figure 3. Such an estimate is the average of the samples obtained for the first column of $A^{\star}$, which corresponds to CMB. We see from the scatter plot and from comparison with Figure 3 that the reconstruction of CMB is very accurate here. The same is true for the other two sources, as shown in Figure 5.

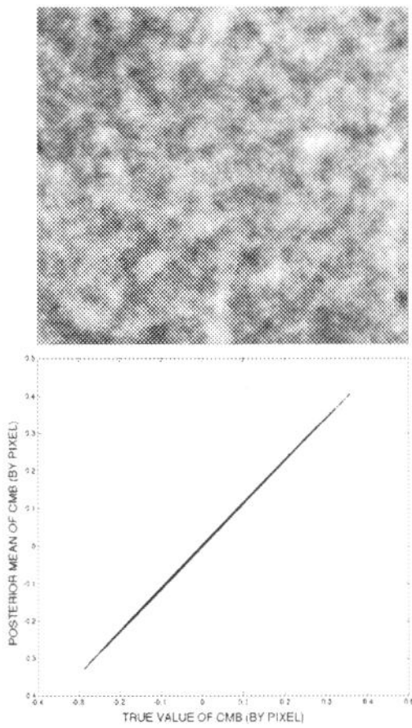

Figure 4: The posterior mean of the reconstruction of the CMB with a scatter plot of true versus posterior mean.

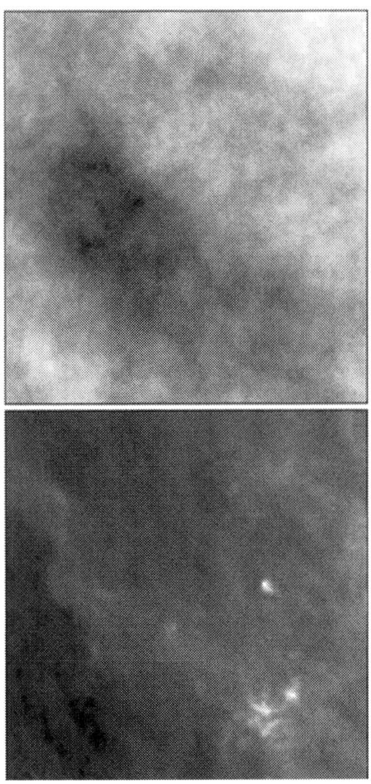

Figure 5: The posterior mean of the reconstruction of synchrotron and galactic dust.

Table 1 shows the mean of the parameters of the model. Regarding the between-sources dependence structure, posterior estimates of $V_{1k}^{-1}, k = 2, 3$ are approximately 0, suggesting independence between CMB and the other sources, as expected. On the other hand, posterior estimate of $V_{23}^{-1} \neq 0$, indicating dependence between synchrotron and galactic dust.

Table 1 Mean estimate of parameters for simulated data

$\hat{\mu} = \begin{pmatrix} 0.020 \\ 0.014 \\ 0.008 \end{pmatrix}$	$\hat{Q} = \begin{pmatrix} 4.665 & 0 & 0 \\ 0 & 0.006 & 0.003 \\ 0 & 0.003 & 0.010 \end{pmatrix} \times 10^8$

Analysis of a WMAP year 5 patch

The WMAP [22] was launched in 2001 and data collection activities finished in 2010. It observes 5 frequencies from 22 to 90 GHz. Figure 6 shows a patch of 5-year WMAP data.

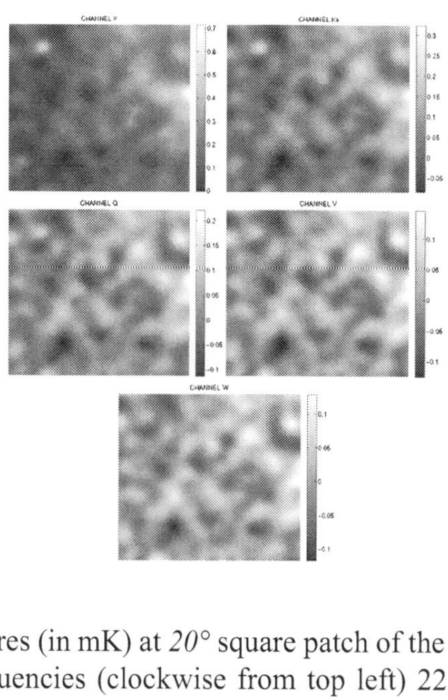

Figure 6: Temperatures (in mK) at *20°* square patch of the sky from WMAP[2] at 5 microwave frequencies (clockwise from top left) 22, 30, 40, 90, and 60 GHz).

The algorithm was implemented with four sources (CMB, synchrotron, dust, and free–free emission). The noise precisions were assumed to be the published values for WMAP detectors. The spectral density for free–free emission was fixed at −2.14 (following [11]) and the synchrotron and dust spectral indices were as in the first example. The number of components in the GMM source model were fixed to be $m=2$, following the same reasoning as in the simulation study. Informative priors were placed on the GMM parameters, based on discussions on the expected marginal properties of the sources. Table 2 shows the mean of the mixture parameters of the model. Figure 7 shows the estimated CMB. The result obtained is in agreement with previous work [3], as can be appreciated in Figure 8 that shows an histogram of the differences between the estimated CMB using the approach presented here and the estimated CMB obtained in [3].

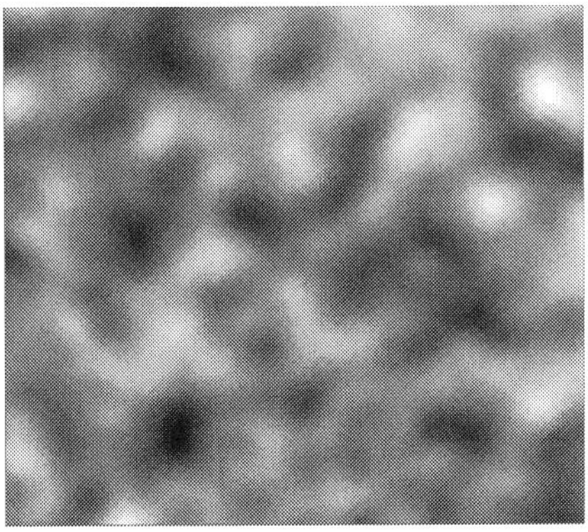

Figure 7: Estimated CMB.

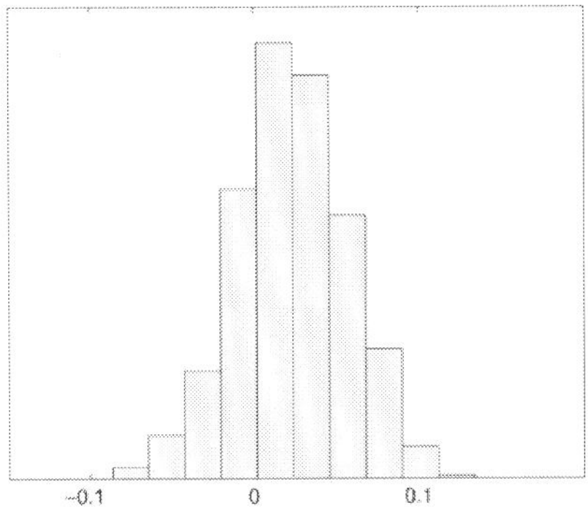

Figure 8: Histogram of the (pixel-by-pixel) differences between the estimated CMB using the approach presented in this article and the estimated CMB obtained in [3]. The result obtained is in agreement with previous work

Table 2 Parameter estimated mean for WMAP data

$a = 1$	$\hat{p}_1 = 0.12$	$\hat{\mu}_1 = \begin{pmatrix} -0.012 \\ 0.051 \\ -0.001 \\ 0.017 \end{pmatrix}$	$\hat{\Omega}_1 = \begin{pmatrix} 5.29 & 0 & 0 & 0 \\ 0 & 1.47 & -0.04 & -1.60 \\ 0 & -0.04 & 0.41 & 0.32 \\ 0 & -1.60 & 0.32 & 3.79 \end{pmatrix} \times 10^5$
$a = 2$	$\hat{p}_2 = 0.88$	$\hat{\mu}_2 = \begin{pmatrix} -0.012 \\ 0.029 \\ 0.003 \\ 0.017 \end{pmatrix}$	$\hat{\Omega}_2 = \begin{pmatrix} 1.45 & 0 & 0 & 0 \\ 0 & 0.54 & 0.08 & -0.20 \\ 0 & 0.08 & 0.28 & 0.26 \\ 0 & -0.20 & 0.26 & 1.09 \end{pmatrix} \times 10^4$

Finally, in order to show the fit of the data to the model, Figure 9 is a scatter plot of the observed value of the d_{jk} with the standardized residuals, with one figure for each frequency $k=1,\ldots,5$.

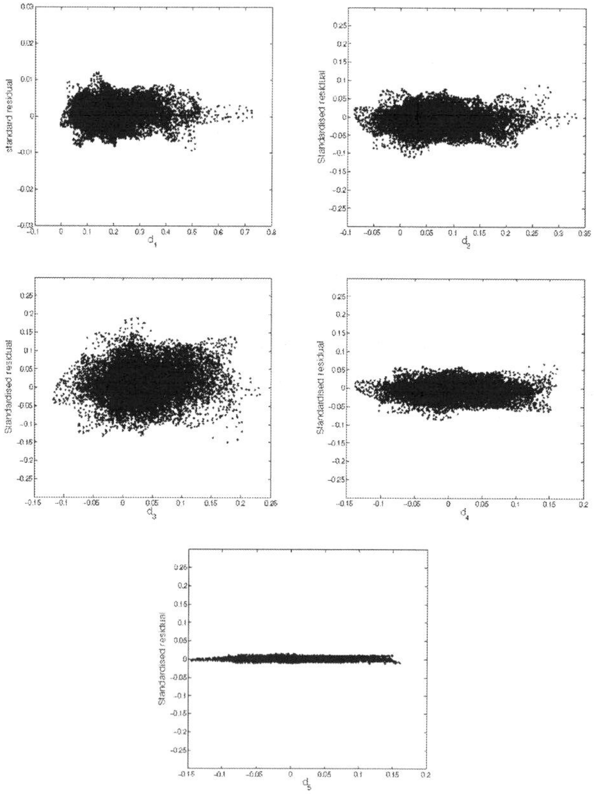

Figure 9: Assessment of model fit. Scatter plot of the posterior predicted values of d_j against the standardized residual over all pixels.

CONCLUSION

A fully Bayesian factor analysis algorithm has been presented and applied to a multi-channel image source separation problem, where dependencies between sources are modeled as a multivariate GMM. The algorithm performs very well on simulated Planck data and has been applied to data from WMAP.

In this study, we extend previous approaches [3] by allowing the source priors to be a mixture of multivariate Gaussian distributions for each pixel.

The development of this methodology is motivated by the need to bring an efficient solution to the separation of components in the microwave radiation maps to be obtained by the satellite mission Planck which has the objective of uncovering CMB radiation. The proposed algorithm successfully incorporates a rich variety of prior information available to us in this problem in contrast to most of the previous work that assumes completely blind separation of the sources. Further, the variational approach presented here overcomes the convergence problems of the MCMC stated in [23], when dealing with large datasets such as that will be provided by the satellite mission Planck.

In the analysis of simulated data, the number of components in the GMM source model turned out to be $m=1$. This means that sources are multivariate Gaussian *a priori*. On the other hand, for real data, the number of components is $m=2$. In blind source separation problem, identifiability relies on the independence of the sources. In this study, in spite of modeling the sources as Gaussians when $m=1$, identifiability is obtained because of the prior information which is incorporated to the model, given structure to the mixing matrix.

Another type of dependence is that a source is spatially correlated. Spatial dependence is most conveniently modeled by a Gaussian MRF and some preliminary work on this idea can be found in [5]. Combining with cross source correlations, one might ultimately consider a mixture of multivariate Gaussian MRF as a prior for the sources. Implementing the analysis with such a prior would be a significant challenge computationally; we hypothesize that it will be difficult to derive a well-behaved MCMC approach. Other functional approximations, such as that of [24], offer feasible alternative to computing the posterior distribution in this case.

Finally, although the technique was developed for the astrophysical source separation problem in mind, it is general and it is applicable to other source separation problems as well.

REFERENCES

1. Kuruoglu EE: Bayesian source separation for cosmology. *IEEE Signal Process. Mag* 2010, 27: 43-54.
2. [http://lambda.gsfc.nasa.gov/product/map/current/m_products.cfm]
3. Wilson SP, Kuruoglu EE, Salerno E: Fully Bayesian source separation of astrophysical images modelled by a mixture of Gaussians. *IEEE J. Sel. Topics Signal Process* 2008, 2(5):685-696.
4. Kayabol K, Kuruoglu EE, Sanz JL, Sankur B, Salerno E, Herranz D: Adaptive Langevin sampler for separation of t-distribution modelled astrophysical maps. *IEEE Trans. Image Process* 2010, 19(9):2357-2368.
5. Kayabol K, Kuruoglu EE, Sankur B: Bayesian separation of images modeled with MRFs using MCMC. *IEEE Trans. Image Process* 2009, 18(5):982-994.
6. Dickinson C, Eriksen HK, Banday AJ, Jewell JB, Gorski KM, Huey G, Lawrence CR, O'Dwyer IJ, Wandelt BD: Bayesian component and separation cosmic microwave background estimation for the five-year WMAP temperature data. *Astrophys. J* 2010, 705: 1607-1623.
7. Bedini L, Herranz D, Salerno E, Baccigalupi C, Kuruoglu EE, Tonazzini A: Separation of correlated astrophysical sources using multiple-lag data covariance matrices. *EURASIP J. Appl. Signal Process* 2005, 2005(15):2400-2412. 10.1155/ASP.2005.2400
8. Bonaldi A, Bedini L, Salerno E, Baccigalupi C, De Zott G: Estimating the spectral indices of correlated astrophysical foregrounds by a second-order statistical approach. *Monthly Notices R. Astron. Soc* 2006, 373: 271-279. 10.1111/j.1365-2966.2006.11025.x
9. Kuruoglu EE: Dependent component analysis for cosmology. *Lecture Notes Comput. Sci* 2010, 6365: 538-545. 10.1007/978-3-642-15995-4_67
10. Ghahramani Z, Beal M: Variational inference for Bayesian mixtures of factor analysers. In *Advances in Neural Information Processing Systems*. Edited by: Solla SA, Leen TK, Muller KR. MIT Press, Cambridge, MA); 2000.
11. Eriksen HK, Dickinson C, Lawrence CR, Baccigalupi C, Banday AJ, Gorski KM, Hansen FK, Lilje PB, Pierpaoli E, Smith KM, Vanderlinde K: C M B component separation by parameter estimation. *Astrophys. J* 2006, 641: 665-682. 10.1086/500499

12. Lee PM: *Bayesian Statistics: An Introduction*. Hodder Arnold H& S, London; 2004.
13. Richardson S, Green P: On Bayesian analysis of mixtures with an unknown number of components (with discussion). *J. R. Stat. Soc. Ser. B* 1997, 59: 731-792. 10.1111/1467-9868.00095
14. Hyvärinen A: Fast and robust fixed-point algorithms for independent component analysis. *IEEE Trans. Neural Netw* 1999, 10(3):626-634. 10.1109/72.761722
15. Leach S, Cardoso JF, Baccigalupi C, Barreiro R, Betoule M, Bobin J, Bonaldi A, Delabrouille J, De Zotti G, Dickinson C, Eriksen HK, González-Nuevo J, Hansen FK, Herranz D, Le Jeune M, López-Caniego M, Martínez-González E, Massardi M, Melin JB, Miville-Deschênes MA, Patanchon G, Prunet S, Ricciardi S, Salerno E, Sanz JL, Starck JL, Stivoli F, Stolyarov V, Stompor R, Vielva P: Component separation methods for the PLANCK mission. *Astron. Astrophys* 2008, 491(2):597-615. 10.1051/0004-6361:200810116
16. Gilks WR, Richardson S, Spiegelhalter DJ: *Markov Chain Monte Carlo in Practice*. CRC Press, Boca Raton, FL; 1996.
17. McKay DJC: *Information Theory, Inference and Learning Algorithms*. Cambridge University Press, Cambridge; 2003.
18. Bishop CM: *Pattern Recognition and Machine Learning*. Springer, New York; 2006.
19. Attias H: *A Variational Bayesian Framework for Graphical Models*. MIT Press, Cambridge, MA; 2000.
20. Collaboration P: The pre-launch plack sky model: a model of sky emission at submillimetre to centimetre wavelengths (in preparation).
21. [http://sci.esa.int/science-e/www/object/index.cfm?fobjectid=34730%26fbodylongid=1595]
22. [http://map.gsfc.nasa.gov/]
23. Wilson SP, Kuruoglu EE, Quirós A: Bayesian factor analysis using Gaussian mixture sources, with application to separation of the cosmic microwave background. *2nd International Workshop on Cognitive Information Processing* 2010.
24. Rue H, Martino S, Chopin N: Approximate Bayesian inference for latent Gaussian models using integrated nested Laplace approximations. *J. R. Stat. Soc. Ser. B* 2008, 71: 319-392.

CHAPTER

6

A Two-Stage Approach Using Gaussian Mixture Models and Higher-Order Statistics for a Classification of Normal and Pathological Voices

Ji Yeoun Lee

MII826, Biomedical Engineering, The Department of Biomedical Engineering, Jungwon University, Donbu-ri #5, Goesan-eup, Goesan-gun, Chungcheongbuk-do 367-805, South Korea

ABSTRACT

A two-stage classifier is used to improve the classification performance between normal and pathological voices. A primary classification between normal and pathological voices is achieved by the Gaussian mixture model (GMM) log-likelihood scores. For samples that do not meet the thresholds for normal or disordered voice in the GMM, the final decision is made by a higher-order statistics (HOS)-based parameter. The normalized skewness

Citation: (APA): Lee, J. Y. (2012). A two-stage approach using Gaussian mixture models and higher-order statistics for a classification of normal and pathological voices. EURASIP journal on advances in signal processing, 2012(1), 252. (8 pages) DOI: https://doi.org/10.1186/1687-6180-2012-252

and kurtosis, and means of the normalized skewness and kurtosis were estimated using a sustained vowel /a/ from 53 normal and 173 pathological voices taken from the Disordered Voice Database. Mel-frequency cepstral coefficients (MFCC)-based GMM, the HOS methods, and a two-stage classifier based on the GMM-HOS were performed for each voice signal. A Mann–Whitney rank sum test was used to detect differences in the means of the HOS-based parameters. A fivefold cross-validation scheme was performed to test the classification method. When 16 Gaussian mixtures were used, the MFCC-based GMM algorithm is performed with 92.0% accuracy. When means of the normalized skewness and kurtosis were used, performances of 82.31 and 83.67% were obtained, respectively. The two-stage classifier with 16 Gaussian mixtures and the mean of the normalized kurtosis classified samples with a 96.96% accuracy were obtained. The proposed two-stage classifier is more accurate than the MFCC-based GMM and HOS methods alone and shows potential for the classification of voices in the clinic.

INTRODUCTION

Speech is integral to day-to-day communication. Speech impediments negatively impact social interactions leading to interest in early detection and treatment of voice disorders. Many researchers have worked towards the goal of automatic and objective classification between normal and pathological voices using minimally invasive methods. A large amount of research has focused on the automatic detection of voice pathologies by means of acoustic analysis, parametric and non-parametric feature extraction, pattern recognition algorithms, and statistical methods [1–15].

Sáenz-Lechón et al. [5] presented an overview of previous classification schemes applied to the Massachusetts Eye & Ear Infirmary (MEEI) Voice Disorders Database [16]. They described some methodological concerns to be considered when designing automatic systems for pathological voice detection. They recommended the use of a commercially well-known database, a cross-validation strategy based on several partitions to obtain averaged classification performances with confidence intervals, a report of the means of a detection error trade-off (DET), and an investigation of the area under receiver operating characteristic (ROC) curves.

The emergence of attractive pattern classification algorithms such as the Gaussian mixture model (GMM), neural network (NN), and hidden Markov model has received attention as a potential means to discriminate between normal and pathological voices [6–13]. The GMM has especially been reported as a very successful classification method [10–13]. Characteristic parameters, such as Mel-frequency cepstral coefficients (MFCC), have also become more popular for voice pathology detection [6, 8, 10–12]. Recently, Wang et al. [12] proposed a GMM supervector kernel-support vector machine (GMM-SVM) classifier which was compared with the GMM classifier as a baseline algorithm. The GMM supervectors were largely effective parameters for the discrimination of normal and pathological voices. A classification accuracy of 96.1% was achieved by SVM classification of the 16 Gaussian GMM supervectors [12].

As an acoustic analysis method, higher-order statistics (HOS) have shown promising results in a number of signal processing applications, and are of particular value when dealing with a mixture of Gaussian and non-Gaussian processes and system nonlinearity [14, 15, 17, 18]. The application of HOS to speech processing has primarily been motivated by its inherent Gaussian suppression and phase preservation properties [17, 18]. Researches in the disordered voice field have been based on the assumption that speech has non-zero HOS that is distinct from that of Gaussian noise [18, 19]. Alonso et al. [19] proposed seven new HOS-based parameters that were obtained directly or indirectly starting from the bispectrum of a voice frame. A success rate of 98.3% was obtained by using both the conventional and the HOS-based parameters with an NN classifier demonstrating the possibility of automatically discriminating pathological from healthy voices using HOS parameters [20]. Further study of how well each HOS parameter can detect pathological voices with the methodological designs recommended by Sáenz-Lechón et al. [5] is merited.

In this article, we propose new HOS-based parameters implemented in the time domain. They are means of the normalized skewness and kurtosis which are calculated from each frame and averaged in a sentence. The HOS-based parameters estimated in the time domain can easily be applied to a real-time environment in contrast to the Fourier series representation of the HOS parameters in frequency domain applied by Alonso et al. [19]. Second, we propose a two-stage approach to further improve the accuracy of the classification between normal and pathological voices. The classification system consists of a MFCC-based GMM algorithm which describes the primary classification achieved by the GMM log-likelihood scores, and an HOS-based parameter as a post-processor.

MATERIAL AND METHODS

Material

Vocal signals were collected from the MEEI Voice Disorders Database [16]. Fifty-three normal and one hundred seventy-three pathological speakers with a wide range of organic, neurological, traumatic, and psychogenic voice disorders were selected. The extracted subset is the same as one described in the study of Wang et al. [12] to compare the result with this study. Voice samples were collected in a controlled environment and sampled with a 50- or 25-kHz sampling rate and 16 bits of resolution. Patients phonated a sustained /a/ (1–3 s). All voice data were down-sampled to 25 kHz and grouped into training (70% of the data) and test (30%) sets to implement all methods. Each set for a fivefold cross-validation scheme was randomly selected from the subset [10, 12].

Statistical analysis

Statistical analysis was conducted using Sigma Stat 3.0 (Jandel Scientific, SanRafael, CA, USA). The Mann–Whitney rank sum test was performed to test the differences between normal and pathological voices for the normalized skewness, the normalized kurtosis, and the means of the normalized skewness and kurtosis. A *p*-value of 0.05 was used for all measures.

MFCC-based GMM method

The voice samples were analyzed with 40-ms interval overlapping 20-ms with thc prcvious frame and multiplied by a hamming window as performed in previous studies [12]. MFCC parameters were extracted and fed into a GMM-based detector enabling a final decision about the absence or presence of pathology. The number of the filter banks was 38, and 36-dimensional MFCC parameters (18 MFCC+18 Delta-MFCC) with 1 derivative were calculated every frame of 18 Mel-cepstral coefficients as in Wang et al. [12]. Cepstral mean subtraction was also used during the extraction to reduce the cepstral bias of the recording channel. The Linde–Buso–Gray algorithm was used for the GMM initialization and the GMM having 8, 16, and 32 mixtures were trained with the expectation-maximization (EM) algorithm to determine the model parameters such as mean vectors, covariance matrices, and mixture weights. For an utterance $X=\{_{x1},x_2,\ldots x_T\}$ where T is the number of frames, the log-likelihood ratio (LLR) applying Bayes' rule, disregarding the constant prior probabilities in the log domain, is presented in Equation (1).

$$\Lambda(X) = \log\left[p\left(\frac{X_N}{\lambda_C}\right)\right] - \log\left[p\left(\frac{X_P}{\lambda_{\bar{C}}}\right)\right] \quad (1)$$

where $\Lambda(X)$ is the LLR, and λ_C and λC^- are GMM models for normal and pathological voices, respectively. Also N and P indicate normal and pathological voices, respectively.

Figure 1 shows the histogram of the LLR estimated from normal and pathological voices in training procedure. The decision threshold, Λ_{NP}, is then set to adjust the tradeoff between rejecting pathological voices (false rejection) and accepting normal voices (false acceptance). In test procedure, the LLR, $\Lambda(X)$, is compared with a threshold $\Lambda_{NP}(X)$ and the voice is said to be pathological if $\Lambda(X)<\Lambda_{NP}(X)$ and normal if $\Lambda(X)>\Lambda_{NP}(X)$.

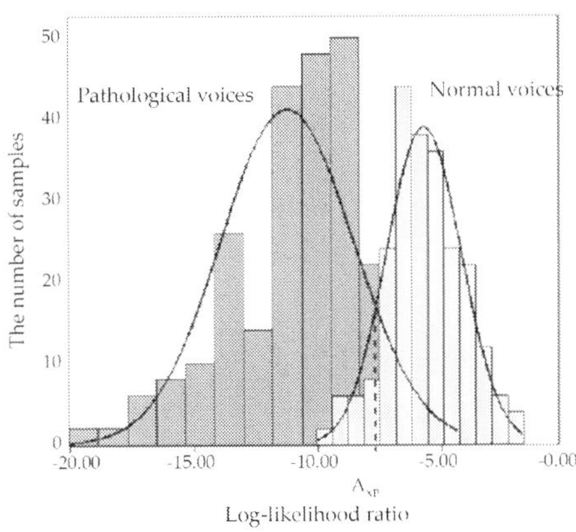

Figure 1: LLR histogram estimated from normal and pathological voices in training procedure. $\Lambda_{NP}(X)$ indicates the decision threshold measured from the MFCC-based GMM algorithm.

HOS method

A speech signal, $x(n)$, which may be normal or pathological, can be expressed as given in Equation (2) [17].

$$x(n) = s(n) + w(n) \quad (2)$$

where $s(n)$ is a non-Gaussian signal generated by the oscillation of the vocal folds and $w(n)$ is Gaussian noise which can be assumed to be zero in normal voices and not to be zero in pathological voices.

Pathological voices are corrupted by noise, $w(n)$, which is directly related to the perceived roughness of the voice [1–3, 20]. If $s(n)$ and $w(n)$ are statistically independent, then the energy of $x(n)$ is the sum of speech and noise energies: $E_x = E_s + E_w$. Second-order statistics are thus directly affected in an additive way by the presence of noise [17, 19]. However, when HOS analysis is applied to pathological voices, unstable and discontinuous statistics of $x(n)$ may be estimated because HOS analysis is blind to Gaussian processes. On the other hand, in a normal voice, the HOS of only non-Gaussian measurements may be extracted because a Gaussian noise can be assumed to be zero. The variation of a non-Gaussian signal which is produced by vibration of the vocal folds can be an important clue for the classification of pathological and normal voices.

If $x(n)$, where $n=0, \pm 1, \pm 2, \ldots$, is a real stationary discrete-time signal and its moments up to order p exist, then its p th-order moment function is given by Equation (3).

$$m_p(\tau_1, \tau_1, \cdots, \tau_{p-1}) \equiv E\{x(n)x(n+\tau_1)\cdots x(n+\tau_{p-1})\} \tag{3}$$

And it depends only on the time differences $\tau_1, \tau_2, \ldots, \tau_{p-1}, \tau_i = 0, \pm 1, \pm 2, \ldots$ for all i. Here, $E\{\cdot\}$ denotes statistical expectation and for a deterministic signal, it is replaced by a time summation over all time samples or time averaging. In addition, if the signal has zero mean, then its cumulant functions (up to order four) are given by Equation (4) [17].

$$\begin{aligned} &\text{second} - \text{order cumulant} : C_2(\tau_1) = m_2(\tau_1) \\ &\text{third} - \text{order cumulant} : C_3(\tau_1, \tau_2) = m_3(\tau_1, \tau_2), \\ &\text{fourth} - \text{order cumulant} : C_4(\tau_1, \tau_2, \tau_3) = m_4(\tau_1, \tau_2, \tau_3) \\ &\quad - m_2(\tau_1)\cdot m_2(\tau_3 - \tau_2) - m_2(\tau_2)\cdot m_2(\tau_3 - \tau_1) \\ &\quad - m_2(\tau_3)\cdot m_2(\tau_2 - \tau_1) \end{aligned} \tag{4}$$

By setting all the lags to zero in the above cumulant expressions, we can obtain the variance, skewness, and kurtosis.

Variance:

$$\begin{aligned} &\text{Variance} : \gamma_2 \equiv C_2(0) = E\{x^2(n)\}, \\ &\text{Skewness} : C_3(0,0) = E\{x^3(n)\}, \\ &\text{Kurtosis} : C_4(0,0,0) = E\{x^4(n)\} - 3\left[E\{x^2(n)\}\right]^2 \end{aligned} \tag{5}$$

When estimating HOS from finite data records, the variance of the estimators is reduced by normalizing the input data to have a unity variance, prior to computing the estimators. Equivalently, the third- and fourth-order statistics are normalized by the appropriate powers of the data variance, thus we define the normalized skewness and kurtosis as shown in Equations (6) and (7) [17].

Normalized skewness :

$$\gamma_3 \equiv \frac{C_3(0,0)}{[C_2(0)]^{1.5}} = \frac{E\{x^3(n)\}}{[E\{x^2(n)\}]^{1.5}} \tag{6}$$

Normalized kurtosis :

$$\gamma_4 \equiv \frac{c_4(0,0,0,)}{[c_2(0)]^2} = \frac{E\{x^4(n)\}}{[E\{x^2(n)\}]^2} - 3.0 \tag{7}$$

In this article, the normalized skewness and kurtosis are extracted in frame as shown in Equation (8).

$$\gamma_{3t} = \frac{\sum_{n=1}^{N} x_t^3(n)}{\left[\sum_{n=1}^{N} x_t^2(n)\right]^{1.5}}, \quad \gamma_{4t} = \frac{\sum_{n=1}^{N} x_t^4(n)}{\left[\sum_{n=1}^{N} x_t^2(n)\right]^{2}} \tag{8}$$

where x_t is the speech sample value of t th frame and N is the number of samples.

The proposed HOS-based parameters are means of the normalized skewness and kurtosis: $\bar{\gamma}_3$ and $\bar{\gamma}_4$. They are estimated in a sentence and have their roots in γ_3 and γ_4 as described in Equation (9). As in the MFCC procedures, voice samples were cut into 40-ms overlapping frames which were shifted by 20 ms and multiplied by a hamming window to extract the HOS-based parameters.

$$\bar{\gamma}_3 = \frac{1}{T}\sum_{t=1}^{T} \gamma_{3t}, \quad \bar{\gamma}_4 = \frac{1}{T}\sum_{t=1}^{T} \gamma_{4t} \tag{9}$$

where γ_{3t} and γ_{4t} are γ_3 and γ_4 extracted in the t th frame, respectively, and T is the number of the frames.

Two-stage classifier based on GMM-HOS

The block diagram of the proposed algorithm is shown in Figure 2. In the training phase, the MFCCs from the voice samples are extracted for each analysis frame. $\bar{\Lambda}_P$ and $\bar{\Lambda}_P$ P indicate the thresholds of the LLR estimated by each GMM for normal and pathological voices. An example is shown in Figure 3 with false acceptance and false rejection plots versus LLR thresholds. Both lines cross over the equal error rate (EER). $\bar{\Lambda}_N$ and $\bar{\Lambda}_P$ are the LLR values determined when the EER is 25.0%. The thresholds, $\bar{\gamma}_{3thre}$ and $\bar{\gamma}_{4thre}$, are determined in advance according to the values to produce the best results when $\bar{\gamma}_3$ or $\bar{\gamma}_4$ are used alone for the pathological voice detection. That is, they are computed for the training data set. In the test phase, the LLR, $\bar{\Lambda}$, is estimated with the feature vector and the pre-trained GMMs. The primary decision is executed by the MFCC-based GMM algorithm. If $\bar{\Lambda}_P \leq \bar{\Lambda} \leq \bar{\Lambda}_N$, the voice samples are processed using the HOS operator. The final decision is realized after calculating $\bar{\gamma}_3$ and $\bar{\gamma}_4$. The values of $\bar{\gamma}_3$ and $\bar{\gamma}_4$ are independently used to classify normal and pathological voice. When γ¯3 is used, the voice is said to be pathological if $\bar{\gamma}_3 < \bar{\gamma}_{3thre}$ thre and normal if $\bar{\gamma}_3 \geq \bar{\gamma}_{3thre}$, thre. When γ¯4 is used, the voice is said to be normal if $\bar{\gamma}_4 < \bar{\gamma}_{4thre}$ thre and pathological if $\bar{\gamma}_4 \geq \bar{\gamma}_{4thre}$ thre.

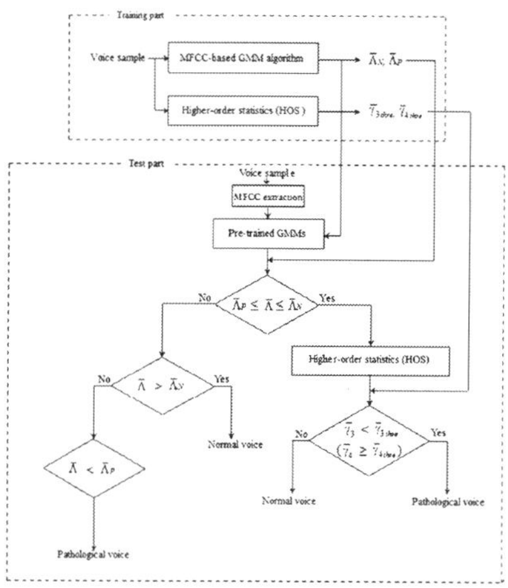

Figure 2: Overall procedure of the two-stage classifier. $\bar{\Lambda}_N$ and $\bar{\Lambda}_P$ are the thresholds of LLR estimated by each GMM for normal and pathological voices in training procedure. $\bar{\gamma}_3$ and $\bar{\gamma}_4$ are means of the normalized skewness and

kurtosis, respectively. $\ddot{\gamma}_3$ and $\ddot{\gamma}_4$ thre are the thresholds optimized from γ_3 and γ_4 in training procedure. Λ is the LLR estimated from the pre-trained GMM in test procedure..

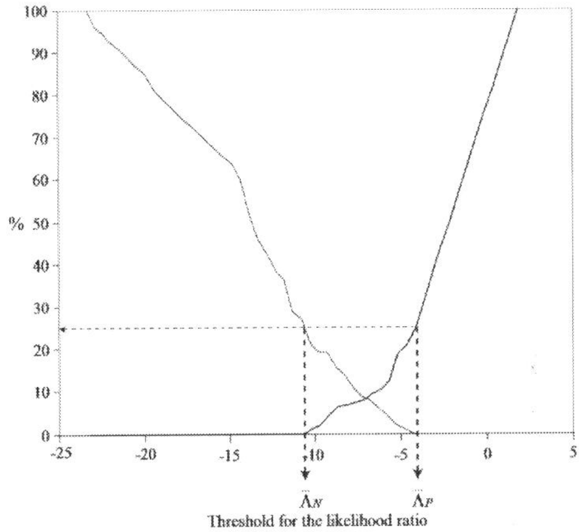

Figure 3: Cumulative false acceptance (right) and false rejection (left). $\ddot{\Lambda}_N$ and $\ddot{\Lambda}_P$ indicate the thresholds of LLR estimated by each GMM for normal and pathological voices in the training procedure. Both lines cross over the EER.

RESULTS

MFCC-based GMM method

The performance was assessed by averaging the results obtained from fivefold cross-validation scheme [10, 12]. Table 1 shows the confusion matrix, accuracy (%) including 95% confidence intervals (CIs), specificity (%), sensitivity (%), and areas under the curve (AUC) according to the number of the Gaussian mixtures. Specificity and sensitivity means the test's ability to identify negative and positive results, respectively. The accuracy is the proportion of true results (both true positives and true negatives) in the population. The GMM models were trained using 8, 16, and 32 mixtures. The average performance was 92.00% when the number of Gaussian mixtures was 16. The result using 32 Gaussian mixtures was also better than one obtained with 8 Gaussian mixtures. Figure 4 shows the area under the ROC curve when MFCC-based GMM method shows the best accuracy. The EER of the MFCC-based GMM method is shown in Figure 5.

Table 1 Performance of the MFCC-based GMM method

Method	Confusion matrix		Accuracy (%)	Specify (%)	Sensitivity (%)	AUC (%)
GMM 8 mixtures	**89.58**	**10.42**	**89.35±3.00**	**89.67**	**89.53**	**95.95**
	10.88	**89.12**				
GMM 16 mixtures	**89.58**	**8.33**	**92.00±4.79**	**92.27**	**91.72**	**98.59**
	7.68	**92.32**				
GMM 32 mixtures	**89.58**	**10.42**	**90.31±4.09**	**90.90**	**89.73**	**96.02**
	8.96	**91.04**				

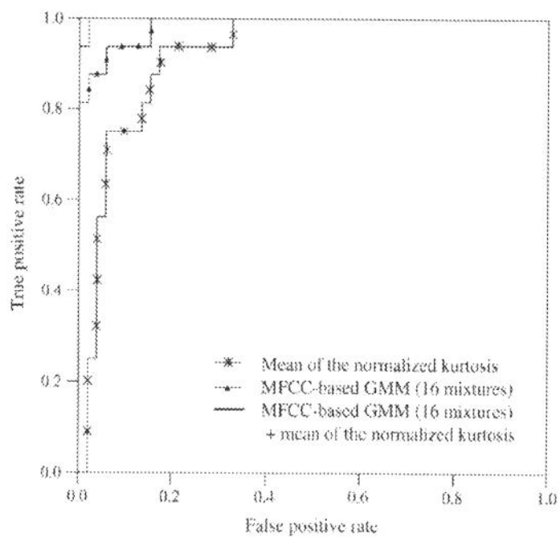

Figure 4: ROC curves estimated in mean of the normalized kurtosis, MFCC- based GMM algorithm (16 Gaussian mixtures), and two- stage classifier.

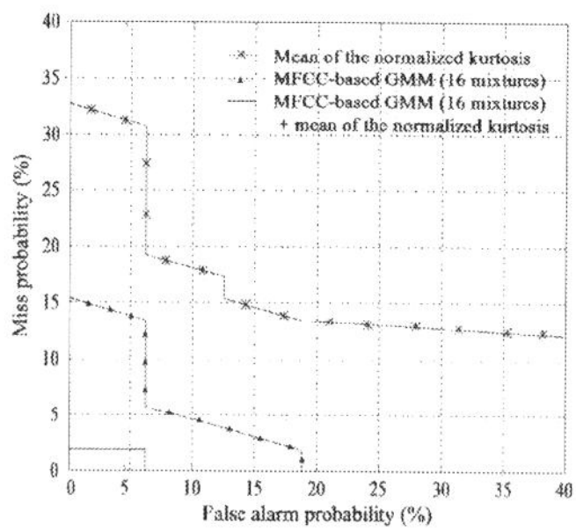

Figure 5: DET curves estimated in mean of the normalized kurtosis, MFCC- based GMM algorithm (16 Gaussian mixtures), and two- stage classifier.

HOS method

Figure 6 presents the distributions of $\bar{\gamma}_3$ and $\bar{\gamma}_4$ of normal and pathological voices. The distributions of $\bar{\gamma}_3$ and $\bar{\gamma}_4$ for pathological voices tended to be skewed to the left and have a leptokurtic distribution $(\bar{\gamma}_4 > 3)$. For normal voices, these distributions tended to be skewed to the right and have a platykurtic distribution $(\bar{\gamma}_4 < 3)$, overall. The distributions of pathological voices had a tendency to show larger variation than those of normal voices. A Mann–Whitney rank sum tests showed a statistically significant difference between normal and disordered voices for $\bar{\gamma}_3$ and $\bar{\gamma}_4$ ($p<0.001$).

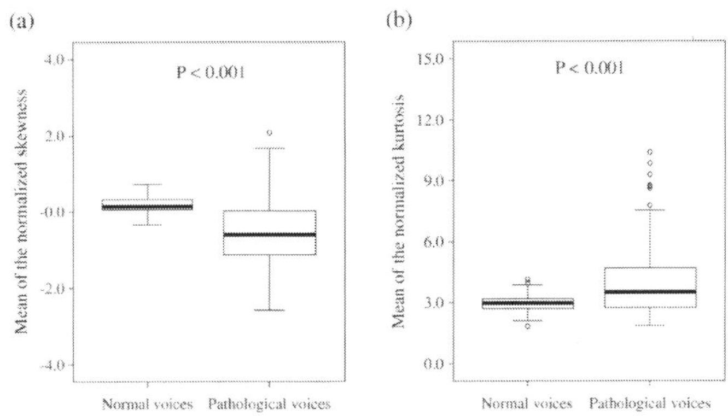

Figure 6: Distributions of the proposed HOS-based parameters. (a) Mean of the normalized skewness. (b) Mean of the normalized kurtosis. The midline represents the median, with the lower and upper boundaries of the box indicating the first and third quartiles, respectively. Whiskers indicate the maximum and minimum values and circles indicate outliers.

The fivefold cross-validation was used to estimate the performances for each parameter. When $\bar{\gamma}_3$ and $\bar{\gamma}_4$ were used to classify normal and pathological voices, the average performances of 82.31 and 83.67% were obtained, respectively. Table 2 shows the confusion matrix, accuracy (%) including 95% CIs, specificity (%), sensitivity (%), and AUC when means of the normalized skewness and kurtosis are used to classify normal and pathological voices. The accuracy of the mean of the normalized kurtosis was higher than that of the mean of the normalized skewness. The ROC curve of the mean of the normalized kurtosis is shown in Figure 4. In Figure 5, the DET curves show the EERs for the mean of the normalized kurtosis. The MFCC-based GMM method outperformed HOS method.

Table 2:- Performance of the means of the normalized skewness and kurtosis

Method	Confusion matrix		Accuracy (%)	Specificity (%)	Sensitivity (%)	AUC (%)
Mean of the normalized skewness	81.25	18.75	82.31±5.67	83.00	81.64	89.36
	16.64	83.36				
Mean of the normalized kurtosis	83.33	16.67	83.67±6.89	83.89	83.49	92.38
	16.00	84.00				

Two-stage classifier based on GMM-HOS

Table 3 shows the confusion matrix, accuracy (%) including 95% CIs, specificity (%), sensitivity (%), and AUC when the two-stage classifier is used. The results were measured in the fivefold cross-validation similar to the MFCC-based GMM and HOS methods. The best performance, 96.96%, was obtained when 16 Gaussian mixtures and the normalized kurtosis were utilized as the classifier. In general, when mean of the normalized kurtosis was used as second classifier, the performance was higher than that of mean of the normalized skewness.

Table 3 Performance of the proposed two-stage classifiers

Method		Confusion matrix		Accuracy (%)	Specificity (%)	Sensitivity (%)	AUC (%)
GMM 8 mixtures	Mean of the normalized skewness $(\bar{\gamma}3\bar{\gamma}3)$	93.75	6.25	94.00±1.67	94.21	93.78	99.69
		5.76	94.24				
	Mean of the normalized kurtosis $(\bar{\gamma}4\bar{\gamma}4)$	93.75	6.25	94.00±1.67	94.21	93.78	99.69
		5.76	94.24				
GMM 16 mixtures	Mean of the normalized skewness$(\bar{\gamma}3\bar{\gamma}3)$	95.83	4.17	96.64±4.09	97.43	95.90	99.69
		2.56	97.44				

Method		Confusion matrix		Accuracy (%)	Specific-ity (%)	Sensi-tivity (%)	AUC (%)
	Mean of the normalize kurto-sis (γ¯4γ¯4)	95. 83	4. 17	96. 96±4. 79	98. 04	95. 92	99. 95
		1.92	98.08				
GMM 32 mix-tures	Mean of the nor-malized skewness (γ¯3γ¯3)	93.75	6.25	94.64±1.92	95.44	93.86	99.69
		4.48	95.52				
	Mean of the nor-malized kurtosis (γ¯4γ¯4)	95.83	4.17	96.00±3.13	96.15	95.84	99.87
		3.84	96.16				

The ROC and DET curves of a two-stage classifier using 16 Gaussian mixtures and mean of the normalized kurtosis are shown in Figures 4 and 5. The AUC of the method was larger than that those of 16 Gaussian mixtures and mean of the normalized kurtosis independently. EER of the two-stage classifier was 3.04% in Figure 5.

CONCLUSION AND DISCUSSION

In this article, we define a two-stage technique to discriminate pathological from normal voices. The newly proposed model is comprised of two parts, an MFCC-based GMM algorithm which describes the primary decision achieved by the LLR scores of the GMM, and post-processing using an HOS analysis block incorporating means of the normalized skewness and kurtosis. The characteristics of the MFCC between normal and pathological voices are presented in the study of Godino-Llorente et al. [10]. A strong correlation between the HOS coefficients (the normalized skewness and kurtosis) and voice classification is demonstrated ($p<0.001$). By introducing these parameters in cases where the MFCC-based GMM algorithm returns uncertain values, the classification can be improved. According to Sáenz-Lechón et al.'s recommendations [5], we utilized a commercially well-known database. Classification performance along with CI was obtained by a cross-validation strategy based on several partitions. Results are also described by DET and AUC.

The two-stage classifier outperformed the individual classification schemes. The best performance, 96.96%, is achieved by combining an

MFCC-based GMM algorithm with 16 Gaussian mixtures and the mean of the normalized kurtosis. The MFCC-based GMM algorithm with 16 Gaussian mixtures performed at 92.0% while the means of the normalized skewness and kurtosis classified correctly 82.31 and 83.67%, respectively. A false decision is occasionally caused by the erroneous EM-based GMM estimation in the intersection regions where they have somewhat low likelihoods. Therefore, it is believed that the performance improvement is mainly due to the fact that our two-stage classifier successfully solves the false decision problem caused by low log-likelihood values.

Many studies have presented a variety of approaches [1–3, 6–13, 19]. Although different datasets and evaluation procedures make it difficult to compare the results of previous studies [5], Godino-Llorente and Gómez-Vilda [8] presented the experimental result using the learning vector quantization methodology, yielding 96% frame accuracy. Afterwards, Godino-Llorente et al. [10] published the article showing an accuracy of 94.07% with 24 MFCC parameters and a GMM of 6 mixtures. A classification performance of 98.3% was obtained by using both the perturbation and seven HOS-based parameters with an NN classifier implemented by Alonso et al. [19]. Recently, Wang et al. [12] proposed a GMM-SVM classifier with a classification performance of 96.1%. Finally, in this article, we combined the MFCC-based GMM method utilized by Godino-Llorente et al. [8, 10, 11] and Wang et al. [12] with HOS parameters to obtain a performance of 96.96%.

The automatic classification between normal and pathological voices remains an open problem that calls for reliable algorithms to aid the clinicians. When the information gathered from simple physically informed GMMs is combined with HOS-based parameters, a valuable classifier can be obtained. This two-stage method can be used for the analysis and assessment of voice quality.

REFERENCES

1. de Oliveira Rosa M, Pereira JC, Grellet M: Adaptive estimation of residual signal for voice pathology diagnosis. *IEEE Trans. Biomed. Eng* 2000, 47(1):96-104. 10.1109/10.817624
2. Shama K, Krishna A, Cholayya NU: Study of harmonics-to-noise ratio and critical band energy spectrum of speech as acoustic indicators of laryngeal and voice pathology. *EURASIP J. Adv. Signal Process.* 2007, 1997(1):50-59.
3. Godino-Llorente JI, Osma-Ruiz V, Sáenz-Lechón N, Gómez-Vilda P, Blanco-Velasco M, Cruz-Roldán F: The effectiveness of the glottal to noise excitation ratio for the screening of voice disorders. *J. Voice* 2010, 24(1):47-56. 10.1016/j.jvoice.2008.04.006
4. Khadivi Heris H, Seyed Aghazadeh B, Nikkhah-Bahrami M: Optimal feature selection for the assessment of vocal fold disorders. *Comput. Biol. Med.* 2009, 39(10):860-868. 10.1016/j.compbiomed.2009.06.014
5. Sáenz-Lechón N, Godino-Llorente JI, Osma-Ruiz V, Gómez-Vilda P: Methodological issues in the development of automatic systems for voice pathology detection. *Biomed. Signal Process. Control* 2006, 1(2):120-128. 10.1016/j.bspc.2006.06.003
6. Godino-Llorente JI, Aguilera-Navarro S, Gómez-Vilda P: Non-supervised neural net applied to the detection of voice impairment. In *Proceeding of IEEE International Conference on Acoustics, Speech, and Signal Processing (ICASSP)*. 6, Istanbul, Turkey; 2000:3594-3597.
7. Hadjitodorov S, Mitev P: A computer system for acoustic analysis of pathological voices and laryngeal disease screening. *Med. Eng. Phys.* 2002, 24(6):419-429. 10.1016/S1350-4533(02)00031-0
8. Godino-Llorente JI, Gómez-Vilda P: Automatic detection of voice impairments by means of short-term cepstral parameters and neural network based detectors. *IEEE Trans. Biomed. Eng.* 2004, 51(2):380-384. 10.1109/TBME.2003.820386
9. Umapathy K, Krishman S, Parsa V, Jamieson DG: Discrimination of pathological voices using a time-frequency approach. *IEEE Trans. Biomed. Eng.* 2005, 52(3):421-430. 10.1109/TBME.2004.842962
10. Godino-Llorente JI, Aguilera-Navarro S, Gómez-Vilda P: Dimensionality reduction of a pathological voice quality assessment system based on Gaussian mixture models and short-term cepstral parameters. *IEEE Trans. Biomed. Eng.* 2006, 53(10):1943-1953.

10.1109/TBME.2006.871883

11. Godino-Llorente JI, Sáenz-Lechón N, Osma-Ruiz V, Aguilera-Navarro S, Gómez-Vilda P: An integrated tool for the diagnosis of voice disorders. *Med. Eng. Phys.* 2006, 28(3):276-289. 10.1016/j.medengphy.2005.04.014
12. Wang X, Zhang J, Yan Y: Discrimination between pathological and normal voices using GMM-SVM approach. *J. Voice* 2010, 25(1):38-43.
13. Das R: A comparison of multiple classification methods for diagnosis of Parkinson disease. *Expert Syst. Appl.* 2010, 37(2):1568-1572. 10.1016/j.eswa.2009.06.040
14. Nikias C, Mendel J: Signal processing with higher-order statistics spectra. *IEEE Signal Process. Mag.* 1993, 10(3):10-37.
15. Mendel JM: Tutorial on higher-order statistics in signal processing and system theory: theoretical results and some applications. *Proc. IEEE* 1991, 79(3):278-305. 10.1109/5.75086
16. Massachusetts Eye and Ear Infirmary: *Voice Disorders Database, Version. 1.03 [CD-ROM]*. Kay Elemetrics Corp, Lincoln Park, NJ; 1994.
17. Nemer E, Goubran R, Mahmoud S: Robust voice activity detection using higher-order statistics in the LPC residual domain. *IEEE Trans. Speech Audio Process.* 2001, 9(3):217-231. 10.1109/89.905996
18. Lee JY, Minsoo H: Automatic assessment of pathological voice quality using higher-order statistics in the LPC residual domain. *EURASIP J. Adv. Signal Process* 2009. Article ID 748207, 8pages
19. Alonso JB, de Leon J, Alonso I, Ferrer MA: Automatic detection of pathologies in the voice by HOS based parameters. *EURASIP J. Appl. Signal Process.* 2001, 2001(4):275-284. 10.1155/S1110865701000336
20. Kent RD, Ball MJ: *Voice Quality Measurement (Thomson Learning).* New York; 2000.

CHAPTER

7

PET Image Segmentation Using a Gaussian Mixture Model and Markov Random Fields

Thomas Layer[1,2], Matthias Blaickner[2], Barbara Knäusl[3], Dietmar Georg[3], Johannes Neuwirth[4], Richard P Baum[5], Christiane Schuchardt[5], Stefan Wiessalla[5] and Gerald Matz[1]

[1]Institute of Telecommunications, Vienna University of Technology, Karlsplatz 13, 1040 Wien Vienna, Austria.

[2]Health & Environment Department, Austrian Institute of Technology, Donau-City-Strasse 1/2, 1220 Wien Vienna, Austria.

[3]Department of Radiation Oncology, Division of Medical Radiation Physics, Christian Doppler Laboratory for Medical Radiation Research for Radiation Oncology, Medical University of Vienna/AKH Vienna, Währinger Gürtel 18-20, 1090 Wien Vienna, Austria.

Citation: (APA): Layer, T., Blaickner, M., Knäusl, B., Georg, D., Neuwirth, J., Baum, R. P., & Matz, G. (2015). PET image segmentation using a Gaussian mixture model and Markov random fields. EJNMMI physics, 2(1), 9. (15 pages) DOI: https://doi.org/10.1186/s40658-015-0110-7

[4]Radiation Safety and Applications, Seibersdorf Labor GmbH, 2444 Seibersdorf, Seibersdorf, Austria.

[5]THERANOSTICS Center for Molecular Radiotherapy and Molecular Imaging (PET/CT) ENETS Center of Excellence, Zentralklinik Bad Berka, Robert-Koch-Allee 9, 99437 Bad Berka, Bad Berka, Germany.

ABSTRACT

Background

Classification algorithms for positron emission tomography (PET) images support computational treatment planning in radiotherapy. Common clinical practice is based on manual delineation and fixed or iterative threshold methods, the latter of which requires regression curves dependent on many parameters.

Methods

An improved statistical approach using a Gaussian mixture model (GMM) is proposed to obtain initial estimates of a target volume, followed by a correction step based on a Markov random field (MRF) and a Gibbs distribution to account for dependencies among neighboring voxels. In order to evaluate the proposed algorithm, phantom measurements of spherical and non-spherical objects with the smallest diameter being 8 mm were performed at signal-to-background ratios (SBRs) between 2.06 and 9.39. Additionally ^{68}Ga-PET data from patients with lesions in the liver and lymph nodes were evaluated.

Results

The proposed algorithm produces stable results for different reconstruction algorithms and different lesion shapes. Furthermore, it outperforms all threshold methods regarding detection rate, determines the spheres' volumes more accurately than fixed threshold methods, and produces similar values as iterative thresholding. In a comparison with other statistical approaches, the algorithm performs equally well for larger volumes and even shows improvements for small volumes and SBRs. The comparison with experts' manual delineations on the clinical data shows the same qualitative behavior as for the phantom measurements.

Conclusions

In conclusion, a generic probabilistic approach that does not require data measured beforehand is presented whose performance, robustness, and swiftness make it a feasible choice for PET segmentation.

Keywords: Expectation maximization, Markov random field, Positron emission tomography, Radiotherapy, Tumor segmentation

BACKGROUND

The determination of the tumor volume is one of the main causes for uncertainties in dosimetry [1]. When trying to assess the volume of a tumor for the sake of treatment planning in external beam radiation therapy (EBRT) or radionuclide therapy, a very common practice is to have an expert manually draw a volume of interest (VOI) on the respective positron emission tomography (PET) or single photon emission computed tomography (SPECT) image. The resulting and inevitable interobserver variations have been reported well enough for different types of cancer [2-4].

Another prevalent approach is the application of a threshold. The simplest choice for the threshold is a fixed percentage of the maximum activity concentration value [5,6]. This thresholding method has been shown to predict well for big volumes but yields large errors in case of small volumes which is attributed to partial volume effects (PVE) and moreover depends on the signal-to-background ratio (SBR).

Despite its questionable scientific meaningfulness, it is still a widespread method even recommended by an international experts' report [7]. Extensions of this method are automatic [8,9] and iterative threshold approaches [10-12]. However, the iterative thresholding method (ITM) requires a regression curve which has to be determined by phantom measurements for every specific imaging setting [13]. Considering the many dependencies of these regression curves such as the (i) manufacturer and the detector of the scanner, (ii) reconstruction algorithm, (iii) nuclide, (iv) SBR, and (v) volume of the lesion, ITM comes at a great expense in terms of work and time. Likewise, the inclusion of physical models of PET images such as a point spread function (PSF) is also dependent on some of the parameters listed above and therefore would require a considerable amount of calibration measurements.

Alternative methods such as watershed and edge detection are also sensitive to noise and different SBRs [14,15].

The aim of this work is to develop an automatic segmentation algorithm for PET images which is as generic as possible, i.e., threshold independent and therefore does not require system-specific regression curves or PSF. Additionally, the volume estimates for small objects shall be improved. Therefore, in this paper, we propose an improved statistical PET image segmentation scheme. Our scheme relies on soft class assignment instead of hard class assignments and fuzzy levels. A Gaussian mixture model (GMM) is established to obtain initial estimates of the volume of the spheres. Subsequently, a MRF is obtained by declaring Markov properties for the unobserved label vector and using a Gibbs distribution to describe neighborhood dependencies. The MRF is then used to obtain the final labeling from the initial GMM labeling vector.

METHODS

Phantom measurements and clinical data

To evaluate the algorithm described below, a NEMA IEC Body Phantom was modified (built in-house at the Medical University of Vienna). The modified phantom differs from the original NEMA IEC Body Phantom only in the substitution of the largest sphere (37 mm) by a sphere of 8-mm diameter. Therefore, the phantom consists of a cylindrical outer body that simulates healthy tissue and six spherical inlays which represent tumor lesions with high tracer uptake. The cylindrical body was homogeneously filled with a water- ^{18}F-FDG solution of low activity concentration (background (BG)) and the spherical inlays were homogeneously filled with a water- ^{18}F-FDG solution of high activity concentration (foreground (FG)). A cylindrical inlay that models the lung was filled with air. The dimensions of all FG objects thus are (diameter [mm]/volume [ml]): 8/0.27, 10/0.52, 13/1.15, 17/2.57, 22/5.58, and 28/11.49. Measurements with different SBRs have been performed and are summarized in Table Table11.

Table 1: Measurements of the modified NEMA sphere phantom

FG	BG	SBR
10.94	5.30	2.06
20.37	5.30	3.84
26.13	5.30	4.90
66.56	9.90	6.72
90.90	9.68	9.39

Activity concentration for the FG objects and for the BG object in kBq/ml and the resulting SBR.

The device in use was a Siemens Biograph 64 TruePoint PET/CT scanner (Siemens, Erlangen, Germany). In accordance with the conditions for NEMA phantom quality assurance measurements in nuclear medicine [16], the average activity concentration never exceeded 10 kBq/ml. This way, the linearity of the scanner's noise equivalent count rate (NECR) is preserved, and the measurements of different SBRs can be compared. The acquisition was performed using emission scans of 10 min. The images were reconstructed with an iterative OSEM2D algorithm (4 iterations on 21 subsets). A preprocessing Gaussian filter of 5 mm was applied. The dimension and volume of the voxels are 4 mm ×4 mm ×3 mm and 0.048 ml, respectively. The more advanced iterative reconstruction algorithm for the Siemens scanner, TrueX (PSF), was not taken into account since recent studies recommended cautiousness with regard to its quantitative meaningfulness [17,18]. The chosen settings correspond to the clinical routine settings at the Medical University of Vienna.

In order to determine the algorithm's performance also with regard to non-spherical objects, measurements of another in-house built phantom containing cylindrical objects have been performed. Cylinders with high activity concentration (FG) and dimensions of (diameter [mm]/volume [ml]): 10/4.08, 15/12.23, 25/37.71, and 38/103.95 have been scanned against a BG with low activity concentration and SBRs according to Table Table2.2. All other settings were identical to the measurements of the modified NEMA IEC Body Phantom (see paragraphs above).

Table 2: Measurements of the cylinder phantom

FG	BG	SBR
34.20	17.40	2
42.70	10.80	4
53.20	8.90	6
75.30	9.40	8

Activity concentration for the FG objects and for the BG object in kBq/ml and the resulting SBR.

Finally, the algorithm was applied on ^{68}Ga-PET data of patients suffering from disseminated neuroendocrine carcinoma that was supplied by the European Neuroendocrine Tumor Society (ENETS) Center of Excellence at the Zentralklinik Bad Berka. Lesions in the liver and lymph nodes from eight patients were segmented, and the resulting volumes were compared to the manual delineation of the experts from the ENETS Center. The image-derived SBRs of the lesions in the lymph nodes are in the order of 15, whereas for the liver they are ≤ 2.

Statistical model

With regard to the smallest object diameter under consideration, which is only twice as large as the voxel size, PVE is a very dominating effect. It is therefore necessary to formulate partial memberships of voxels, as provided by the probability theory.

PET voxel values correspond to the activity concentration. Therefore, a PET data set comprising N voxels can be represented by an $N\times 1$ vector $\mathbf{x}=(x_1 \ldots x_N)$ whose elements are real-valued and positive, $x_n \geq 0$, $n=1,\ldots,N$. In order to describe the membership of the PET voxels in K distinct objects, we introduce the $K\times N$ label matrix:

$$Z = \begin{pmatrix} z_{11} & \ldots & z_{1N} \\ \vdots & \ddots & \vdots \\ z_{K1} & \ldots & z_{KN} \end{pmatrix} = (\mathbf{z}_1 \ldots \mathbf{z}_N). \tag{1}$$

The elements z_{nk} are binary, $z_{nk} \in \{0,1\}$, with $z_{nk}=1$ indicating that voxel n belongs object k (note that $z_{nl}=0$ for $l \neq k$, i.e., the row sums of $\mathbf{Z}$ equal 1).

PET image segmentation amounts to estimating the unknown label matrix $\mathbf{Z}$ given the data $\mathbf{x}$. The proposed algorithm consists of two consecutive steps: the *coarse estimation step* fits a basic model, yielding fairly good initial estimates. These estimates are then improved in the *correction step.* The coarse estimation is performed via a modified expectation-maximization algorithm for a Gaussian mixture model (EMGMM) using information from the analysis of phantom data. The correction step compensates for overestimation of small volumes by sampling from a Gaussian MRF, using Gibbs distributions to obtain the final labeling. The Gibbs interaction parameters are chosen to act only on voxels at the boundary of two objects, i.e., voxels whose neighbors are attributed to different objects in the coarse estimation step.

For a voxel belonging to object k, we model x_n as Gaussian with mean μ_k and standard deviation σ_k, $x_n \sim (x_n; \mu_k, \sigma_k)$. Since the label specifying the object is not known, the joint likelihood function of **x** and **Z** can be written as a GMM [19,20], which is used in the coarse estimation step:

$$p(\mathbf{x}, \mathbf{Z}; \Theta) = \prod_{k=1}^{K} \prod_{n=1}^{N} [\tau_k \mathcal{N}(x_n; \mu_k, \sigma_k)]^{z_{nk}} \quad (2)$$

Here, τ_k denotes the prior probabilities (normalized suchlike that $\sum_{k=1}^{K} \tau_k = 1$)and $\Theta = \begin{pmatrix} \mu_1 \ldots \mu_K \\ \sigma_1 \ldots \sigma_K \end{pmatrix}$ is a $2 \times K$ is a 2×K matrix containing all the Gaussian means and standard deviations.

Note that (2) does not model any statistical dependence (interaction) of different voxels. Such an interaction will be modeled only in the correction step with a Gibbs distribution [21-29] that imposes specific Markov properties on the label matrix **Z**, thus resulting in a Gaussian Markov random field (GMRF). More specifically, we use the model:

$$p(\mathbf{x}, \mathbf{Z}; \Theta, \Gamma) = \prod_{n=1}^{N} p(x_n|\mathbf{z}_n)\, p(\mathbf{Z}; \Gamma) \quad (3)$$

where $p(x_n|z_n) = \mathcal{N}(x_n; \boldsymbol{\Theta}\mathbf{z}_n)$ is Gaussian and models the voxel activity via the mean and standard deviation $(\mu_k,\sigma_k)=\boldsymbol{\Theta}\mathbf{z}_n$ associated to the kth object. Furthermore, $p(\mathbf{Z};\boldsymbol{\Gamma})$ is a Gibbs distribution [27,28] describing the interaction of the labels z_{nk}:

$$p(\mathbf{Z}; \Gamma) = \frac{1}{\mathcal{Z}} \prod_{n=1}^{N} \exp\left(- \sum_{m \in \mathcal{M}_n} \mathbf{z}_n^T \Gamma \mathbf{z}_m \right). \quad (4)$$

Here, $\mathcal{M}_n$ denotes the (first-order) neighborhood of the nth voxel, i.e., all voxels that share a surface with the nth voxel, hence $|\mathcal{M}_n| = 6$. Furthermore, the coupling matrix is given by $\Gamma = \begin{pmatrix} 0 & \gamma \\ \gamma & 0 \end{pmatrix}$.

This implies that statistical interaction is modeled only for the voxels with different labels, i.e., belonging to different objects. Finally, $\mathcal{Z}$ is the partition function [27,28]:

$$\mathcal{Z} = \sum_{\mathbf{Z}} \prod_{n=1}^{N} \exp\left(- \sum_{m \in \mathcal{M}_n} \mathbf{z}_n^T \Gamma \mathbf{z}_m \right) \quad (5)$$

The specific choice of the coupling matrix $\boldsymbol{\Gamma}$ involves only voxels on the boundary between two objects. Since the voxels at the boundary of a

FG object predominantly are part of the BG, the probability of labeling a boundary voxel as FG is decreased.

Coarse estimation step

The FG objects of the phantom are analyzed by defining VOIs. Within the VOI, only the FG (*k*=1) and the BG (*k*=2) need to be distinguished (hence, *K*=2, *cf.* (1)). The problem of simultaneously estimating the labeling matrix **Z** and the parameter matrix $\boldsymbol{\Theta}$ can be formulated in terms of the expectation-maximization (EM) algorithm (see [19-21,30-37] for further details and application):

$$\hat{\Theta}^{(i+1)} = \arg\max_{\Theta} E_{\mathbf{Z}|\mathbf{X};\hat{\Theta}^{(i)}} \left\{ \log p(\mathbf{x}, \mathbf{Z}; \Theta) \right\}, \tag{6}$$

where $p(\mathbf{x},\mathbf{Z};\boldsymbol{\Theta})$ is defined in (2).

Unfortunately, small FG objects consist only of a few voxels and hence represent poor statistical ensembles for the EM algorithm, resulting in inaccurate parameter estimates. Furthermore, in an FG object, the number of PVE voxels tends to be equal to or even larger than the number of pure FG voxels, causing the PVE voxels to be classified as FG (note that the statistical ensemble for the BG is rather large and therefore reliable). Consequently, we observe an SBR-dependent overestimation of the FG volumes which in many cases are not even morphologically connected objects. For the BG, the normalized standard deviation $\tilde{\sigma}_2 = \frac{\sigma_2}{\mu_2}$ amounts to 10% uniformly over all object sizes and SBRs. In contrast for the FG, $\tilde{\sigma}_1$ depends significantly on the object size and SBR as can be seen in Figure1.

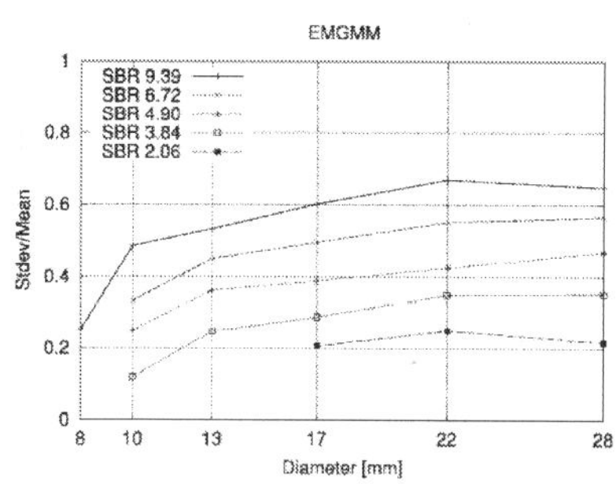

Figure 1: Normalized standard deviation versus object diameter. Plot of normalized standard deviation σ˜2 for BG (estimated using the conventional

EM algorithm) versus object diameter at several SBRs with the remodeled NEMA phantom (including morphologically not connected objects).

In order to avoid the problem with the increased FG standard deviation, we use a modified EM algorithm in which σ_1 is set equal to σ_2. The resulting EM starts with an initial, reasonable guess for the means and standard deviations of FG as well as BG and iterates Bayesian estimation and maximization as follows:

- The conditional expectation $\bar{z}_{nk}^{(i)} = E_{z_{nk}|x_n;\theta_k^{(i)}}\{z_{nk}\}$ is updated as:

$$\bar{z}_{nk}^{(i)} = \frac{\tau_k^{(i)} \mathcal{N}\left(x_n; \mu_k^{(i)}, \sigma_k^{(i)}\right)}{\sum_{k=1}^{2} \tau_k^{(i)} \mathcal{N}\left(x_n; \mu_k^{(i)}, \sigma_k^{(i)}\right)}. \quad (7)$$

- The parameter updates read:

$$\tau_k^{(i+1)} = \sum_{n=1}^{N} \frac{\bar{z}_{nk}^{(i)}}{N},$$

$$\mu_k^{(i+1)} = \frac{\sum_{n=1}^{N} x_n \bar{z}_{nk}^{(i)}}{\sum_{n=1}^{N} \bar{z}_{nk}^{(i)}},$$

$$\sigma_2^{(i+1)} = \sqrt{\frac{\sum_{n=1}^{N} (x_n - \mu_2^{(i+1)})^2 \bar{z}_{n2}^{(i)}}{\sum_{n=1}^{N} \bar{z}_{n2}^{(i)}}},$$

$$\sigma_1^{(i+1)} = \sigma_2^{(i+1)}. \quad (8)$$

- A soft labeling is achieved by using the conditional probabilities $\bar{z}_{nk}^{(i)}$:

$$\hat{z}_{nk}^{(i)} = \bar{z}_{nk}^{(i)}. \quad (9)$$

Due to the usage of a prior probability for **Z**, the resulting parameter estimations are done according to weighted averages.

Correction step

To improve the coarse estimates obtained via the modified EM algorithm operating on the Gaussian mixture model, a second labeling problem for **Z** is formulated on the GMRF (3) via a maximum *a posteriori* (xMAP) estimate:

$$\hat{\mathbf{Z}} = \arg\max_{\mathbf{Z}} \left\{ p(\mathbf{x}, \mathbf{Z}; \boldsymbol{\Theta}, \gamma) \right\}. \tag{10}$$

We solve (10) numerically by Metropolis sampling according to the procedure described below using the labeling obtained in the first coarse estimation step as initialization with its parameters held fixed to obtain a refined segmentation.

New labeling proposals are generated by changing local values of **Z** (i.e., changing a FG voxel to BG or vice versa). Assume $\mathbf{z}_n^{(i+1)}$ is the voxel label that has been changed and let $(\mu_k, \sigma_k) = \boldsymbol{\Theta}\mathbf{z}_n^{(i+1)}$; furthermore,

$$P_n^{(i+1)} \propto \exp\left(- \sum_{m \in \mathcal{M}_n} \mathbf{z}_n \boldsymbol{\Gamma} \mathbf{z}_m - \frac{(x_n - \mu_k)^2}{2\sigma_k^2} \right) \tag{11}$$

denotes the local conditional probability of the labeling proposal. If $P_n^{i+1} \geq P_{n'}^{i}$, the new label is accepted. If $P_n^{i+1} < P_{n'}^{i}$, the proposal labeling is accepted with probability Pi+1n/Pin and rejected with probability $1 - P_n^{i+1}/P_n^{i}$.

After a predefined number L of full samples (labeling configurations where each voxel is processed once) have been obtained in this manner, the final labeling is computed by forming the arithmetic mean of all accepted full proposal labeling, i.e.,

$$\hat{\mathbf{Z}} = \frac{1}{L} \sum_{l=1} \mathbf{Z}^{(l)}. \tag{12}$$

Threshold methods

For clinical applications of volume determination in PET, values of 36% to 42% of the maximum value have been proposed as local threshold [5]. In this context 'local' means that the threshold is calculated using the VOI and not the whole image.

With ITM, regression curves are needed, i.e. measurements of a phantom at different SBRs. The percentage threshold yielding the true volume is calculated as function of the volume and the SBR. Using ITM for automatic segmentation, an initial threshold is applied to a VOI followed by the determination of the resulting volume (V) and SBR. With those values, the corresponding threshold from the regress function %Thr=$f(V,\text{SBR})$ is further applied to the VOI, resulting in an iterative update scheme. The algorithm stops when the deviation between two iterations is ≤0.1%.

For purposes of evaluation and comparison, we applied the EMGMM, the proposed GMRF algorithm, and the aforementioned thresholding methods to the phantom measurements as well as the clinical data discussed in the 'Phantom measurements and clinical data' section. A graphical user interface was built by the object-oriented programming language IDL to visualize and process the data and to draw VOIs around each FG object. To investigate whether the results depend on the VOI size, we used VOIs consisting of 14×14×20, 14×14×40, 22×14×20, and 22×14×40 voxels.

RESULTS

Figures 2a,b,c,d,e,f, 3a,b, and 4a,b,c in this section show the relative volume error (in percent) for the FG objects versus their diameter. For each diameter, several bars are drawn to account for the different SBRs. FG objects are considered as detected if the segmentation yields two morphological connected objects, i.e., a FG object surrounded by BG volume. If a FG objects is not detected according to this definition, this is indicated graphically by a volume error of −5%. Furthermore, the dice similarity coefficient (DSC) is presented for the comparison of GMRF vs. ground truth of the CT as well as for the clinical Query: Please shorten the title for Figure 4 up to 15 words only to comply with the journal instruction. data.

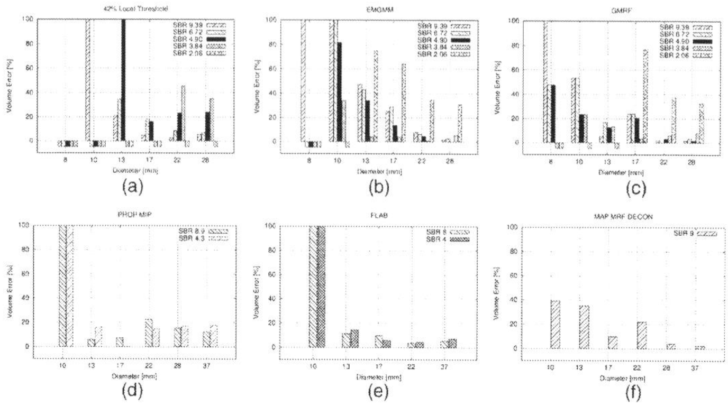

Figure 2: Volume error achieved by various methods for different NEMA sphere diameters, and SBRs. (a) 42% thresholding, (b) EMGMM, (c) GMRF, (d) statistical approach acting on MIP [40], (e) fuzzy locally adaptive Bayesian approach [38], and (f) maximum *a posteriori* GMRF approach with subsequent deconvolution [39]. (a), (b), and (c) use VOIs of 14×14×20 voxels; (d), (e), and (f) were copied visually from [38,40] and [39]. Note that the diameter ranges vary in (a-f).

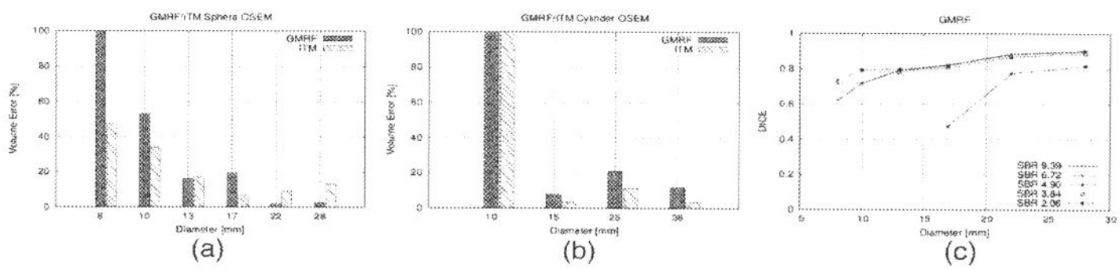

Figure 3: Volume error achieved by GMRF versus ITM averaged over different SBRs. (a) NEMA sphere OSEM reconstructions. (b) Cylinder phantom OSEM reconstructions. **(c)** DSC between GMRF and ground truth from CT scan for spherical objects.

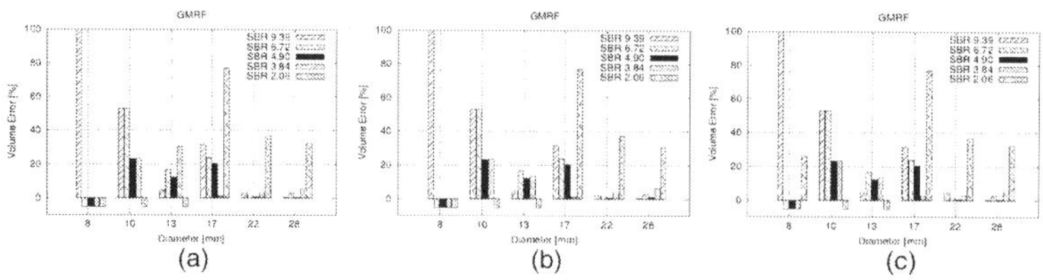

Figure 4: Volume error for different NEMA sphere diameters and SBRs obtained with GMRF on VOIs of size (a) *14×14×40* , (b) *22×14×20* , and (c) *22×14×40* voxels.

Furthermore, Tables Tables33 and and44 give an overview of the number of detected FG objects with regard to each segmentation strategy and SBR.

Table 3: Number of FG objects detected by 36% and 42% thresholding, ITM, EMGMM and GMRF

SBR	36%	42%	ITM	EMGMM	GMRF
2.06	0	0	0	4	3
3.84	0	2	4	5	5
4.90	3	4	4	5	6
6.72	4	4	5	5	6
9.39	5	5	6	6	6

The maximum amount of detectable FG objects is 6.

Table 4: Number of FG objects (cylinder phantom) detected at different SBRs by the ITM and the GMRF

SBR	ITM	GMRF
2	0	0
4	2	3
6	3	3
8	4	4

Threshold methods

Figure 2a shows the results for local thresholding with a threshold of 42%. Clearly, for small FG objects, the activity concentration is underestimated due to PVE which leads to large volume errors. It is seen that for all FG objects the error increases with decreasing SBR. This also applies to ITM whose results are depicted in Figure 3a,b. Here, the volume error is averaged over the SBR in order to keep the amount of results readily comprehensible. No threshold method can detect FG objects at an SBR of 2.06. Furthermore only the ITM is able to detect the smallest FG object of 8-mm diameter and only at the biggest SBR of 9.39.

EMGMM AND GMRF

The correction step is performed with γ=1,000 since the GMRF algorithm produces stable results in this region (see Figure5a).5a). Since the coarse estimation step provides initial values that are already close to the energetic minimum, the label configurations attain equilibrium after the first ten samples which get discarded. For $L > 70$, no change in results is observed. Figure 5b shows the GMRF segmentation result of the 28-mm sphere at SBR=9.39 as compared to the ground truth derived from the CT. For this, the center of the sphere was identified in the high-resolution CT and the known radius of the real phantom was plotted.

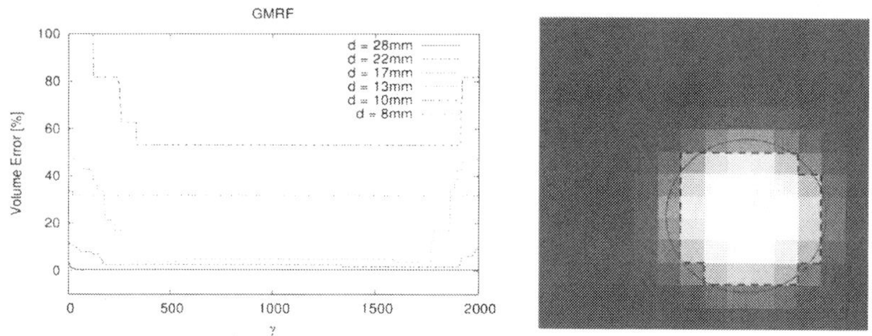

(a) (b)

Figure 5: Volume error and segmentation result. (a) Volume error achieved by GMRF for different NEMA sphere diameters at SBR=9.39 drawn over the parameter γ. (b) Segmentation result of the 28-mm sphere at SBR=9.39 after application of the GMRF (dashed lines) and ground truth from CT (black circle).

The results of the proposed GMRF method and its EMGMM initialization are shown in Figure,2c,b, respectively, for VOIs comprising 14×14×20 voxels. With regard to both spheres and cylinders, EMGMM as well as GMRF achieve a much better detection rate of FG objects than the threshold methods (see Tables Tables3 and 4).4). In particular, GMRF detects all six spherical FG objects at SBRs above 4.90 as well as the three larger spheres at SBR 2.03. The total detection rate score of the GMRF with 26/30 and 10/16 for spheres and cylinders, respectively, clearly supersedes the ITM with 19/30 and 9/16.

Concerning volume segmentation, GMRF achieves lower errors than the fixed threshold approaches for almost all combinations of SBR and diameter. In a direct comparison of GMRF and ITM (see Figure 3a,b), ITM performs better on smaller spheres while GRMF is better on larger ones. For cylindrical objects, ITM shows slightly smaller errors for diameters ≥15 mm. Figure 3c shows the DSC which was calculated for the phantom with spherical objects as a comparison of GMRF segmentation vs. ground truth from the CT. For objects with diameters ≥13 mm and SBR ≥3.84, the DSC ≥0.8.

Comparison to threshold-independent algorithms

Figure2e,f shows the results obtained by Hatt and co-workers using a fuzzy locally adaptive Bayesian (FLAB) algorithm [38] and by the maximum *a posteriori* estimation for a GMRF with deconvolution (MAP-MRF-DECON) designed by Gribben et al. [39], respectively (the data has been manually copied from the above cited publications). For further comparison, a statistical approach by Dewalle-Vignion and co-workers acting on maximum intensity projections (PROP MIP) [40] including the fuzzy c-means technique is shown in Figure 2d. While those papers also performed NEMA phantom measurements, FG objects with different diameter ranges were used. In particular, none of the previous works used a FG object as small as 8 mm of diameter or an SBR as low as 2.06.

Comparing GMRF with the MAP-MRF-DECON algorithm, the former performs better for some diameters and worse on others. However, a systematic comparison is not possible since [39] provides results only for the rather large SBR of 9. PROP MIP and FLAB yield quite accurate volumes for spheres greater than 10 mm but performs poorly when it comes to spheres smaller than 13-mm diameter (the smallest sphere used in [40] and [38] is 10 mm). At such diameters, GMRF approaches have to be preferred due to their better performance.

VOI dependencies of GMRF

Finally, Figure 4 shows the results obtained with GMRF for different VOI sizes. As can be seen for all spheres with diameter ≥10 mm, the volume errors are constant over the chosen VOI range with regard to all SBRs. Concerning the 8-mm sphere, no correlation can be found between the detection rate and VOI size. However, if detected, the volume error is also constant over the VOI range.

Clinical data

With regard to the clinical data, GMRF's segmentation of the metastases in the lymph nodes yields volumes which are 15% to 20% bigger than the manual delineation. This value is constant over a volume range down to 3 ml wherefrom the overestimation gets bigger. Taking into account that the SBR for the lesions in the lymph nodes is 15, this behavior is very similar to the measurements of the NEMA spheres at higher SBRs. Likewise, the GMRF's segmentation of lesions in the liver with very small SBRs (≤2) shows a volume which is constantly smaller by 30% and therefore reflects very well the behavior of phantom measurements at very low SBRs.

Figure 6 shows the DSC between the manual experts' delineation and the resulting volumes when GMRF is applied to the clinical data. For metastases in the lymph nodes, the mean DSC is 0.89 and for the liver 0.80.

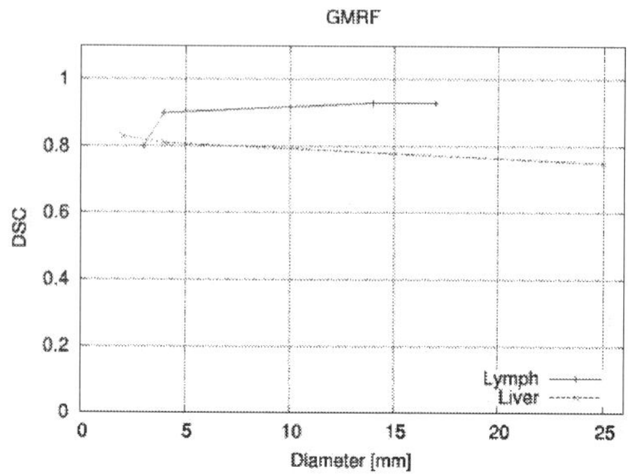

Figure 6: DSC between manual delineation and GMRF for [68] Ga-PET of neuroendocrine metastases in the liver and lymph node.

DISCUSSION

Forcing equal standard deviations for BG and FG as discussed in the 'Coarse estimation step' section helped to reduce the segmentation error. While this approach appears somehow arbitrary, it can be qualitatively understood as follows: Figure 1 shows the SBR diversification for the normalized standard deviation $\tilde{\sigma}1$ of FG. For higher SBR, $\tilde{\sigma}1$ and therewith the volume estimates increase because the PVE voxels get encompassed by the FG object. By setting $\sigma_1=\sigma_2$, the PVE voxels are interpreted as BG such that in the subsequent estimate σ_2 increases. As a result, also σ_1 increases and parts of the PVE voxels are re-assigned to FG, in turn lowering σ_1. This interplay causes a conversion towards the solutions shown in the 'Results' section.

Likewise, the value γ=1,000 was introduced without further explanation. Numerical calculations have shown that values for γ between 500 and 1,500 yield stable non-zero volume estimates for the FG objects instead of downgrading the volumes until they vanish (increasing the amount of neighborhood voxels $|M_n|$ leads to zero volume solutions). Since the coarse estimation step already yields good initial label estimates with a large value of γ as defined above, the system reaches equilibrium after the first three full samples have been obtained. Storing more than 70 subsequent configurations at equilibrium according to (12) does not change the solutions presented here.

When comparing the proposed GMRF method and its initializing EMGMM (see Figure Figure2c,b),2c,b), it becomes apparent that the proposed correction step enforced in the GMRF decreases the volume error for small FG objects as intended by the choices described in the 'Correction step' section. This emphasizes the usefulness of MRFs as a powerful tool regarding PET image segmentation, especially in the case of small objects. In this sense, the findings presented here confirm the work of Gribben et al. [39] that also uses MRFs. Nonetheless, it can be seen in Figure 2 that the initialization step is not fully corrected for SBR diversification, especially for small FG objects. Acting with Gibbs distributions on this solution as described above is not sensible to different SBRs. Therefore, future work should aim for compensating SBR dependencies in the correction step.

The results of the GMRF also show that for spheres with diameter ≥13 mm, the cold-wall effect can be accounted for. For smaller spheres, the cold-wall effect gets noticeable as for all segmentation approaches in the literature [8-40].

The choice of the comparison methods shall be briefly outlined. Despite the existence of iterative threshold approaches (see the 'Background' section), fixed thresholds are very common in clinical practice. As we can see in the 'Results' section, GMRF clearly outreaches the fixed threshold approaches both in the detection rate and volume error, a result not very surprising. However, the fact that GMRF also outperforms ITM on detection rate and produces similar results with regard to the volume error and DSC is remarkable all the more, as GMRF uses no *a priori* knowledge whatsoever whereas ITM has a the luxury of a validated calibration curve. Taking into account the dependencies of these regression curves (see the 'Background' section) and as a consequence thereof the need of not only one but many phantom calibration measurements, GMRF seems a much more practical approach for clinics since it is equally reliable but can be directly applied for different systems, nuclide, reconstructions, etc. Furthermore, an ITM only can realize a segmentation of whole voxels whereas the inclusion of probability theory as in our approach can assign partial classifications to voxels which is inevitable in the case of small volumes. The comparison with the threshold-independent algorithms also shows that the algorithm in this work performs equally well for larger volumes and even shows improvements for small volumes and SBRs.

When discussing the segmentation of the clinical data, one has always to recall that unlike the real volume of the phantoms the manual delineation does not represent the ground truth, i.e., it can only be a comparative analysis. Systematic overestimation as well as underestimation of the tumor volume has an extensive impact on radiation treatment planning. In practice, the so-called planning target volume (PTV) is an extension of the visible tumor taking into account microscopic spread as well as uncertainties in dose delivery and patient positioning. Therefore, a systematic overestimation will partially coincide with the PTV and be of no severe consequences as long as the overlap is not too big and thus causes widespread damage to the surrounding healthy tissue. In contrast, a systematic underestimation involves the danger of not irradiating the entire tumor and thus enhances the chances for relapse.

A mean DSC of 0.9 for high SBRs, the robustness of the results over a large volume range, and the resemblance to the results of the phantom measurements concerning the SBR dependency make the GMRF a suitable candidate for future studies that encompass patient data of higher quantity and larger diversity. Moreover, to study realistic images with inhomogeneous activity distributions and be in possession of the ground truth, simulated images have to be included as also done by various authors [11,40-43].

The proposed GMRF produces stable results for different lesion shapes (see Figure 3) and therefore underlines its potential for future clinical use. Another benefit of the proposed algorithm is its fast convergence. With the termination condition:

$$|\mu_k^{(i+1)} - \mu_k^{(i)}| < 0.01, \quad k = 1, 2, \tag{13}$$

the number of EMGMM iterations in the case of detection did not exceed 50. Given a standard laptop (dual core 2×1.8 GHz), the overall processing time stays well below 1 s even for the case of large VOI. In this time frame, it is even feasible to use the algorithm several times to repeat the processing with subVOIs to improve the detection rate for small objects.

CONCLUSIONS

Segmentation of PET data remains a very challenging issue since pertinent algorithms are very sensitive to a variety of parameters. The aim of this work was to investigate the aspects that were not sufficiently covered in the literature so far, namely the impact of the SBR, the segmentation of objects with a diameter smaller than 10 mm, and the waiver of any *a priori* knowledge such as a regression curve. Therefore, phantom measurements with spherical as well as non-spherical objects with SBRs ranging from 2 to 9.36 have been evaluated, including a FG sphere with a diameter of 8 mm. Additionally, lesions from clinical data have been segmented.

Combining an EMGMM with MRFs, taking advantage of Gibbs distributions to describe neighbor dependencies, results in a significant decrease of the overestimation of small volumes on the one hand and on the other hand yields vanishing volume errors in the case of bigger objects and high SBRs. The proposed algorithm has advantages over threshold methods and can be applied to any PET data, not requiring any system-specific regression curves in order to account for the given nuclide, manufacturer, or reconstruction algorithm. The comparison with experts' manual delineations on clinical images shows the same qualitative behavior as for the phantom measurements. In connection with its rapidness and the insensitivity towards reconstruction algorithm and lesion shape, the proposed algorithm is a suitable choice for PET segmentation, even though there is still room for improvement in future work.

ACKNOWLEDGEMENTS

This work was co-funded by the Austrian Federal Ministry for Transport, Innovation and Technology within the program ModSim Computational Mathematics which is part of the program Research, Innovation, Technology and Information Technology.

REFERENCES

1. Hamilton CS, Ebert MA. Volumetric uncertainty in radiotherapy. Clin Oncol. 2005;17:456–64. doi: 10.1016/j.clon.2005.03.014.
2. Greco C, Rosenzweig K, Cascini GL, Tamburrini O. Current status of PET/CT for tumour volume definition in radiotherapy treatment planning for non-small cell lung cancer (NSCLC) Lung Cancer. 2007;57:125–34. doi: 10.1016/j.lungcan.2007.03.020.
3. Breen SL, Publicover J, De Silva S, Pond G, Brock K, O'Sullivan B, et al. Intraobserver and interobserver variability in GTV delineation on FDG-PET-CT images of head and neck cancers. Int J Radiat Oncol Biol Phys. 2007;68:763–70. doi: 10.1016/j.ijrobp.2006.12.039.
4. Morarji K, Fowler A, Vinod SK, Ho Shon I, Laurence JM. Impact of FDG-PET on lung cancer delineation for radiotherapy. J Med Imaging Radiat Oncol. 2012;56:195–203. doi: 10.1111/j.1754-9485.2012.02356.x.
5. Erdi YE, Mawlawi O, Larson SM, Imbriaco M, Yeung H, Finn R, et al. Segmentation of lung lesion volume by adaptive positron emission tomography image thresholding. Cancer. 1997;80:2505–9. doi: 10.1002/(SICI)1097-0142(19971215)80:12+<2505::AID-CNCR24>3.0.CO;2-F.
6. Erdi YE, Rosenzweig K, Erdi AK, Macapinlac HA, Hu YC, Braban LE, et al. Radiotherapy treatment planning for patients with non-small cell lung cancer using positron emission tomography (PET) Radiother Oncol. 2002;62:51–60. doi: 10.1016/S0167-8140(01)00470-4.
7. MacManus M, Nestle U, Rosenzweig KE, Carrio I, Messa C, Belohlavek O, et al. Use of PET and PET/CT for radiation therapy planning: IAEA expert report 2006-2007. Radiother Oncol. 2009;91:85–94. doi: 10.1016/j.radonc.2008.11.008.
8. Daisne JF, Sibomana M, Bol A, Doumont T, Lonneux M, Grégoire V. Tri-dimensional automatic segmentation of PET volumes based on measured source-to-background ratios: influence of reconstruction algorithms. Radiother Oncol. 2003;69:247–50. doi: 10.1016/S0167-8140(03)00270-6.
9. Brambilla M, Matheoud R, Secco C, Loi G, Krengli M, Inglese E. Threshold segmentation for PET target volume delineation in radiation treatment planning: the role of target-to-background ratio and target size. Med Phys. 2008;35:1207–13. doi: 10.1118/1.2870215.

10. Jentzen W, Freudenberg L, Eising EG, Heinze M, Brandau W, Bockisch A. Segmentation of PET volumes by iterative image thresholding. J Nucl Med. 2007;48:108–14. [PubMed]
11. Tylski P, Stute S, Grotus N, Doyeux K, Hapdey S, Gardin I, et al. Comparative assessment of methods for estimating tumor volume and standardized uptake value in (18)F-FDG PET. J Nucl Med. 2010;51:268–76. doi: 10.2967/jnumed.109.066241.
12. Vauclin S, Doyeux K, Hapdey S, Edet-Sanson A, Vera P, Gardin I. Development of a generic thresholding algorithm for the delineation of 18FDG-PET-positive tissue: application to the comparison of three thresholding models. Phys Med Biol. 2009;54:6901–16. doi: 10.1088/0031-9155/54/22/010.
13. Lee JA. Segmentation of positron emission tomography images: some recommendations for target delineation in radiation oncology. Radiother Oncol. 2010;96:302–7. doi: 10.1016/j.radonc.2010.07.003.
14. Riddell C, Brigger P, Carson RE, Bacharach SL. The watershed algorithm: a method to segment noisy PET transmission images. IEEE Trans Nucl Sci. 1999;46:713–9. doi: 10.1109/23.775604.
15. Reutter B, Klein GJ, Huesman RH. Automated 3-D segmentation of respiratory-gated pet transmission images. IEEE Trans Nucl Sci. 1997;44:2473–6. doi: 10.1109/23.656454.
16. Herzog H, Tellmann L, Hocke C, Pietrzyk U, Casey ME, Kuwert T. NEMA NU2-2001 guided performance evaluation of four Siemens ECAT PET-scanners. IEEE Trans Nucl Sci. 2004;51:2662–9. doi: 10.1109/TNS.2004.835778.
17. Knäusl B, Hirtl A, Dobrozemsky G, Bergmann H, Kletter K, Dudczak R, et al. PET based volume segmentation with emphasis on the iterative TrueX algorithm. Zeitschrift für medizinische Physik. 2011;22:29–39. doi: 10.1016/j.zemedi.2010.12.003.
18. Knäusl B, Rausch IF, Bergmann H, Dudczak R, Hirtl A, Georg D. Influence of PET reconstruction parameters on the TrueX algorithm. a combined phantom and patient study. Nuklearmedizin. 2013;52:28–35. doi: 10.3413/Nukmed-0523-12-07.
19. Bishop CM. Mixture models and EM. In: Jordan M, Kleinberg J, Schölkopf B, editors. Pattern recognition and machine learning. New York: Springer; 2006.
20. McLachlan GJ, Krishnan T. The EM algorithm and extensions. New

York: Wiley; 1997.

21. Blekas K, Likas A, Galatsanos NP, Lagaris IE. A spatially constrained mixture model for image segmentation. IEEE Trans Neural Netw. 2005;16(2):494–8. doi: 10.1109/TNN.2004.841773.
22. Kato Z, Zerubia J, Berthold M, Pieczyński W. Unsupervised adaptive image segmentation. IEEE Int Conf Acoustics, Speech and Signal Processing. 1995;4:2399–2402.
23. Pieczyński W. Statistical image segmentation. Mach Graph Vis. 1992;1(1/2):261–8.
24. Braathen B, Pieczyński W. Global and local methods of unsupervised Bayesian segmentation of images. Mach Graph Vis. 1993;2(1):39–52.
25. Lakshmanan S, Derin H. Simultaneous parameter estimation and segmentation of Gibbs random fields using simulated annealing. IEEE Trans Pattern Anal Mach Intell. 1989;11:799–813. doi: 10.1109/34.31443.
26. Montgomery DWG, Amira A, Zaidi H. Fully automated segmentation of oncological PET volumes using a combined multiscale and statistical model. Med Phys. 2007;34:722–36. doi: 10.1118/1.2432404.
27. Bishop CM. Graphical models. In: Jordan M, Kleinberg J, Schölkopf B, editors. Pattern recognition and machine learning. New York: Springer; 2006.
28. Li SZ. Markov random field modeling in image analysis. Beijing: Springer; 2009.
29. MacKay D. Information theory, inference, and learning algorithms. Cambridge: Cambridge University Press; 2005. Maximum likelihood and clustering.
30. MacKay D. Information theory, inference, and learning algorithms. Cambridge: Cambridge University Press; 2005. Ising models.
31. Pokrić M, Thacker NA, Scott MLJ, Jackson A. Medical image computing and computer-assisted intervention, vol. 2208. Utrecht, The Netherlands: Springer; 2001. Multidimensional medical image segmentation with partial voluming.
32. Moon TK. The expectation-maximization algorithm. IEEE Signal Processing Mag. 1996;13:47–60. doi: 10.1109/79.543975.
33. Fessler JA, Hero AO. Space-alternating generalized expectation-maximization algorithm. IEEE Trans Signal Process. 1994;42:2664–

77. doi: 10.1109/78.324732.

34. Fessler JA, Hero AO. Penalized maximum-likelihood image reconstruction using space-alternating generalized EM algorithms. IEEE Trans Image Process. 1995;4:1417–29. doi: 10.1109/83.465106.
35. Dempster AP, Laird NM, Rubin DB. Maximum likelihood from incomplete data via the EM algorithm. J Roy Stat Soc. 1977;39(1):1–38.
36. Kay SM. Maximum likelihood estimation. In: Oppenheim AV, editor. Fundamentals of statistical signal processing, volume I: estimation theory. New Jersey: Prentice Hall; 1993.
37. Su KH, Chen JS, Lee JS, Hu CM, Chang CW, Chou YH, et al. Image segmentation and activity estimation for microPET 11C-raclopride images using an expectation-maximum algorithm with a mixture of Poisson distributions. Comput Med Imaging Graph. 2011;35:417–26. doi: 10.1016/j.compmedimag.2011.01.004.
38. Hatt M, Cheze le Rest C, Turzo A, Roux C, Visvikis D. A fuzzy locally adaptive Bayesian segmentation approach for volume determination in PET. IEEE Trans Med Imag. 2009;28:881–93. doi: 10.1109/TMI.2008.2012036.
39. Gribben H, Miller P, Wang H, Carson K, Hounsell A, Zatari A. 5th IEEE int symp biomedical imaging: from nano macro. Paris, France: The institute of electrical and electronic engineers engineering in medicine and biology society signal processing society;; 2008.. Automated MAP-MRF EM labelling for volume determination in PET.
40. Dewalle-Vignion AS, Betrouni N, Lopes R, Huglo D, Stute S, Vermandel M. A new method for volume segmentation of PET images, based on possibility theory. IEEE Trans Med Imag. 2011;30:409–23. doi: 10.1109/TMI.2010.2083681.
41. Hatt M, Lamare F, Boussion N, Roux C, Turzo A, Cheze Lerest C, et al. Fuzzy hidden Markov chains segmentation for volume detemination and quantitation in PET. Phys Med Biol. 2007;52:3467–91. doi: 10.1088/0031-9155/52/12/010.
42. Hatt M, Boussion N, Lamare F, Collet C, Salzenstein F, Roux C, et al. 3rd IEEE international symposium on biomedical imaging: nano to macro. Arlington, Virginia, USA: The institute of electrical and electronic engineers engineering in medicine and biology society signal processing society;; 2006. Fuzzy versus hard hidden Markov

chains segmentation for volume determination and quantitation in noisy PET images.

43. Ballangan C, Wang X, Fulham M, Eberl S, Yin Y, Feng D. Automated delineation of lung tumors in PET images based on monotonicity and a tumor-customized criterion. IEEE Trans Inf Technol Biomed. 2011;15:691–702. doi: 10.1109/TITB.2011.2159307.

CHAPTER

8

Kernel Analysis Based on Dirichlet Processes Mixture Models

Jinkai Tian [1,2], Peifeng Yan [1,2] and Da Huang [1,2],

[1] State Key Laboratory of High Performance Computing, National University of Defense Technology, Changsha 410073, China

[2] College of Computer, National University of Defense Technology, Changsha 410073, China

ABSTRACT

Kernels play a crucial role in Gaussian process regression. Analyzing kernels from their spectral domain has attracted extensive attention in recent years. Gaussian mixture models (GMM) are used to model the spectrum of kernels. However, the number of components in a GMM is fixed. Thus, this model suffers from overfitting or underfitting. In this paper, we try to combine the

Citation: (APA): Tian, J., Yan, P., & Huang, D. (2019). Kernel Analysis Based on Dirichlet Processes Mixture Models. Entropy, 21(9), 857. (18 pages) DOI: https://doi.org/10.3390/e21090857 URL: https://www.mdpi.com/1099-4300/21/9/857/htm

spectral domain of kernels with nonparametric Bayesian models. Dirichlet processes mixture models are used to resolve this problem by changing the number of components according to the data size. Multiple experiments have been conducted on this model and it shows competitive performance.

Keywords: Dirichlet processes; Dirichlet processes mixture models; Gaussian mixture models; kernel; spectral analysis

INTRODUCTION

Probabilistic models are essential for machine learning to have a correct understanding of problems. While frequentist models are considered a traditional method for statisticians, Bayesian models are more widely used in machine learning fields. A Bayesian model is a statistical model where probability is used to represent all uncertainties in the model. Uncertainty of both input and output are under consideration in a Bayesian model. A parametric Bayesian model assumes a finite set of parameters which is not flexible because the complexity of the model is bounded while the amount of data is unbounded. A Bayesian nonparametric model is a Bayesian model on infinite-dimensional parameter space. The amount of information captured by the model from the data can grow as the amount of the data grows. These models are always connected with some kinds of stochastic processes to describe an infinite number of parameters. For example, the Gaussian process (GP) and Dirichlet process mixture models (DPMM) are two cornerstones of Bayesian nonparametric models.

Dirichlet processes were first introduced in Reference [**1**]. The stick-breaking methods [**2**], Pólya urn scheme [**3**] and the Chinese restaurant process (CRP) [**4**,**5**] are several different definitions and constructions of Dirichlet processes. Dirichlet processes are the basis for many Bayesian nonparametric models and are used to process spatial data [**6**,**7**], tree-structured data [**8**,**9**], relational data [**10**,**11**] and sequential data [**12**,**13**,**14**,**15**].

The recent development of deep learning has led to the extensive use of the neural network in a variety of domains and a higher performance than conventional methods is achieved. However, this is less than satisfactory for statisticians because of the uninterpretability of black-box function approximation. Neal [**16**] shows that as the number of hidden unions approaches infinity, Bayesian neural networks converge to Gaussian processes. Thus Gaussian processes are regarded as powerful and interpretable alternatives to neural networks. It provides a principled, practical and probabilistic approach to learning in kernel machines [**17**].

The property of a Gaussian process is closely related to its kernel (or covariance function). Kernels provide a depiction of smoothness and periodicity for models and parameter selection of kernel is essential for model training. There are many widely used kernels, for example, the squared exponential (SE) kernels, rational quadratic kernels, polynomial kernels and the Matérn kernels family. The expressiveness of a kernel is defined as the ability to discover the inherent patterns of the data.

Recent research [**18**, **19**, **20**] has proposed some expressive kernels to discover patterns without human intervention. Deep architectures are used in designing kernels. Other previous work [**21**, **22**] sought to combine the Gaussian process into a Bayesian neural network framework. However, these approaches usually do not have closed forms and are thus less interpretable. Meanwhile, sophisticated approximate inference techniques, for example, variational inference and expectation propagation, are required in the training step. Combining and modifying existing kernels to make new kernels is straightforward because the sum, the product or the convolution of two kernels is also a kernel. There are many challenges [**23**, **24**] to designing kernels according to this principle for specialized applications. Specific restrictions are enforced on these kernels to avoid overfitting and complicated inference. However, it is hard to generalize these kernels to other applications for their specialized kernel structures.

We use Gaussian mixture models (GMM) to analyze the spectral density of popular kernels and find that traditional GMM suffer from overfitting and underfitting by a fixed number of components. Therefore, we seek to combine an infinite mixture model in designing kernels. This innovation introduces more flexibility and expressiveness into designed kernels.

Our work is distinct in that we combine Dirichlet process mixture models into analyzing and designing kernels from the spectral domain. Using traditional Gaussian mixture models in kernel analysis [**25**] imposes restraint on the expressiveness of kernels. Dirichlet process mixture models are flexible models to fit data because the complexity of models grows as more data are observed. By using it to our advantage, we propose the infinite mixture (IM) kernel. The combination of these two traditional Bayesian nonparametric models is one of the most innovative contributions of this work. Theoretically, we prove that with possibly infinite components in mixture models, designed kernels can fit any linear or nonlinear function by arbitrary accuracy. Besides, a Markov Chain Monte Carlo (MCMC) implementation of a hierarchical Bayesian model for spectral analysis is presented and experiments are conducted on real-world temporal data which show the IM kernel's competitiveness in comparison with other kernels and its superiority in extrapolation.

To the best of our knowledge, this is the first trial to combine the Dirichlet process mixture model in designing kernels. The result is promising on account of the computational convenience and expressiveness of Dirichlet process mixture models. Our infinite version has several advantages. (1) We use MCMC methods (collapsed Gibbs sampling specifically) to avoid the local minimum problem which frequently occurs in derivative-based methods like EM used in conventional GMM. (2) The number of Gaussian components K is automatically determined by the provided data, which alleviates the requirement for expertise. On the other hand, it is beneficial for the customization of a kernel for a specific scene. (3) Setting K in advance limits the expressiveness of kernels, while an infinite version makes more details available.

The main contribution of the paper can be summarized in two parts:

- We try to combine a hierarchical Bayesian model with the spectral mixture kernel. Our method shows better and more robust performance.
- The original spectral mixture kernel initializes the mean of each component uniformly between 0 and the Nyquist frequency. We analyze the robustness of the spectral mixture kernel and find this strategy to have fluctuant performance. One of the strongest advantages of our method is that the infinite version has a better solution for initialization and has more stable performance.

In **Section 2**, we have a brief review of the Dirichlet process and the Gaussian process. We introduce a hierarchical Dirichlet mixture model in **Section 3** and use this model to analyze kernels in the spectral domain and put forward the concept of *the infinite mixture kernel* in **Section 4**. In **Section 5**, we conduct multiple experiments to show the robustness and good performance of the IM kernel.

BACKGROUND

Dirichlet Distribution

Dirichlet distribution is an extension of Beta distribution from bivariate to multivariate and can be expressed as:

$$f(x_1,\ldots,x_K;\alpha_1,\ldots,\alpha_K) = \frac{1}{\mathrm{B}(\alpha)}\prod_{i=1}^{K} x_i^{\alpha_i - 1}, \quad (1)$$

Where $x_1,\ldots,x_K > 0$ and $\sum_{i=1}^{K} = 1.$. The Beta function B(α) is the normalizer of the distribution and can be expressed in the form of Gamma function:

$$\mathrm{B}(\alpha) = \frac{\prod_{i=1}^{K}\Gamma(\alpha_i)}{\Gamma\left(\sum_{i=1}^{K}\alpha_i\right)}, \quad \alpha = (\alpha_1,\ldots,\alpha_K). \quad (2)$$

Thus we can regard a sample from the Dirichlet distribution $(x_1,\ldots,x_K)$ as a distribution itself.

The probability density function (PDF) of the gamma distribution is:

$$p(x|\alpha,\theta) = \mathcal{G}(\alpha,\theta) = \frac{x^{\alpha-1}e^{-x/\theta}}{\Gamma(\alpha)\theta^{\alpha}}. \quad (3)$$

There is another version of the PDF of the gamma distribution which is also widely used:

$$p(x|\alpha,\theta) = \mathcal{G}_R(\alpha,\theta) = \frac{x^{\alpha/2-1}e^{-\alpha x/2\theta}}{\Gamma(\alpha/2)(2\theta/\alpha)^{\alpha/2}}. \quad (4)$$

The transformation between these two versions is:

$$\mathcal{G}(\alpha,\theta) = \mathcal{G}_R(2\alpha,\alpha\theta) \quad (5)$$

$$\mathcal{G}_R(\alpha,\theta) = \mathcal{G}(\frac{\alpha}{2},\frac{2\theta}{\alpha}). \quad (6)$$

Dirichlet Process

Dirichlet processes (DP) are a family of stochastic processes whose observations are probability distributions. A Dirichlet process has two parameters, a base distribution *H* and a positive real-valued scalar α called scaling or concentration parameter. Given these two parameters and a measurable set *S*, the Dirichlet process DP(H,α) is a stochastic process whose sample path is a probability distribution over *S*, such that the following holds:

For any finite partition of *S*, denoted $B_{i=1}^{n}$, if

$$G \sim \mathrm{DP}(H,\alpha), \quad (7)$$

then

$$((G(B_1),\ldots,G(B_n)) \sim \mathrm{Dir}(\alpha H(B_1),\ldots,\alpha H(B_n)). \quad (8)$$

The concentration parameter α is considered the pseudo count number and the base distribution H can be considered the prior knowledge of the distribution [**26**].

Gaussian Process

The Gaussian process is widely used as a statistical model for modeling functions. In the field of machine learning, we are not especially interested in drawing random function from the prior. There always exist raw training data or evaluation results of a model from a series of experiments iterations. These data provide extra knowledge of the model and incorporating this knowledge into the model is of primary interest. In the Gaussian process regression, we seek to predict f(x) based on limited observations.

We express Gaussian process as:

$$f(x) \sim \mathcal{GP}(m(x), k(x,x')). \quad (9)$$

Suppose f(x) is a real process, we can define the mean function m(x) and the covariance function k(x,x′) as:

$$\begin{aligned} m(x) &= \mathbb{E}[f(x)], \\ k(x,x') &= \mathbb{E}[(f(x)-m(x))(f(x')-m(x'))]. \end{aligned} \quad (10)$$

A HIERARCHICAL DIRICHLET PROCESS MIXTURE MODEL

A Gaussian mixture model can be written as:

$$p(y|\mu_1,\ldots,\mu_K,\Lambda_1,\ldots,\Lambda_K,\pi_1,\ldots,\pi_K) = \sum_{i=1}^{K} \pi_i \mathcal{N}(\mu_i,\Lambda_i^{-1}), \quad (11)$$

where μ_i denotes the mean of the *i*-th component, Λ_i denotes the precision of the *i*-th component and πi denotes the weight or mixture coefficient of the *i*-th component which satisfies $\pi_i \geq 0$ for $i = 1,\ldots,K$. A GMM model can be learned using expectation–maximization (EM) algorithm. However, the number of components K is set manually here which is inconvenient in practical applications because K should vary with the complexity of data.

Given hyperparameters for means, precisions and weights respectively, a hierarchical model in high dimension with a potentially infinite number of Gaussian components is implemented. We first consider inference in a finite version with *K* components and then extend it into infinite hierarchical models as the limiting case of the finite one.

We outline our model in a probabilistic graphical model for clarity. **Figure 1** illustrates the hierarchical structure embedded in the model. Nine nodes represent different collections of random variables and one node represents observations. We classify these random variables into hyperparameters, global variables and local variables. The first column contains hyperparameters with priors set by sufficient statistics from the data. The second column contains means, precisions and weights of *K* components which we call global variables. The node tagged by ci represents the local variable that indicates which component generates this data point.

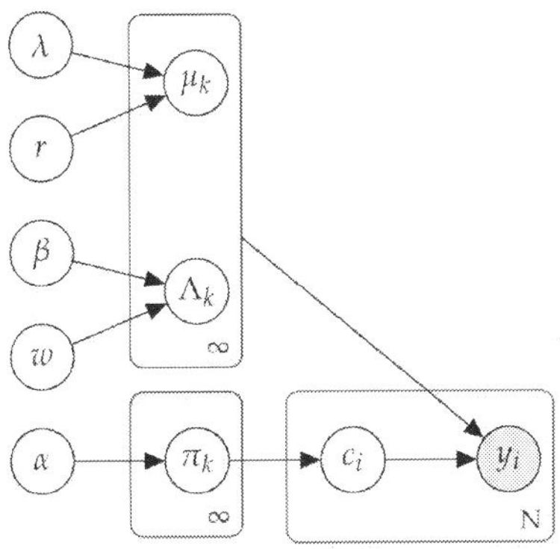

Figure 1. Graphical model representation of an infinite hierarchical mixture model. Nodes denote random variables. Under the condition of high dimensionality, each node can be a collection of multiple random variables. Edges denote dependence. Two blue plates denote replication of global hidden variables and the red plate denotes replication of local hidden variables and observations. Notice that blue plates have potentially infinite number of replications, while the red plate has only *N* replications corresponding to each data point.

In Gaussian mixture models, the data $y_{j\,j=1}^{N}$ is assumed to come from a generative model:

$$y_j|c_j \sim \mathcal{N}\left(\mu_{c_j}, \Lambda_{c_j}^{-1}\right). \quad (12)$$

To establish a hierarchical model, we put priors on μi and Λi:

$$p(\mu_i|\lambda, r) \sim \mathcal{N}\left(\lambda, r^{-1}\right), \quad p(\Lambda_i|\beta, w) \sim \mathcal{G}\left(\beta, w^{-1}\right), \quad (13)$$

where G denotes the gamma distribution. First we have a deeper insight into μ. We give hyperparameters λ and r vague priors as:

$$p(\lambda) \sim \mathcal{N}(\mu_y, \Sigma_y), \tag{14}$$

$$p(r) \sim \mathcal{G}\left(\frac{1}{2}, 2\Sigma_y^{-1}\right) \propto r^{-1/2} \exp\left(\frac{-r\Sigma_y}{2}\right), \tag{15}$$

where $\mu_y = \frac{1}{N}\sum_{j=1}^{N} y_j$ and $\Sigma_y = \frac{1}{N}\sum_{j=1}^{N}(y_j - \mu_y)^2$ denote the mean and variance of all observations.

When considering prior in Bayesian inference, the usage of conjugate priors allows the results to be derived in closed form. It is conventional to use Gaussian, gamma and inverse-gamma distribution as conjugate priors for scalar means, precisions and variances, respectively [**27**]. Considering the Wishart distribution is a generalization of the gamma distribution to multiple dimensions, we use Gaussian, Wishart and inverse-Wishart distribution as conjugate priors for high-dimensional hyperparameters. The posterior distribution of μ given other variables is:

$$p(\mu_i | \lambda, r, \Lambda_i, \mathbf{c}, \mathbf{y}) \sim \mathcal{N}\left(\lambda + \frac{(\bar{y}_i - \lambda)}{n_i\Lambda_i + r} n_i\Lambda_i, (n_i\Lambda_i + r)^{-1}\right), \tag{16}$$

Where $n_i = \sum_{j=1}^{N} \delta(c_j, i)$ denotes the number of data points generated by the i-th component and y¯i=1ni∑j:cj=iyj denotes the mean of observations in this component. As to hyperparameters, taking Equation (**14**) and Equation (**15**) as priors and Equation (**13**) as likelihood, we get the posteriors:

$$p(\lambda | \mu_1, \ldots, \mu_K, r) \sim \mathcal{N}\left(\mu_y + \frac{\sum_{i=1}^{K} \mu_i - K\mu_y}{\Sigma_y^{-1} + Kr} r, (\Sigma_y^{-1} + Kr)^{-1}\right),$$

$$p(r | \mu_1, \ldots, \mu_K, \lambda) \sim \mathcal{W}\left(\frac{1}{D}\left(\Sigma_y + \sum_{i=1}^{K}(\mu_i - \lambda)^{\top}(\mu_i - \lambda)\right)^{-1}, \frac{K+1}{D}\right), \tag{17}$$

where W denotes the Wishart distribution and D denotes the number of dimensions. The posterior of precision and its hyperparameters can be derived in the same way.

Theweightofcomponentsπ1,…,πKhaveahyperparameterα.Wecansimply set the prior of $p(\alpha^{-1}) \sim \mathcal{G}(\frac{1}{2}, 2)$, which means $p(\alpha) \propto \alpha^{-3/2} \exp(-\frac{1}{2\alpha})$. The joint distribution of N category variables cj is:

$$p(c_1, \ldots, c_N | \pi_1, \ldots, \pi_K) = \prod_{i=1}^{K} \pi_i^{n_i}. \tag{18}$$

Notice that this distribution is a multinomial distribution with permutation rather than combination and the Dirichlet distribution is the conjugate prior of the multinomial distribution in Bayesian statistics. In order to bring as less prior information as possible, we give a symmetric Dirichlet distribution prior for πi:

$$p(\pi_1, \ldots, \pi_K | \alpha) \sim \text{Dirichlet}(\alpha/K, \ldots, \alpha/K) = \frac{\Gamma(\alpha)}{\Gamma(\alpha/K)^K} \prod_{i=1}^{K} \pi_i^{\alpha/K-1}. \quad (19)$$

By combining Equations (**18**) and (**19**) we can integrate out the weight:

$$p(c_1, \ldots, c_N | \alpha) = \frac{\Gamma(\alpha)}{\Gamma(n+\alpha)} \prod_{i=1}^{K} \frac{\Gamma(\alpha/K + n_i)}{\Gamma(\alpha/K)}. \quad (20)$$

Collapsed Gibbs sampling requires updating one variable each time, so cj for all j=1,…,N should be considered separately. We can derive from Equation (**20**) directly:

$$p(c_j = i | \mathbf{c}_{-j}, \alpha) = \frac{n_{-j,i} + \alpha/K}{n - 1 + \alpha}, \quad (21)$$

where c_{-j} denotes cj for i=1,…,j−1,j+1,…,N and $n_{-j,i}$ denotes the number of data points generated by the *i*-th component without the *j*-th data point.

When *K* rounds towards infinity, Equation (**21**) become $p(c_j = i | \mathbf{c}_{-j}, \alpha) = n_{-j,i}/(n - 1 + \alpha)$Notice that the probability of the *j*-th data point in all existing components is $\sum_{i=1}^{K} p(c_j = i | \mathbf{c}_{-j}, \alpha) = (n-1)/(n-1+\alpha)$, which is not equal to 1. Thus the probability of this data point belonging to a new component is $p(c_j \neq c_{j'} \text{ for all } j' \neq j | \mathbf{c}_{-j}, \alpha) = \alpha/(n-1+\alpha)$. The mean and precision of the new component are sampled from the prior. In the meantime, components are removed when they get empty. After enough iterations, this model tends to be stabilized. The detailed derivation of all formulas in this section can be found in **Appendix A**.

KERNEL ANALYSIS IN THE SPECTRAL DOMAIN

In this section, we analyze kernels in their spectral domain and propose a new kernel called the infinite mixture (IM) kernel. The IM kernel is defined as a kernel with infinite Gaussian mixture components in its spectral domain. Considering the symmetric property of spectral density for stationary kernels, the spectral density of the infinite mixture kernel is designed as:

$$\theta(s_i) \sim \mathcal{N}(\mu_i, \Lambda_i^{-1}),$$
$$S(s) = \sum_{i=1}^{\infty} \frac{1}{2} w_i \left(\theta(s_i) + \theta(-s_i)\right), \tag{22}$$

where μi, Λi and wi are the mean, precision and weight for the *i*-th component respectively.

The design of a kernel is of essential importance in Gaussian process. When considering interpolation, we need the covariance function to be smooth enough, which means if two input points are close to each other, we expect the value of the function at those points to be similar. This requires the covariance matrix to have the attribute that if two points are closer, the corresponding element in the matrix will be larger. The squared exponential (SE) kernel, also called the radial basis function kernel, is the most widely used kernel and has the form:

$$k(x, x') = \exp\left(-\frac{\|x - x'\|^2}{2\ell^2}\right). \tag{23}$$

It is infinitely differentiable, which means the Gaussian process fitted with this kernel has mean square derivatives in any order and is thus very smooth. A stationary kernel is a function of τ=x−x'. According to the definition, the SE kernel is a typical stationary kernel and can be reformulated as $k(\tau) = \exp(-\|\tau\|^2/2\ell^2)$ The application of the Bochner's theorem (Theorem 1) [28,29] enables us to analyze a stationary kernel from its spectral domain.

Theorem 1.

A complex-valued function k on RD *is the function of a weakly stationary mean square continuous complex-valued random process on* RD *if and only if it can be represented as:*

$$k(\tau) = \int_{\mathbb{R}^D} e^{2\pi i s \cdot \tau} d\mu(s), \tag{24}$$

where μ is a positive finite measure.

If μ has a density $S(s)$, that is, $d\mu(s) = S(s)ds,$, then *S* is called the spectral density or power spectral of kernel *k*. Furthermore, when the spectral density exists, the Wiener-Khintchine theorem shows that the covariance function and the spectral density are Fourier duals of each other:

$$k(\tau) = \int S(s)e^{2\pi i s^\top \tau} ds, \qquad S(s) = \int k(\tau)e^{-2\pi i s^\top \tau} d\tau. \tag{25}$$

By substituting the SE kernel in Equation (**23**) into (**25**), we can get the corresponding spectral density of the SE kernel as a Gaussian function $S(s) = (2\pi\ell^2)^{D/2}\exp(-2\pi^2\ell^2 s^\top s)$, where D denotes the dimensionality of s. Note that the variance of process is $k(\mathbf{0}) = \int S(s)ds$, which is exactly the integral in spectral domain.

Analyzing kernel from a spectral (or frequency) perspective, the characteristic of zero-centered Gaussian in the spectral density of the SE kernel restricts its expressiveness. Wilson and Adams [**25**] proposed the spectral mixture (SM) kernel as an extension of traditional kernels by using several Gaussian components instead of one zero-centered Gaussian.

The number of components of the SM kernel is set to a fixed value. However, without additional expertise, setting K in advance is not suitable for practical application which gives rise to underfitting or overfitting. When the inherent pattern of data is simple, setting K with a large value will take noise in frequency domain into consideration and result in overfitting. Meanwhile, when the inherent pattern is sophisticated, setting K with a small value will lead to the ignorance of details and result in underfitting.

In order to address this problem, we combine an infinite mixture model into kernel design and propose the infinite mixture kernel under a Bayesian framework. By combining Equations (**25**) and (**22**), we get the analytic expression of the IM kernel:

$$k(\tau) = \sum_{i=1}^{\infty} w_i \exp(-2\pi^2\tau^2\Lambda_i^{-1})\cos(2\pi\tau\mu_i), \tag{26}$$

where $\sum_{i=1}^{\infty} w_i = 1$ and $w_i \geq 0$

The IM kernel can assign K automatically as the distribution of data changes. The Wiener's Tauberian theorem (Theorem 2) gives the theory ground to the assertion that an infinite mixture of Gaussians is dense in the set of all possible distributions.

Theorem 2.

Let f∈L1(R) *be an integrable function. The span of translations* $f_a(x) = f(x+a)$ *is dense in* L_1(R) *if and only if the Fourier transform of f has no real zeros.*

This theorem is stronger because it only uses translation transformation to make the span dense in L1(R) function space. Various popular kernels can be reconstructed by the IM kernel via analyzing their spectral density which shows the expressiveness of the IM kernels from another point of view.

Before applying infinite mixture models, we need to get the empirical spectral density of the data first. From Theorem 1, the spectral density is proportional to the discrete Fourier transform (DFT) of the squared data. The time complexity of DFT is $O(n^2)$, which can be reduced to O(nlogn) using fast Fourier transform (FFT). FFT is used here to calculate the spectral density. Notice that if the observations are not equidistant, data should be preprocessed. A simple way to preprocess the data is using interpolation methods like linear interpolation, polynomial interpolation and spline interpolation. A more precise and complex approach is using a compressed sensing algorithm [**30**]. Further experiments show that when the input has a grid structure except for a lack of input in a few ranges, it works well without preprocessing. We refer to it as a quasi-grid structure. The empirical spectral density is taken as an unnormalized distribution and we get observations by slice sampling [**31**].

EXPERIMENTS

In this section, we conduct a series of experiments to verify the robustness and good performance of the IM kernel on multiple datasets. We first introduce details of the implementation of the hierarchical mixture model and analyze the independence of samples. Then we perform robustness test on the SE kernel and the IM kernel, finding that performance of the SE kernel fluctuates wildly with different random seeds while IM kernel maintains good performance. Finally, we compare IM kernel with other widely used kernels in several datasets. The IM kernel shows smaller negative log-likelihood and mean square error (MSE) than others. Matlab (version 9.1) code to perform all experiments is available on GitHub (https://github.com/deeperKernelInsight/infinitemixturekernel). All experiments are conducted on a PC with a 3.20 GHz Hexa-core CPU and 16 GB RAM.

An MCMC Implementation of the Hierarchical Model

A collapsed Gibbs sampler marginalizes over a part of dependent variables while sampling from other variables. This strategy is widely used in hierarchical Bayesian models like latent Dirichlet allocation (LDA). Collapsing out the Dirichlet distribution in this model helps to accelerate sample steps. We use collapsed Gibbs sampling as the specific MCMC method to sample hyperparameters, local variables and global variables. Meanwhile, the exact order of sampling parameters influences the performance. In the implementation, we sample means and precisions for all components first, with hyperparameters next and indicator variables at last.

We use the airline passenger dataset which records monthly passenger numbers from 1949 to 1961 here to look into this MCMC process. Before the Markov chain becomes stationary, there are a few iterations for "burn-in". In all experiments below, we set 1000 iterations for "burn-in". Since collapsed Gibbs sampling draws only one parameter each time, observations close to each other tend to be correlated. **Figure 2** plots the autocorrelation function of hyperparameters taking lag as *x*-axis. From this figure, the parameters are deemed independent of itself after 200 iterations. After the "burn-in" iterations, we record observations every 100 iterations and choose the most suitable one to refine.

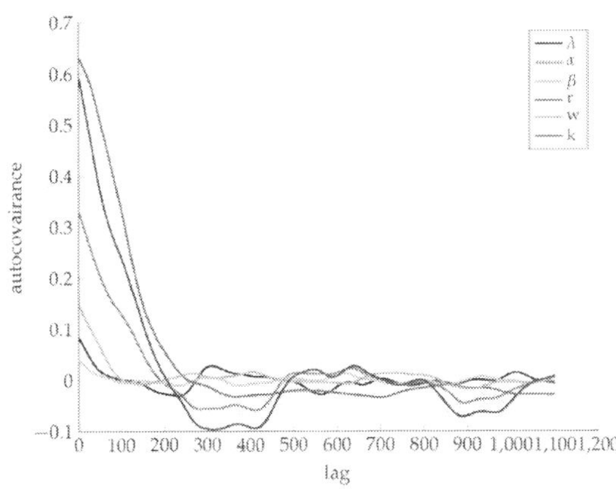

Figure 2. Autocorrelation analysis for hyperparameters.

Figure 3a shows the number of components in the process of training with a different number of sampling points. For universality, we set only one component in the beginning. It increases rapidly in the first few iterations and stays stable with slight fluctuations. With the number of sampling points increases, the number of components used to depict the model grows. The primary cause is that models become complex with the growth of sampling points and this is where the flexibility emerges. **Figure 3**b plots histogram of 100 observations for α in the condition of 10,000 sampling points. The concentrated parameter α takes values around 2.5, which means only $\alpha/(n+\alpha)\approx 0.025\%$ of data points belonging to a new component.

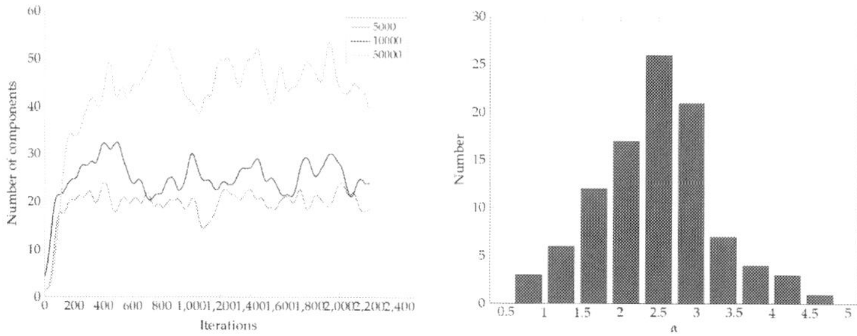

(a) Number of components

(b) Number of α

Figure 3. Parameters observed in the process of Markov Chain Monte Carlo (MCMC).

Robustness Test

Since the learning surface for a spectral mixture kernel is often high multimodal, the original model of the SE kernel is vulnerable for its stern demand for initialization.

The SE kernel takes a rather simple and greedy strategy by setting initial weights as k(0)/K, initial variances sampled from a truncated Gaussian distribution with a mean equal to the range of the data and initial frequencies sampled from a uniform distribution from 0 to its Nyquist frequency. Figure 4 shows the performance of the SM kernel with different random seeds. It performs well when the random seed takes the value 3,141,879.

However, taking 579 and 1457 as random seeds results in bad performance. Notice that there are obvious performance gaps between random seeds which indicates that the simple strategy used by the SM kernel is not stable and is thus not a good option. Performances of the IM kernel with different random seeds are not plotted because results remain stable and overlap in the chart.

By switching the number of sampling points, the performances are slightly different. With a small number of sampling points, kernel tends to show secular trends at the expense of details. Meanwhile, with a large number of sampling points, kernel tends to show more details while secular trend flattens out.

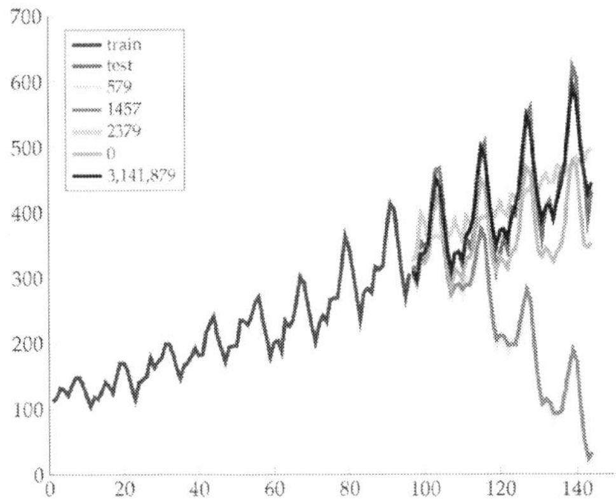

Figure 4. Performance of the spectral mixture (SM) kernel with different random seeds.

Performance on Multiple Datasets

The datasets we employ are publicly available and has been used in previous research [**24**]: (a) monthly gasoline demand in Ontario; (b) monthly death tolls from bronchitis in the UK; (c) mean monthly air temperature; (d) weekday bus ridership in Iowa city. We divide these datasets each into two parts with 2/3 data as the training set and others as the testing set. We attempt to give fair treatment to these kernels by initializing hyperparameters with high marginal log-likelihoods and fully training kernels to convergence. The marginal log-likelihood with noise observations can be expressed as:

$$\log p(\mathbf{y}|X,\theta) = -\frac{1}{2}\mathbf{y}^{\top}\left(K_{\theta}+\sigma_n^2 I\right)^{-1}\mathbf{y} - \frac{1}{2}\log\left|K_{\theta}+\sigma_n^2 I\right| - \frac{n}{2}\log 2\pi. \quad (27)$$

We use GPML toolbox (**http://www.gaussianprocess.org**) to refine parameters and evaluate other popular kernels.

Figure 5 shows the predictive distribution of the IM kernel and its spectral density. The top row displays predicting results on the testing set with the training and testing set in blue and black respectively. The mean of the predictive distribution is in red. The IM kernel shows good predictive ability with the testing set located in 95% of the predictive mass shown in shadow. The bottom row shows the log spectral density of the IM kernel in red and the log empirical spectral density in blue. The learned kernel fitted well with the empirical spectral. The log spectral density of kernel in the Temperature dataset shows the ability to filter noise.

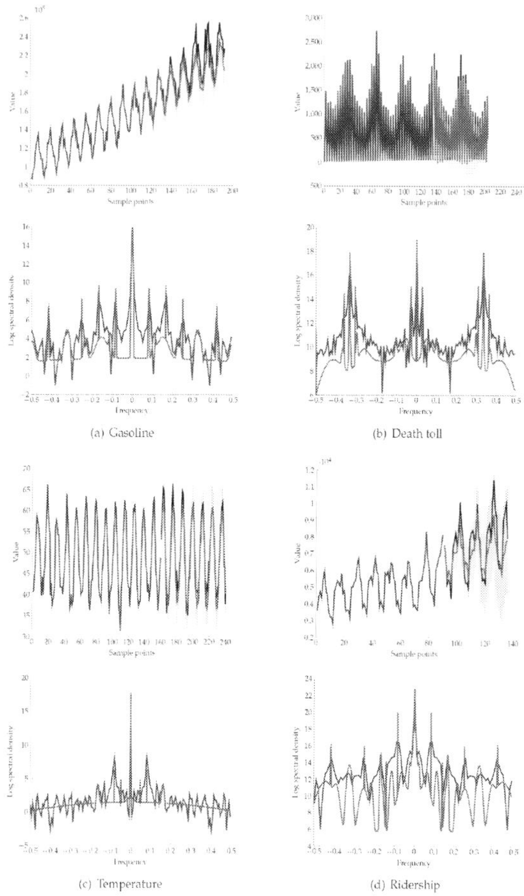

Figure 5. Performance of the infinite mixture (IM) kernel on a bunch of datasets.

Table 1 and Table 2 show negative log-likelihood and MSE of the SE kernel, the Matérn kernel, the squared exponential (SE) kernel and the IM kernel with different numbers of sampling points. In experiments, we set numbers of sampling points as 5000, 10,000 and 50,000. Only the standard deviations of the IM and SM kernels are recorded in the brackets due to the space limitation. The IM kernel shows better performance than others both in negative log-likelihood and in MSE. Meanwhile, when the number of sampling performance increases, the result turns out better. While a larger number of sampling points needs a longer time to sample and inference, the trade-off between accuracy and time cost is concerned in practical application.

Table 1. Negative Log-Likelihood. (The results with the best performance are shown in bold).

Dataset	Unit	IM			SM	Matérn	SE
		5×10^3	1×10^4	5×10^4			
Gasoline	1×10^3	1.31(0.09)	1.31(0.07)	**1.27(0.06)**	14.3(0.27)	1.49	1.44
Death toll	1×10^2	8.55(0.14)	8.49(0.20)	**8.48(0.10)**	9.66(0.43)	28,500,000	11.2
Temperature	1×10^2	3.80(0.12)	3.70(0.15)	3.71(0.15)	**3.68(0.29)**	5.71	4.74
Ridership	1×10^2	6.57(0.23)	6.96(0.26)	**6.48(0.14)**	7.01(0.33)	44,100,000	7.66

Table 2. Mean Square Error (MSE). (The results with the best performance are shown in bold).

Dataset	Unit	IM			SM	Matérn	SE
		5×10^3	1×10^4	5×10^4			
Gasoline	1×10^8	3.50(0.21)	3.27(0.19)	**2.94(0.13)**	6.18(0.76)	335,000,000	394
Death toll	1×10^4	1.74(0.06)	**1.67(0.03)**	1.83(0.03)	6.35(0.16)	1,040,000	80.7
Temperature	1	7.05(0.10)	6.75(0.32)	**5.89(0.10)**	6.85(0.69)	81.1	1850
Ridership	1×10^6	2.32(0.08)	3.12(0.11)	**1.14(0.13)**	2.64(0.18)	750,000	52.0

CONCLUSIONS

We introduced the infinite mixture kernel by combining a hierarchical Bayesian model into the analyze of the spectral domain. The extensive experiments have shown the robustness and better performance of our method than the finite version. The results are competitive with popularly used kernels and the spectral density of the IM kernel provides an extra interpretation of the inherent pattern of the data.

In future work, one can combine the IM kernel with approximate inference algorithms. What's more, in this paper we only analyze the kernel of a specific stochastic process. Although the Gaussian distribution and the Gaussian processes are widely used in many different fields, there are many other distributions and their corresponding stochastic processes, for example, the q-Gaussian distribution and the distributions which emerge in the anomalous diffusion contexts and their corresponding stochastic processes. Analyzing the performance of these unconventional distributions and their corresponding stochastic processes can result in interesting discoveries. We believe that kernel analysis from a spectral perspective helps to design more powerful kernels while a deeper investigation is still needed for future research.

Author Contributions

Conceptualization, J.T.; methodology, J.T.; software, J.T.; validation, J.T.; formal analysis, P.Y.; investigation, P.Y.; writing–original draft preparation, J.T.; writing–review and editing, J.T. and P.Y.; supervision, D.H.

Funding

This work was partially supported by the National Natural Science Foundation of China (grants 61803375 and 91648204), by the National Key Research and Development Program of China (grants 2017YFB1001900 and 2017YFB1301104) and by the National Science and Technology Major Project.

Conflicts of Interest

The authors declare no conflict of interest. The funders had no role in the design of the study; in the collection, analyses, or interpretation of data; in the writing of the manuscript, or in the decision to publish the results.

Abbreviations

The following abbreviations are used in this manuscript:

GP	Gaussian process
GMM	Gaussian mixture models
DPMM	Dirichlet processes mixture model
PDF	Probability density function
CDF	Cumulative distribution function
SM	Spectral mixture kernel
SE	Squared exponential kernel
IM	Infinite mixture kernel

Appendix A. Hierarchical Bayesian Models

A traditional Gaussian mixture model is expressed as:

$$p(y|\mu_1,\ldots,\mu_k,s_1,\ldots,s_k,\pi_1,\ldots,\pi_k) = \sum_{j=1}^{k} \pi_j \mathcal{N}\left(\mu_j, s_j^{-1}\right), \tag{A1}$$

where μ_j, s_j, and π_j are the mean, the precision, and the proportional hyperparameters for the j-th components.

The mean μ_j is given Gaussain prior:

$$p(\mu_j|\lambda, r) \sim \mathcal{N}\left(\lambda, r^{-1}\right), \quad \text{(A2)}$$

where λ and r are hyperparameters.

The hyperparameters λ and r are also given Gaussian and Gamma priors:

$$p(\lambda) \sim \mathcal{N}\left(\mu_y, \sigma_y^2\right), \quad \text{(A3)}$$

$$p(r) \sim \mathcal{G}\left(1, \sigma_y^{-2}\right) \propto r^{-1/2} \exp\left(-r\sigma_y^2/2\right), \quad \text{(A4)}$$

where μ_y and σ^2y are the mean and variance of all observations.

Take Equation (**A2**) as the prior and Equation (**A1**) as the likelihood function, we can get the posterior distribution for the means μj:

$$\begin{aligned} p(\mu_j|y_j, s_j, \lambda, r) &\propto p(\mu_j|s_j, \lambda, r) p(y_j|\mu_j, s_j, \lambda, r) \\ &\propto p(\mu_j|\lambda, r) p(y_j|\mu_j, s_j) \\ &\propto \mathcal{N}(\mu_j|\lambda, r^{-1}) \prod_{i=1}^{n_j} \mathcal{N}(y_{j,i}|\mu_j, s_j^{-1}) \\ &\propto \exp\left(-\frac{r}{2}(\mu_j - \lambda)^2 - \frac{s_j}{2}\sum_{i=1}^{n_j}(y_{j,i} - \mu_j)^2\right) \\ &\propto \exp\left(-\frac{s_j n_j + r}{2}\left(\mu_j - \frac{r\lambda + s_j n_j \overline{y_j}}{s_j n_j + r}\right)^2\right) \\ &\sim \mathcal{N}\left(\frac{r\lambda + s_j n_j \overline{y_j}}{s_j n_j + r}, (s_j n_j + r)^{-1}\right), \end{aligned} \quad \text{(A5)}$$

where yj is all observations belonging to class j, nj is the number of these observations, and $\overline{y}_j = \frac{1}{n_j}\sum y_{j,i}$.

Take Equation (**A3**) as the prior and Equation (**A2**) as the likelihood function, we get the posterior distribution for the hyperparameter λ:

$$\begin{aligned} p(\lambda|\mu_1, \ldots, \mu_k, r) &\propto p(\lambda|r) p(\mu_1, \ldots, \mu_k|\lambda, r) \\ &\propto p(\lambda) \prod_{j=i}^{k} p(\mu_j|\lambda, r) \\ &\propto \mathcal{N}(\lambda|\mu_y, \sigma_y^2) \prod_{j=1}^{k} \mathcal{N}(\mu_j|\lambda, r^{-1}) \\ &\propto \exp\left(-\frac{1}{2\sigma_y^2}(\lambda - \mu_y)^2 - \frac{r}{2}\sum_{j=1}^{k}(\mu_j - \lambda)^2\right) \\ &\propto \exp\left(-\frac{kr + \sigma_y^{-2}}{2}\left(\lambda - \frac{\mu_y \sigma_y^{-2} + r\sum_{j=1}^{k}\mu_j}{kr + \sigma_y^{-2}}\right)\right) \\ &\sim \mathcal{N}\left(\frac{\mu_y \sigma_y^{-2} + r\sum_{j=1}^{k}\mu_j}{kr + \sigma_y^{-2}}, (kr + \sigma_y^{-2})^{-1}\right). \end{aligned} \quad \text{(A6)}$$

Take Equation (**A4**) as the prior and Equation (**A2**) as the likelihood function, we get the posterior distribution for the hyperparameter *r*:

$$
\begin{aligned}
p(r|\mu_1,\ldots,\mu_k,\lambda) &\propto p(r)\prod_j p(\mu_{j=1}^k|\lambda,r) \\
&\propto \mathcal{G}_R\left(r|1,\sigma_y^{-2}\right)\prod_{j=1}^{k}\mathcal{N}(\mu_j|\lambda,r^{-1}) \\
&\propto r^{-\frac{1}{2}}\exp\left(-\frac{\sigma_y^2}{2}r\right)\prod_{j=1}^{k}r^{\frac{1}{2}}\exp\left(-\frac{r}{2}(\mu_j-\lambda)^2\right) \\
&\propto r^{\frac{k-1}{2}}\exp\left(-\frac{1}{2}\left(\sigma_y^2+\sum_{j=1}^{k}(\mu_j-\lambda)^2\right)r\right) \\
&\sim \mathcal{G}_R\left(k+1,(k+1)\left(\sigma_y^2+\sum_{j=1}^{k}(\mu_j-\lambda)^2\right)^{-1}\right) \\
&\sim \mathcal{G}\left(\frac{k+1}{2},2\left(\sigma_y^2+\sum_{j=1}^{k}(\mu_j-\lambda)^2\right)^{-1}\right).
\end{aligned}
\tag{A7}
$$

The precision of each component sj is given Gamma prior:

$$
p\left(s_j|\beta,w\right)\sim\mathcal{G}\left(\beta,w^{-1}\right),
\tag{A8}
$$

where β and *w* are hyperparameters with inverse Gamma and Gamma priors:

$$
p\left(\beta^{-1}\right)\sim\mathcal{G}(1,1)\Longrightarrow p(\beta)\propto\beta^{-3/2}\exp(-1/(2\beta)),
\tag{A9}
$$

$$
p(w)\sim\mathcal{G}\left(1,\sigma_y^2\right).
\tag{A10}
$$

Take Equation (**A8**) as the prior and Equation (**A1**) as the likelihood function, we can get the posterior distribution for the means s_j:

$$
\begin{aligned}
p(s_j|\boldsymbol{y}_j,\mu_j,\beta,w) &\propto p(s_j|\mu_j,\beta,w)p(\boldsymbol{y}_j|\mu_j,s_j,\beta,w) \\
&\propto p(s_j|\beta,w)p(\boldsymbol{y}_j|\mu_j,s_j) \\
&\propto \mathcal{G}_R(s_j|\beta,w^{-1})\prod_{i=1}^{n_j}\mathcal{N}(y_{j,i}|\mu_j,s_j^{-1}) \\
&\propto s_j^{\frac{\beta}{2}-1}\exp\left(-\frac{\beta w}{2}w_j\right)\prod_{i=1}^{n_j}s_j^{\frac{1}{2}}\left(-\frac{s_j}{2}(y_{j,i}-\mu_j)^2\right) \\
&\propto s_j^{\frac{\beta}{2}-1+\frac{n_j}{2}}\exp\left(-\frac{w\beta}{2}s_j-\frac{1}{2}\sum_{i=1}^{n_j}(y_{j,i}-\mu_j)^2 s_j\right) \\
&\sim \mathcal{G}_R\left(\beta+n_j,(\beta+n_j)\left(w\beta+\sum_{i=1}^{n_j}\left(y_{j,i}-\mu_j\right)^2\right)^{-1}\right).
\end{aligned}
\tag{A11}
$$

Take Equation (**A10**) as the prior and Equation (**A8**) as the likelihood function, we get the posterior distribution for the hyperparameter *w*:

$$\begin{aligned} p(w|s_1,\ldots,s_k,\beta) &\propto p(w|\beta)p(s_1,\ldots,s_k|w,\beta) \\ &\propto p(w)\prod_{j=1}^{k} p(s_j|w,\beta) \\ &\propto \mathcal{G}_R(w|1,\sigma_y^2)\prod_{j=1}^{k}\mathcal{G}_R(s_j|\beta,w^{-1}) \\ &\propto w^{-\frac{1}{2}}\exp\left(-\frac{w}{2\sigma_y^2}\right)\prod_{j=1}^{k} w^{\frac{\beta}{2}}\exp\left(-\frac{\beta s_j}{2}w\right) \\ &\propto w^{\frac{k\beta-1}{2}}\exp\left(-\frac{1}{2}(\sigma_y^{-2}+\beta\sum_{j=1}^{k}s_j)w\right) \\ &\sim \mathcal{G}_R\left(k\beta+1,(k\beta+1)(\sigma_y^{-2}+\beta\sum_{j=1}^{k}s_j)^{-1}\right). \end{aligned} \quad \text{(A12)}$$

Take Equation (**A9**) as the prior and Equation (**A8**) as the likelihood function, we get the posterior distribution for the hyperparameter β:

$$\begin{aligned} p(\beta|s_1,\ldots,s_k,w) &\propto p(\beta|w)p(s_1,\ldots,s_k|\beta,w) \\ &\propto p(\beta)\prod_{j=1}^{k} p(s_j|\beta,w) \\ &\propto \beta^{-\frac{3}{2}}\exp\left(-\frac{1}{2\beta}\right)\prod_{j=1}^{k}\mathcal{G}_R(s_j|\beta,w^{-1}) \\ &\propto \beta^{-\frac{3}{2}}\exp\left(-\frac{1}{2\beta}\right)\prod_{j=1}^{k}\frac{s_j^{\frac{\beta}{2}-1}e^{-\frac{\beta w s_j}{2}}}{\Gamma\left(\frac{\beta}{2}\right)\left(\frac{2}{w\beta}\right)^{\frac{\beta}{2}}} \\ &\propto \Gamma\left(\frac{\beta}{2}\right)^{-k}\beta^{-\frac{3}{2}}\left(\frac{w\beta}{2}\right)^{\frac{k\beta}{2}}\exp\left(-\frac{1}{2\beta}-\frac{\beta w}{2}\sum_j s_j\right)\prod_{j=1}^{k}s_j^{\frac{\beta}{2}-1}. \end{aligned} \quad \text{(A13)}$$

The posterior distribution of β is in analytical form but not a stantard form. It can be proved that $p\left(\log(\beta)|s_1,\ldots,s_k,w\right)$ is log-concave. Thus we can use adaptive rejection sampling method [**32**] to sample this distribution.

The proportional hyperparameters π1,…,πk are given a symmetric Dirichlet prior:

$$p\left(\pi_1,\ldots,\pi_k|\alpha\right) \sim \text{Dirichlet}(\alpha/k,\ldots,\alpha/k) = \frac{\Gamma(\alpha)}{\Gamma(\frac{\alpha}{k})^k}\prod_{j=1}^{k}\pi_j^{\frac{\alpha}{k}-1}. \quad \text{(A14)}$$

The probability of all indicators cj under the condition where π1,… ,πk are fixed is:

$$p\left(c_1,\ldots,c_n|\pi_1,\ldots,\pi_k\right) = \prod_{j=1}^{k}\pi_j^{n_j}. \quad \text{(A15)}$$

We can integrate out the proportional hyperparameters and get the distribution of each indicator conditioning on α:

$$\begin{aligned}p(c_1,\ldots,c_n|\alpha) &= \int p(c_1,\ldots,c_k|\pi_1,\ldots,\pi_k)\,p(\pi_1,\ldots,\pi_k)\,d\pi\\ &= \int \frac{\Gamma(\alpha)}{\Gamma(\frac{\alpha}{k})^k}\prod_{j=1}^{k}\pi_j^{\frac{\alpha}{k}-1}\prod_{j=1}^{k}\pi_j^{n_j}\,d\pi\\ &= \frac{\Gamma(\alpha)}{\Gamma(\frac{\alpha}{k})^k}\int\prod_{j=1}^{k}\pi_j^{n_j+\frac{\alpha}{k}-1}\,d\pi\\ &= \frac{\Gamma(\alpha)}{\Gamma(\frac{\alpha}{k})^k}\frac{\prod_{j=1}^{k}\Gamma(n_j+\frac{\alpha}{k})}{\Gamma(n+\alpha)}\\ &= \frac{\Gamma(\alpha)}{\Gamma(n+\alpha)}\prod_{j=1}^{k}\frac{\Gamma(n_j+\frac{\alpha}{k})}{\Gamma(\frac{\alpha}{k})}.\end{aligned}$$

(A16)

When using Gibbs sampling methods to sample c, we have to sample each ci conditioning on others ci. So we need to get the conditional distribution of ci:

$$\begin{aligned}p(c_i=j|\mathbf{c}_{-i},\alpha) &= \frac{p(\mathbf{c}|\alpha)}{p(\mathbf{c}_{-i}|\alpha)}\\ &= \frac{\frac{\Gamma(\alpha)}{\Gamma(n+\alpha)}\prod_{j=1}^{k}\frac{\Gamma(n_j+\frac{\alpha}{k})}{\Gamma(\frac{\alpha}{k})}}{\frac{\Gamma(\alpha)}{\Gamma(n-1+\alpha)}\frac{\Gamma(n_{-i,j}+\frac{\alpha}{k})}{\Gamma(n_j+\frac{\alpha}{k})}\prod_{j=1}^{k}\frac{\Gamma(n_j+\frac{\alpha}{k})}{\Gamma(\frac{\alpha}{k})}}\\ &= \frac{n_{-i,j}+\alpha/k}{n-1+\alpha},\end{aligned}$$

(A17)

where $n_{-i,j}$ is the number of the *j*-th components except the *i*-th point.

α is given an inverse gamma distribution:

$$p\left(\alpha^{-1}\right)\sim\mathcal{G}_R(1,1)\Longrightarrow p(\alpha)\propto\alpha^{-\frac{3}{2}}\exp\left(-\frac{1}{2\alpha}\right).$$

(A18)

Given α, the probability of the number of points for all components, n1,… ,nk is:

$$p(n_1,\ldots,n_k|\alpha)=\frac{\alpha^k\Gamma(\alpha)}{\Gamma(n+\alpha)}.$$

(A19)

Taking Equation (**A18**) as the prior and Equation (**A19**) as the likelihood function, we can get the posterior distribution of α:

$$\begin{aligned}p(\alpha|n_1,\ldots,n_k) &\propto p(\alpha)p(n_1,\ldots,n_k|\alpha)\\ &\propto \alpha^{-\frac{3}{2}}\exp\left(-\frac{1}{2\alpha}\right)\frac{\alpha^k\Gamma(\alpha)}{\Gamma(n+\alpha)}.\end{aligned}$$

(A20)

REFERENCES

1. Ferguson, T.S. A Bayesian analysis of some nonparametric problems. *Ann. Stat.* **1973**, *1*, 209–230.
2. Ishwaran, H.; James, L.F. Gibbs sampling methods for stick-breaking priors. *J. Am. Stat. Assoc.* **2001**, *96*, 161–173.
3. Blackwell, D.; MacQueen, J.B. Ferguson distributions via Pólya urn schemes. *Ann. Stat.* **1973**, *1*, 353–355.
4. Pitman, J. *Combinatorial Stochastic Processes*; Technical Report 621; Department Statistics, University of California: Berkeley, CA, USA, July 2002.
5. Aldous, D.J. Exchangeability and related topics. *École d'Été de Probabilités de Saint-Flour XIII—1983* **1985**, *1117*, 1–198.
6. Sudderth, E.B.; Jordan, M.I. Shared segmentation of natural scenes using dependent Pitman-Yor processes. In *Advances in Neural Information Processing Systems 21*; Curran Associates, Inc.: New York, NY, USA, 2009; pp. 1585–1592.
7. Gelfand, A.E.; Kottas, A.; MacEachern, S.N. Bayesian nonparametric spatial modeling with Dirichlet process mixing. *J. Am. Stat. Assoc.* **2005**, *100*, 1021–1035.
8. Liang, P.; Petrov, S.; Jordan, M.; Klein, D. The infinite PCFG using hierarchical Dirichlet processes. In Proceedings of the 2007 Joint Conference on Empirical Methods In Natural Language Processing and Computational Natural Language Learning (EMNLP-CoNLL), Prague, Czech Republic, 28–30 June 2007; pp. 688–697.
9. Johnson, M.; Griffiths, T.L.; Goldwater, S. Adaptor grammars: A framework for specifying compositional nonparametric Bayesian models. In *Advances in Neural Information Processing Systems*; Curran Associates, Inc.: New York, NY, USA, 2007; pp. 641–648.
10. Navarro, D.J.; Griffiths, T.L. Latent features in similarity judgments: A nonparametric Bayesian approach. *Neural Comput.* **2008**, *20*, 2597–2628.
11. Kemp, C.; Tenenbaum, J.B.; Griffiths, T.L.; Yamada, T.; Ueda, N. Learning systems of concepts with an infinite relational model. In Proceedings of the AAAI'06 21st National Conference on Artificial Intelligence, Boston, MA, USA, 16–20 July 2006; pp. 381–388.
12. Fox, E.; Sudderth, E.B.; Jordan, M.I.; Willsky, A.S. Nonparametric Bayesian learning of switching linear dynamical systems. In *Advances*

in Neural Information Processing Systems; Curran Associates, Inc.: New York, NY, USA, 2009; pp. 457–464.

13. Fox, E.; Jordan, M.I.; Sudderth, E.B.; Willsky, A.S. Sharing features among dynamical systems with beta processes. In *Advances in Neural Information Processing Systems*; Curran Associates, Inc.: New York, NY, USA, 2009; pp. 549–557.
14. Paisley, J.; Carin, L. Hidden Markov models with stick-breaking priors. *IEEE Trans. Signal Process.* **2009**, *57*, 3905–3917.
15. Beal, M.J.; Ghahramani, Z.; Rasmussen, C.E. The infinite hidden Markov model. In *Advances in Neural Information Processing Systems*; Curran Associates, Inc.: New York, NY, USA, 2002; pp. 577–584.
16. Neal, R.M. *Bayesian Learning for Neural Networks*; Springer Science & Business Media: Berlin, Germany, 2012; Volume 118.
17. Rasmussen, C.E. Gaussian processes in machine learning. *Adv. Lect. Mach. Learn.* **2003**, *3176*, 63–71.
18. Salimbeni, H.; Deisenroth, M. Doubly stochastic variational inference for deep Gaussian processes. In *Advances in Neural Information Processing Systems*; Curran Associates, Inc.: New York, NY, USA, 2017; pp. 4588–4599.
19. Wilson, A.G.; Hu, Z.; Salakhutdinov, R.R.; Xing, E.P. Stochastic Variational Deep Kernel Learning. In Proceedings of the NIPS'16 30th International Conference on Neural Information Processing Systems, Barcelona, Spain, 5–10 December 2016; pp. 2594–2602.
20. Yang, Z.; Wilson, A.; Smola, A.; Song, L. A La Carte–Learning Fast Kernels. In Proceedings of the 18th International Conference on Artificial Intelligence and Statistics, San Diego, CA, USA, 9–12 May 2015; pp. 1098–1106.
21. Damianou, A.; Lawrence, N. Deep gaussian processes. In Proceedings of the 16th International Conference on Artificial Intelligence and Statistics, Scottsdale, AZ, USA, 29 April–1 May 2013; pp. 207–215.
22. Hinton, G.E.; Salakhutdinov, R.R. Using deep belief nets to learn covariance kernels for Gaussian processes. In NIPS'07 Proceedings of the 20th International Conference on Neural Information Processing Systems, Vancouver, BC, Canda, 3–6 December 2007; pp. 1249–1256.
23. Gönen, M.; Alpaydın, E. Multiple kernel learning algorithms. *J. Mach. Learn. Res.* **2011**, *12*, 2211–2268.
24. Duvenaud, D.; Lloyd, J.R.; Grosse, R.; Tenenbaum, J.B.; Ghahramani,

Z. Structure discovery in nonparametric regression through compositional kernel search. *arXiv* **2013**, arXiv:1302.4922.

25. Wilson, A.; Adams, R. Gaussian process kernels for pattern discovery and extrapolation. In Proceedings of the 30th International Conference on Machine Learning, Atlanta, GA, USA, 16–21 June 2013; pp. 1067–1075.
26. Antoniak, C.E. Mixtures of Dirichlet processes with applications to Bayesian nonparametric problems. *Ann. Stat.* **1974**, *2*, 1152–1174.
27. Murphy, K.P. Conjugate Bayesian analysis of the Gaussian distribution. *def* **2007**, *1*, 16. Available online: **http://citeseerx.ist.psu.edu/viewdoc/summary?doi=10.1.1.126.4603** (accessed on 31 August 2019).
28. Stein, M.L. *Interpolation of Spatial Data: Some Theory for Kriging*; Springer Science & Business Media: Berlin, Germany, 2012.
29. Cox, D.R. *The Theory of Stochastic Processes*; Routledge: Abingdon-on-Thames, UK, 2017.
30. Donoho, D.L. Compressed sensing. *IEEE Trans. Inf. Theory* **2006**, *52*, 1289–1306.
31. Neal, R.M. Slice sampling. *Ann. Stat.* **2003**, *31*, 705–767.
32. Gilks, W.R.; Wild, P. Adaptive rejection sampling for Gibbs sampling. *J. R. Stat. Soc. C* **1992**, *41*, 337–348.

CHAPTER

9

A Mixture of Generalized Tukey's *g* Distributions

José Alfredo Jiménez[1] and Viswanathan Arunachalam[1]

[1] Department of Statistics, Universidad Nacional de Colombia, Carrera 45 No. 26-85, Bogota, Colombia

ABSTRACT

Mixtures of symmetric distributions, in particular normal mixtures as a tool in statistical modeling, have been widely studied. In recent years, mixtures of asymmetric distributions have emerged as a top contender for analyzing statistical data. Tukey's g family of generalized distributions depend on the parameters, namely, g, which controls the skewness. This paper presents the probability density function (pdf) associated with a mixture of Tukey's

Citation: (APA): Jiménez, J. A., & Arunachalam, V. (2016). A Mixture of Generalized Tukey's g-Distributions. Journal of Probability and Statistics, 2016. (7 pages) DOI: https://doi.org/10.1155/2016/3509139 URL: https://www.hindawi.com/journals/jps/2016/3509139/

g family of generalized distributions. The mixture of this class of skewed distributions is a generalization of Tukey's g family of distributions. In this paper, we calculate a closed form expression for the density and distribution of the mixture of two Tukey's g families of generalized distributions, which allows us to easily compute probabilities, moments, and related measures. This class of distributions contains the mixture of Log-symmetric distributions as a special case

INTRODUCTION

The main focus of interest in financial economics is the distribution of stock market returns. Mandelbrot [1] suggested the family of stable Paretian distributions for stock market returns. Fama [2] established that the normality assumption of the empirical data does not hold as the distribution is fat tailed. Kon [3] and Tse [4] used a mixture of normal distributions for stock return. Fielitz and Rozelle [5] proposed a mixture of nonnormal stable distributions for stock price. Consequently, greater emphasis has been placed on using distributions which have asymmetry and leptokurtic properties. Recently Jimenez et al. [6] proposed option pricing ´ based mixture of log-skew-normal distributions. If extreme events tend to occur more frequently than normal events, then skewness and kurtosis of nonnormal distributions play an essential role for the volatility smile. The most important and useful characteristic of Tukey's g-h family of distributions introduced by Tukey [7] is that it covers most of the Pearsonian family of distributions. It can also generate several known distributions, for example, lognormal, Cauchy, exponential, and Chi-squared (see Mart´ınez and Iglewicz [8], page 363). From Tukey's g-h family of distribution, we obtain g distribution, which is closely related to lognormal distribution and possesses similar properties of moments. Tukey's g-h family of distributions have been used to study financial markets. Badrinath and Chatterjee [9, 10] and Mills [11] used g-h to model the return on a stock index, as well as the return on shares in several markets. Dutta and Babbel [12] found that the skewness and leptokurtic behavior of LIBOR were modeled effectively using g-h distribution. Dutta and Babbel [13] used g-h to model interest rates and options on interest rates, while Tang and Wu [14] proposed a new method for the Decomposition of Portfolio VaR. Dutta and Perry [15] and recently Jimenez and Arunachalam ´ [16] used g-h distribution to study the operational risk for heavy tailed severity models. Jimenez and Arunachalam [17] ´ provided explicit expressions for VaR and CVaR calculations using the family of Tukey's

g-h distributions. Currently Jimenez et al. [18] studied generalization of Tukey's ′ g-h family of distributions, when the standard normal random variable is replaced by a continuous random variable U with mean 0 and variance 1. The subfamily of g distributions exhibits skewness and has great importance in the study of asymmetric distributions for analyzing data. This kind of distribution allows us to obtain scaled Log-symmetric distributions. Vitiello and Poon [19] considered a simple mixture of two g distributions for option pricing data. The purpose of this paper is to present a mixture of Tukey's g distributions and derive some statistical properties including the pdf and moment generating function and its properties. The paper is organized as follows: Section 2 presents Tukey's g-h family of generalized distributions and its pdf, as well as the cumulative distribution function (cdf). In Section 3, some theoretical results of the mixture of two Tukey's g families of generalized distributions are presented and Section 4 explains the methodology of calculating estimation of parameters by the method of moments. Section 5 discusses the adjustment methodology of our proposed model to real data of Heating-Degree-Days (HDD) indices and finally, in Section 6, we conclude.

TUKEY'S g FAMILY OF GENERALIZED DISTRIBUTIONS

Tukey [7] introduced the family g-h distributions by means of two nonlinear transformations given by

$$Y = T_{g,h}(Z) = \frac{1}{g}\left(\exp\{gZ\} - 1\right)\exp\left\{\frac{hZ^2}{2}\right\} \tag{1}$$

with $g = 0$, h ∈/ R, where the distribution of Z is standard normal. When these transformations are applied to a continuous random variable U with mean 0 and variance 1 such that its pdf $f_U(\cdot)$ is symmetric about the origin and cdf $F_U(\cdot)$, the transformation $T_{g,h}(U)$ is obtained, which henceforth will be termed Tukey's g-h generalized distribution. If h=0, Tukey's g-h generalized distribution reduces to

$$T_{g,0}(U) = \frac{1}{g}\left(\exp\{gU\} - 1\right) \tag{2}$$

which is known as Tukey's g generalized distribution. In order to model an arbitrary random variable X using the transformation given in (2), Hoaglin and Peters [20] introduced two new parameters, A (location) and B (scale), and proposed the following linear transformation:

$$X = A + BY \quad \text{with } Y = T_{g,0}(U). \tag{3}$$

The following properties for pdf, cdf, and quantile functions of Tukey's g generalized distribution were established by Jimenez et al. [18] in terms of the pdf and cdf of ′ X as follows:

$$\text{pdf: } f_X(x;\lambda,g,\theta) = \frac{1}{g(x-\theta)} f_U\left(\frac{1}{g}\ln\left(\frac{x-\theta}{B/g}\right)\right) \quad \text{if } g(x-\theta) > 0,$$

$$\text{cdf: } F_X(x;\lambda,g,\theta) = F_U\left(\frac{1}{g}\ln\left(\frac{x-\theta}{B/g}\right)\right) \quad \text{if } g(x-\theta) > 0,$$

$$\text{qf: } F_U^{-1}(q) = u_q = \frac{1}{g}\ln\left(\frac{x_q-\theta}{B/g}\right) \quad \text{if } g\left(x_q-\theta\right) > 0, \tag{4}$$

where $\lambda = \ln(B/|g|)$ and $\theta = A - B/g$. We say that the random variable X has a Log-symmetric distribution (such distributions are all asymmetric; see for reference Johnson et al. [21] and Stuart and Ord [22]) with three parameters: threshold (θ), scale (λ), and shape (g), denoted by $X \sim \text{LS}(\lambda, g, \theta)$.

The first expression of (4) allows us to obtain the following pdf associated with Tukey's g distribution. Table 1 shows the parameters of the pdf of X that we obtain using a selected set of well known symmetrical distributions (from Jimenez et al. ′ [18]).

The nth moment of the random variable $Y=T_{g,0}(U)$ is given by

$$\mu_n'(Y) = \frac{1}{g^n}\sum_{k=0}^{n}(-1)^k\binom{n}{k}M_U(\tilde{g}), \quad \text{if } g \neq 0, \tag{5}$$

Where $\tilde{g} = (n-k)g$ and (t) is the moment generating function of the random variable U, which are even function; that is, $M(t) = M_U(-t)$.

Table 2 shows parameters of the pdf and the moment generating function for a random variable U, using a selected set of well-known symmetrical distributions. Expression (5) allows us to obtain the moments of Tukey's g generalized distribution. The nth moment of the random variable X given by (3) can be obtained using the formula

$$\mathbb{E}\left[(X-\mathbb{E}[X])^n\right] = \mu_n(X)$$
$$= \left(\frac{B}{g}\right)^n \sum_{k=0}^{n} (-1)^k \binom{n}{k} M_U(\tilde{g}) M_U^k(g), \quad (6)$$

Where $B/g = \operatorname{sgn}(g)e^{\lambda}$ Note that the above expression of the nth moment does not depend on the parameter θ. Formulas for calculating the standardized skewness $\beta_1(X)$, and standardized excess kurtosis, $\beta_2(X)$, are given by

$$\beta_1(X) = \operatorname{sgn}(g) \cdot \frac{\left(M_U(3g) - M_U^3(g)\right) - 3\left(M_U(2g) - M_U^2(g)\right)M_U(g)}{\left[M_U(2g) - M_U^2(g)\right]^{3/2}},$$

$$\beta_2(X) = \frac{M_U(4g) - 4M_U(3g)M_U(g) + 3M_U^2(2g)}{\left[M_U(2g) - M_U^2(g)\right]^2} - 3, \quad (7)$$

where sgn(·) denote the signum function. Note that these expressions only depend on the parameter g and its sign, respectively. Any LS distribution should satisfy the following test given in Stuart and Ord [22]:

$$\beta_2(X) - \beta_1^2(X) - 1 \geq 0. \quad (8)$$

THE MIXTURE OF TWO g DISTRIBUTIONS

We assume that Y follows a Log-Symmetric Mixture (LSMIX) distribution. Let us assume that (y) is the weighted sum of m-component LS densities; that is,

$$f_Y(y;\Lambda) = \sum_{j=1}^{m} \omega_j f_U\left(y;\lambda_j, g_j, \theta_j\right). \quad (9)$$

Table 1: Parameters of the pdf of the random variable $Z = \ln(X)$.

Distribution of the r.v. U	Parameters μ, a	σ, b	$g \neq 0$	Distribution of the r.v. Z	Parameters μ, a	σ, b
Laplace	0	$\frac{\sqrt{2}}{2}$	$\lvert g \rvert < \frac{\sqrt{2}}{n}$	Log-Laplace	$\ln\left(\frac{B}{\lvert g \rvert}\right)$	$\frac{\sqrt{2}}{2}\lvert g \rvert$
Logistic	0	$\frac{\sqrt{3}}{\pi}$	$\lvert g \rvert < \frac{\pi}{\sqrt{3}n}$	Log-Logistic	$\ln\left(\frac{B}{\lvert g \rvert}\right)$	$\frac{\sqrt{3}}{\pi}\lvert g \rvert$
Normal	0	1	$g \in \mathbb{R}$	Lognormal	$\ln\left(\frac{B}{\lvert g \rvert}\right)$	$\lvert g \rvert$
HyperSec	0	$\frac{2}{\pi}$	$\lvert g \rvert < \frac{\pi}{2n}$	LoghyperSec	$\ln\left(\frac{B}{\lvert g \rvert}\right)$	$\frac{2}{\pi}\lvert g \rvert$
HyperCsc	0	$\frac{\sqrt{2}}{\pi}$	$\lvert g \rvert < \frac{\pi}{\sqrt{2}n}$	LoghyperCsc	$\ln\left(\frac{B}{\lvert g \rvert}\right)$	$\frac{\sqrt{2}}{\pi}\lvert g \rvert$

Table 2: Parameters of the pdf and moment generating functions of the random variable U

Distribution of the r.v. U	Parameters μ, a	σ, b	$g \neq 0$	$M_U(g)$
Laplace	0	$\frac{\sqrt{2}}{2}$	$\lvert g \rvert < \frac{\sqrt{2}}{n}$	$\frac{2}{2-g^2}$
Logistic	0	$\frac{\sqrt{3}}{\pi}$	$\lvert g \rvert < \frac{\pi}{\sqrt{3}n}$	$\sqrt{3}g\csc\left(\sqrt{3}g\right)$
Normal	0	1	$g \in \mathbb{R}$	$\exp\left\{\frac{1}{2}g^2\right\}$
HyperSec	0	$\frac{2}{\pi}$	$\lvert g \rvert < \frac{\pi}{2n}$	$\sec(g)$
HyperCsc	0	$\frac{\sqrt{2}}{\pi}$	$\lvert g \rvert < \frac{\pi}{\sqrt{2}n}$	$\sec^2\left(\frac{g}{\sqrt{2}}\right)$

We use the notation $Y \sim$ LSMIX(Λ)**, where $\Lambda = (\xi 1,..., \xi m)$**, and each element $\xi_j = (\omega_j, \lambda_j, g_j, \theta_j)$ is the parameter vector that defines the jth component and probability weights, ωj, satisfying the conditions

$$\sum_{j=1}^{m} \omega_j = 1, \quad 0 < \omega_j < 1, \text{ for each } j. \tag{10}$$

According to Titterington et al. [23] the two-component mixture of known distributions is set by two weights. Let

$$X = A + BY \text{ with } Y \sim \text{LSMIX}(\Lambda). \tag{11}$$

Then we can assume that (x) is the weighted sum of two Tukey's g mixture densities such that $g1g2 > 0$. Thus

$$f_X(x) = \begin{cases} 0, & \text{if } x \le \theta_1, \\ \dfrac{\omega_1}{g_1(x-\theta_1)} f_U(z_1), & \text{if } \theta_1 < x \le \theta_2, \\ \dfrac{\omega_1}{g_1(x-\theta_1)} f_U(z_1) + \dfrac{1-\omega_1}{g_2(x-\theta_2)} f_U(z_2), & \text{if } x > \theta_2, \end{cases} \tag{12}$$

where, without loss of generality, we let $\theta_1 < \theta_2$, $0 \le \omega_1 \le 1$ and for $j = 1, 2$

$$z_j = \frac{1}{g_j} \ln\left(\frac{x-\theta_j}{B/g_j}\right) \tag{13}$$

with $\theta j = A - (B/gj)$, $\lambda j = (B/|g_j|)$. We use the notation $X \sim$ GTMIX(A, B, g_1, g_2, $\omega 1$). Vitiello and Poon [19] did not provide the piecewise nature of the mixture density function above in (12). In this case the cdf of X is given by

$$F_X(x) = \begin{cases} 0, & \text{if } x \le \theta_1, \\ \omega_1 F_U\left(\dfrac{\ln(x-\theta_1)-\lambda_1}{g_1}\right), & \text{if } \theta_1 < x \le \theta_2, \\ \omega_1 F_U\left(\dfrac{\ln(x-\theta_1)-\lambda_1}{g_1}\right) + \omega_2 F_U\left(\dfrac{\ln(x-\theta_2)-\lambda_2}{g_2}\right), & \text{if } x > \theta_2, \end{cases} \tag{14}$$

where $\omega_2 = 1-\omega_1$. Begin with the fact that the quartile function is the inverse of the cdf. Thus, replacing $x_q > \theta_2$ in (14), we obtain

$$F_X(x_q) = \omega_1 F_U\left(\frac{\ln(x_q-\theta_1)-\lambda_1}{g_1}\right) + \omega_2 F_U\left(\frac{\ln(x_q-\theta_2)-\lambda_2}{g_2}\right). \tag{15}$$

If we assume that $U \sim (0, 1)$, (12) can be written as

$$f_X(x) = \begin{cases} 0, & \text{if } x \le \theta_1, \\ \dfrac{\omega_1}{g_1(x-\theta_1)} \varphi(z_1), & \text{if } \theta_1 < x \le \theta_2, \\ \dfrac{\omega_1}{g_1(x-\theta_1)} \varphi(z_1) + \dfrac{1-\omega_1}{g_2(x-\theta_2)} \varphi(z_2), & \text{if } x > \theta_2, \end{cases} \tag{16}$$

where $\varphi(z)$ is the standard normal pdf. Note that the expression above matches the pdf of a mixture of three-parameter lognormal distributions. Letting $\theta_1 = \theta_2 = 0$, the above pdf reduces to that of a mixture of two-parameter lognormal distributions. Given that every normal pdf is a version of the standard normal pdf then if $U \sim (\mu, \sigma_2)$ we have

$$f_U(u, \mu, \sigma) = \frac{1}{\sigma}\varphi\left(\frac{u-\mu}{\sigma}\right), \quad \text{with } \mu \in \mathbb{R}, \ \sigma > 0, \tag{17}$$

and (12) can be written as

$$f_X(x) = \begin{cases} 0, & \text{if } x \le \theta_1^*, \\ \dfrac{\omega_1}{g_1^*\left(x-\theta_1^*\right)}\varphi(z_1), & \text{if } \theta_1^* < x \le \theta_2^*, \\ \dfrac{\omega_1}{g_1^*\left(x-\theta_1^*\right)}\varphi(z_1) + \dfrac{1-\omega_1}{g_2^*\left(x-\theta_2^*\right)}\varphi(z_2), & \text{if } x > \theta_2^*. \end{cases} \tag{18}$$

If the parameters g_j are scaled by σ, that is, $g*_j = \sigma g_j$, then

$$z_j = \frac{\ln\left(x-\theta_j^*\right) - \lambda_j^*}{g_j^*} \tag{19}$$

With $\theta_j^* = A - (B\sigma/g_j^*)$, $\lambda_j^* = \ln(B\sigma/|g_j^*|) + (\mu/\sigma)g_j^*$. Note that the expression above matches the pdf of a mixture of three-parameter lognormal distributions, which is a generalization of the pdf given in (16), and we use the notation $X \sim \text{LSMIX}(\lambda*_1, \lambda*_2, g*_1, g*_2, \theta*_1, \theta*_2, \omega_1)$. Similarly, we can obtain pdf of a mixture of distributions for the random variables listed in Table 1.

ESTIMATION OF THE MIXTURES OF TWO TUKEY'S g DISTRIBUTIONS

In this section, we explain the estimation of the mixture of two Tukey's g distributions. The expected value of X is given by

$$\mathbb{E}[X] = \mu_1' = \sum_{j=1}^{2} \omega_j \left(\theta_j + \frac{B}{g_j} M_U(g_j) \right) \tag{20}$$

The nth raw moment of the random variable X is given by

$$\mathbb{E}[X^n] = \sum_{j=1}^{2} \omega_j \sum_{k=0}^{n} \binom{n}{k} \left(\frac{B}{g_j} \right)^{n-k} \theta_j^k M_U(\widetilde{g_j}), \tag{21}$$

where $g1g2 = 0,/ \tilde{g}j = (n-k)gj$ and $MU(t)$ is the moment generating function of the random variable U. The central moments μn of the random variable X are given by

$$\mathbb{E}\left[(X - \mu_1')^n \right] = \mu_n(X)$$
$$= \sum_{j=1}^{2} \omega_j \sum_{k=0}^{n} \binom{n}{k} \left(\frac{B}{g_j} \right)^{n-k} (\theta_j - \mu_1')^k M_U(\widetilde{g_j}). \tag{22}$$

The first five central moments are as follows:

$$\begin{aligned}
\mu_1 &= \omega_1 \eta_1 + \omega_2 \eta_2 = 0, \\
\mu_2 &= \omega_1 \left(\sigma_1^2 + \eta_1^2 \right) + \omega_2 \left(\sigma_2^2 + \eta_2^2 \right), \\
\mu_3 &= \omega_1 \left(3\sigma_1^2 + \eta_1^2 \right) \eta_1 + \omega_2 \left(3\sigma_2^2 + \eta_2^2 \right) \eta_2, \\
\mu_4 &= \omega_1 \left(3\sigma_1^4 + 6\eta_1^2 \sigma_1^2 + \eta_1^4 \right) \\
&\quad + \omega_2 \left(3\sigma_2^4 + 6\eta_2^2 \sigma_2^2 + \eta_2^4 \right), \\
\mu_5 &= \omega_1 \left(15\sigma_1^4 + 10\eta_1^2 \sigma_1^2 + \eta_1^4 \right) \eta_1 \\
&\quad + \omega_2 \left(15\sigma_2^4 + 10\eta_2^2 \sigma_2^2 + \eta_2^4 \right) \eta_2,
\end{aligned} \tag{23}$$

where for $j = 1, 2$

$$\eta_j = \frac{B}{g_j} M_U\left(g_j\right) + \theta_j - \mu_1',$$

$$\sigma_j^2 = \left(\frac{B}{g_j}\right)^2 \left[M_U\left(2g_j\right) - M_U^2\left(g_j\right)\right].$$

(24)

Because $\theta_1 < \theta_2$, upon equating population moments to the corresponding sample moments, it follows from (23) that

$$\omega_1\left(\frac{B}{g_1} M_U\left(g_1\right) + \theta_1 - m_1\right)$$
$$+ \omega_2\left(\frac{B}{g_2} M_U\left(g_2\right) + \theta_2 - m_1\right) = 0.$$

(25)

Left-hand side of system (23) is multiplied by $\omega_1 + \omega_2 = 1$; the equations take the following form:

$$\omega_1\eta_1 + \omega_2\eta_2 = 0,$$
$$\omega_1\left(\sigma_1^2 + \eta_1^2 - m_2\right) + \omega_2\left(\sigma_2^2 + \eta_2^2 - m_2\right) = 0,$$
$$\omega_1\left(3\eta_1\sigma_1^2 + \eta_1^3 - m_3\right) + \omega_2\left(3\eta_2\sigma_2^2 + \eta_2^3 - m_3\right) = 0,$$
$$\omega_1\left(3\sigma_1^4 + 6\eta_1^2\sigma_1^2 + \eta_1^4 - m_4\right)$$
$$+ \omega_2\left(3\sigma_2^4 + 6\eta_2^2\sigma_2^2 + \eta_2^4 - m_4\right) = 0,$$
$$\omega_1\left(15\eta_1\sigma_1^4 + 10\eta_1^3\sigma_1^2 + \eta_1^5 - m_5\right)$$
$$+ \omega_2\left(15\eta_2\sigma_2^4 + 10\eta_2^3\sigma_2^2 + \eta_2^5 - m_5\right) = 0,$$

(26)

where mi ($i = 1, 2, \ldots$) denote the ith central moment of the sample. Equations (26) accordingly constitute a system of five equations to be solved simultaneously for the estimates of the five parameters A, B, g_1, g_2, and ω_1.

Note that, from the first equation of system of (26), it follows that

$$\omega_1 = \frac{\eta_2}{\eta_2 - \eta_1}.$$

(27)

We eliminate $\omega 1$ between the first and the subsequent equations of (26) in turn and thereby reduce the system to the following four equations in four unknowns A, B, g_1, and g_2:

$$\begin{aligned} &\eta_1^{-1}\left(\sigma_1^2+\eta_1^2-m_2\right)=\eta_2^{-1}\left(\sigma_2^2+\eta_2^2-m_2\right), \\ &\eta_1^{-1}\left(3\eta_1\sigma_1^2+\eta_1^3-m_3\right)=\eta_2^{-1}\left(3\eta_2\sigma_2^2+\eta_2^3-m_3\right), \\ &\eta_1^{-1}\left(3\sigma_1^4+6\eta_1^2\sigma_1^2+\eta_1^4-m_4\right) \\ &\quad=\eta_2^{-1}\left(3\sigma_2^4+6\eta_2^2\sigma_2^2+\eta_2^4-m_4\right), \\ &\eta_1^{-1}\left(15\eta_1\sigma_1^4+10\eta_1^3\sigma_1^2+\eta_1^5-m_5\right) \\ &\quad=\eta_2^{-1}\left(15\eta_2\sigma_2^4+10\eta_2^3\sigma_2^2+\eta_2^5-m_5\right). \end{aligned} \tag{28}$$

These systems of equations are solved computationally by using scientific software package and we do not need to verify the unique solution of the system as the parameter estimates. We skip further details and numerical illustration owing to space constraint.

ILLUSTRATION

In this section we discuss some examples and applications of the results derived in Section 3 with two examples. In the first example, we discuss the pricing of a call option using a mixture of two Tukey's g-generalized distributions as an example to illustrate the results of Section 3. In the second example, we examine the empirical real data of HeatingDegree-Day to demonstrate usefulness of our approach of mixture of LS distributions.

Jimenez et al. [24] derived the option price of an European ′ option assuming that the terminal price distribution follows a g-generalized distribution. Instead if we use a mixture of two Tukey's classes of g-generalized distributions, then the price of the call option denoted by $C(t, \tau; K)$ with a strike price K and maturity date $T=t+\tau$ can be expressed as follows:

$$\begin{aligned} C(t,\tau;K) &= \sum_{j=1}^{2}\omega_j e^{-r\tau} \\ &\cdot\left[\int_{-\delta_j}^{\infty} e^{\lambda_j+g_j u} f_U(u)\,du-\left(K-\theta_j\right)F_U\left(\delta_j\right)\right], \end{aligned} \tag{29}$$

where $K>\theta_2$ and

$$\delta_j = \frac{\lambda_j - \ln\left(K - \theta_j\right)}{g_j} \quad \text{for } j = 1, 2.$$

(30)

When $U \sim (0, 1)$, (29) reduces to

$$C(t, \tau; K) = \sum_{j=1}^{2} \omega_j e^{-r\tau} \cdot \left[e^{\lambda_j + (1/2)g_j^2} \Phi\left(\delta_j - g_j\right) - \left(K - \theta_j\right) \Phi\left(\delta_j\right)\right],$$

where $\Phi(\cdot)$ denotes the cdf of a standard univariate normal variable. If we assume that $\theta_1 = \theta_2 = 0$, then (31) reduces to

$$C(t, \tau; K) = \sum_{j=1}^{2} \omega_j e^{-r\tau} \left[e^{\lambda_j + (1/2)g_j^2} \Phi\left(\delta_j - g_j\right) - K\Phi\left(\delta_j\right)\right].$$

(32)

Note that when $gj = \sigma j\sqrt{\tau}$, these expressions coincide with the option pricing formula given in Bahra [25]. The authors also established closed form formula for the calculation of the sensitives measures of option pricing (Greek parameters of the option). Here we wish to observe that our mixture model uses less unknown parameters for calculating the option pricing, whereas Vitiello and Poon [19] used nine unknown parameters to obtain the same for the mixture of two g- distributions. It has been known that when we increase the number of parameters, we lose degrees of freedom and it is no longer acceptable for the best fit of data. This gives an advantage of our approach for the mixture of two g- generalized distributions. We now present, as an example, the use of HeatingDegree-Days (HDD) in relation to winter temperature risk as a substitute for gas demand. HDD based contracts are listed on the Chicago Mercantile Fxchange (CME). We consider an example that consists of monthly aggregate Heating-DegreeDay (HDD) data values at the Chicago O'Hare International Airport from December 1979 to December 2000 given in Wang [26] and explored also by Vitiello and Poon [19]. We describe first a LS distribution with three parameters based method to infer the implied risk-neutral probability density (RND). In Table 3, we present the estimated values of the three parameters of lognormal and Log-Logistic distributions; our interest is to compare with Vitiello and Poon [19] risk-neutral densities with our proposed mixture model. The smaller value of the Kolmogorov-Smirnov (KS) test confirms that the data obeys the LS distributions with three parameters. We wish to observe that Anderson-Darling (AD) test is more sensitive to the tails of the LS distributions in comparison with KS test. In this case, we choose the LogLogistic distribution as the best fit for the HDD data. The implicit risk-neutral densities (RND) of LS distributions are shown in Figure 1 and compared with Figure 6 of Vitiello and Poon [19]. We have obtained a similar plot by our method with less unknown parameters than method given by Vitiello and Poon [19]. Furthermore, their KS test value of 13.6326% which is higher than the KS test values of Table 3 favors the best fit for the frequency of the LS distributions. Therefore, finite mixtures are attractive from the application viewpoint because of its flexibility and permit us to model various kinds of shaped distributions. In Table 4,

Table 3: Estimates for adjusting the LS(**Λ**).

Distribution of three parameters	Parameters			Test of adjusted	
	λ	g	θ	AD (%)	KS (%)
Lognormal	5.7407	0.5265	798.2540	37.53	12.3041
Log-Logistic	3.0225	283.8185	824.1814	32.68	11.1367

Table 4: Estimates for adjusting the mixture of LS(**Λ**).

Mixture of distributions	Parameters				KS (%) test
	$(\lambda_1;\lambda_2)$	$(g_1;g_2)$	$(\theta_1;\theta_2)$	ω	
Lognormal	6.5749	−0.1268	1797.9995	0.8188	8.0866
	11.0933	5.7574×10^{-4}	−64213.6284		
Log-Logistic	10.1338	−538.2433	1620.7711	0.8182	8.5928
	303.7429	6818.2894	−5309.4256		

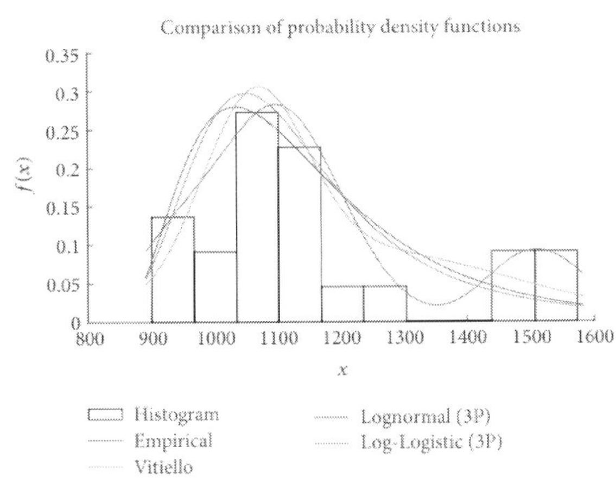

Figure 1: Empirical and LS(**Λ**) **densities estimated from HDD.**

we give the estimate values of the parameters of the mixture LS distributions. These parameters are estimated using (28). The estimated two g-densities and the implied risk-neutral densities (RND) are shown in Figure

2. We observe that the bimodal LS mixture distribution has same fitting performance of the empirical distribution function (EDF) and lognormal mixture distribution gives best goodness of fit using the KS test.

CONCLUSIONS

This paper presents a mixture of Tukey's g-generalized distributions and its properties. The methodology of estimating the unknown parameters by the method of moments is also presented.The proposed model has the advantage that it provides flexibility, when skewness, kurtosis, or other moments of the underlying distribution do not follow a normal distribution.

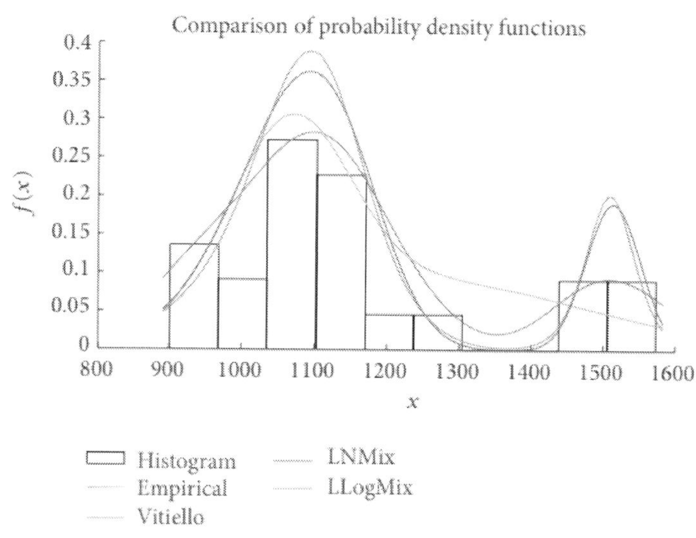

Figure 2: Empirical and two-g densities estimated from HDD.

Some special cases of well-known distributions are obtained from the proposed model.

Competing Interests

The authors declare that they have no competing interests.

REFERENCES

1. B. Mandelbrot, "The variation of certain speculative prices," *The Journal of Business*, vol. 36, no. 4, pp. 394–419, 1963.
2. E. F. Fama, "The behavior of stock-market prices," *The Journal of Business*, vol. 38, no. 1, pp. 34–105, 1965.
3. S. J. Kon, "Models of stock returns—a comparison," *The Journal of Finance*, vol. 39, no. 1, pp. 147–165, 1984.
4. Y. K. Tse, "Price and volume in the tokyo stock exchange: an exploratory study," *Working Paper* 1573, University of Illinois at Urbana-Champaign, 1989.
5. B. D. Fielitz and J. P. Rozelle, "Stable distributions and the mixtures of distributions hypotheses for common stock returns," *Journal of the American Statistical Association*, vol. 78, no. 381, pp. 28–36, 1983.
6. J. A. Jiménez, V. Arunachalam, and G. M. Serna, "Option pricing based on a log-skew-normal mixture," *International Journal of Theoretical and Applied Finance*, vol. 18, no. 8, Article ID 1550051, 22 pages, 2015.
7. J. W. Tukey, "Modern techniques in data analysis," in *Proceedings of the NSF Sponsored Regional Research Conference*, Southeastern Massachusetts University, North Dartmouth, Mass, USA, 1977.
8. J. Martínez and B. Iglewicz, "Some properties of the tukey g and h family of distributions," *Communications in Statistics—Theory and Methods*, vol. 13, no. 3, pp. 353–369, 1984.
9. S. G. Badrinath and S. Chatterjee, "On measuring skewness and elongation in common stock return distributions: the case of the market index," *The Journal of Business*, vol. 61, no. 4, pp. 451–472, 1988.
10. S. G. Badrinath and S. Chatterjee, "A data-analytic look at skewness and elongation in common-stock-return distributions," *Journal of Business & Economic Statistics*, vol. 9, no. 2, pp. 223–233, 1991.
11. T. C. Mills, "Modelling skewness and kurtosis in the london stock exchange ft-se index return distributions," *The Statistician*, vol. 44, no. 3, pp. 323–332, 1995.
12. K. K. Dutta and D. F. Babbel, *On Measuring Skewness and Kurtosis in Short Rate Distributions: The Case of the US Dollar London Inter Bank Offer Rates*, The Warton School, University of Pennsylvania, Philadelphia, Pa, USA, 2004.

13. K. K. Dutta and D. F. Babbel, "Extracting probabilistic information from the prices of interest rate options: tests of distributional assumptions," *Journal of Business*, vol. 78, no. 3, pp. 841–870, 2005.
14. X. Tang and X. Wu, "A new method for the decomposition of portfolio VaR," *Journal of Systems Science and Information*, vol. 4, no. 4, pp. 721–727, 2006.
15. K. K. Dutta and J. Perry, "A tale of tails: an empirical analysis of loss distribution models for estimating operational risk capital," *Working Paper* 06-13, Federal Reserve Bank of Boston, 2007.
16. J. A. Jiménez and V. Arunachalam, "Evaluating operational risk by an inhomogeneous counting process based on Panjer recursion," *The Journal of Operational Risk*, vol. 11, no. 1, pp. 1–21, 2016.
17. J. A. Jiménez and V. Arunachalam, "Using Tukey's g and h family of distributions to calculate value at risk and conditional value at risk," *Journal of Risk*, vol. 13, no. 4, pp. 95–116, 2011.
18. J. A. Jiménez, V. Arunachalam, and G. M. Serna, "A generalization of Tukey's g-h family of distributions," *Journal of Statistical Theory and Applications*, vol. 14, no. 1, pp. 28–44, 2015.
19. L. Vitiello and S.-H. Poon, "General equilibrium and risk neutral framework for option pricing with a mixture of distributions," *The Journal of Derivatives*, vol. 15, no. 4, pp. 48–60, 2008.
20. D. C. Hoaglin and S. C. Peters, *Software for Exploring Distribution Shape*, Laboratory for Information and Decision Systems, Massachusetts Institute of Technology, 1979.
21. N. L. Johnson, S. Kotz, and N. Balakrishnan, *Continuous Univariate Distributions*, vol. 1, John Wiley & Sons, New York, NY, USA, 1994.
22. A. Stuart and J. K. Ord, *Kendall's Advanced Theory of Statistics: Distribution Theory*, vol. 1, Edward Arnold, London, UK; Halsted Press, New York, NY, USA, 6th edition, 1994.
23. D. M. Titterington, A. F. Smith, and U. E. Makov, *Statistical Analysis of Finite Mixture Distributions*, vol. 7, John Wiley & Sons, New York, NY, USA, 1985.
24. J. A. Jiménez, V. Arunachalam, and G. M. Serna, "Option pricing based on the generalised Tukey distribution," *International Journal of Financial Markets and Derivatives*, vol. 3, no. 3, pp. 191–221, 2014.
25. B. Bahra, "Implied risk-neutral probability density functions from option prices: a central bank perspective," in *Forecasting Volatility in*

the Financial Markets, J. Knight and S. Satchell, Eds., pp. 201–226, Elsevier, 3rd edition, 1997.

26. S. S. Wang, "A universal framework for pricing financial and insurance risks," *ASTIN Bulletin*, vol. 32, no. 2, pp. 213–234, 2002.

CHAPTER 10

Poisson Mixture Regression Models for Heart Disease Prediction

Chipo Mufudza and Hamza Erol

Statistics Department, Cukurova University, 01330 Adana, Turkey

ABSTRACT

As heart disease becomes the global leading cause of deaths, high efficiency in heart disease diagnosis can help alleviate heart failure dangers. Model based clustering techniques via Poisson mixture regression models are therefore here applied and analysed to seek efficiency in heart disease prediction. Although extensive work has been done in heart disease prediction via supervised classifications in data mining, reality doesn't always provide the required information hence the unsupervised classification approach.

Citation: (APA): Mufudza, C., & Erol, H. (2016). Poisson mixture regression models for heart disease prediction. Computational and mathematical methods in medicine, 2016. (10 pages) DOI: https://doi.org/10.1155/2016/4083089 URL: https://www.hindawi.com/journals/cmmm/2016/4083089/

Poisson mixture regression models were analysed under two classes: standard and concomitant variable mixture regression models on heart disease count data. Heart disease diagnosis counts were modeled via by a Poisson process and prediction via mixture regression models in a generalised linear model (GLM) context on heart disease data. Results show that a two component concomitant variable Poisson mixture regression model predicts heart disease better than both the standard Poisson mixture regression model and the ordinary GLM Poisson regression model. This was determined by its low BIC value a criteria which shows that it fits the data far much better as it clearly identifies the risk factors. Furthermore a Zero Inflated Poisson mixture regression (ZIPMR) model was also applied on heart disease data to cater for zeros and it turned out the best model over all the Poisson mixture regression models. Thus heart disease prediction was effectively done componentwise with males, chest pain, non anginal pain, lack of exercise, and resting electrocardiographs identified as major risks for heart failure among other risks with cholestrol, resting blood pressure and age less likely to affect heart failure.

INTRODUCTION

Heart disease encloses a number of conditions that influence the heart and not just heart attacks. These may include functional problems of the heart such as heart-valve abnormalities, high blood pressure (BP), smoking, diet, cholesterol, or irregular heart rhythms. Problems like these can lead to heart failure, arrhythmia, and a host of other problems. The work in [1] claims that heart diseases have become the leading global death accounting for 17.3 million deaths per year and killing 1 in 7 in the US alone. Therefore effective and efficient automated heart disease prediction systems can be beneficial to both the patient and cardiologist. Although there has been increasing interests on heart disease problems especially with the use of data mining techniques and algorithms, most of them concentrated on a supervised classification approach through different classifications and algorithms.

In a comparative approach research by Bagirov et al. [2] they showed that it is possible to classify heart disease problems using either supervised or unsupervised classifications. Supervised classification on different patients status by data mining algorithms to predict heart disease has been explored by various authors in machine learning. H. D. Masethe and M. A. Masethe [3] identified that the predictive accuracy of J48, over REPTREE and CART, was reliable for heart disease prediction in South Africa. Weighted fuzzy models based on supporting have also been studied and analysed showing an improvement against the network based models [4]. Supervised classification algorithms can improve the efficiency to cardiologist as shown by Taneja [5] on PGI data where over 7300 observations were classified using J48 and Naive Bayes in WEKA. In a general review on the heart disease using data mining techniques done by Kaur and Singh [6], they summarised that most researches show that the main risk factors are cholesterol, lack of exercise, obesity, and high blood pressure whilst the best algorithms seem to be dominated by decision trees.

Supervised classifications predictions can also be improved in some cases by incorporating unsupervised classification techniques like clustering as a preprocessing procedure. However, it may not always be the case as shown by Soni et al. [7] that decision trees can still outperform the Bayesian classification even when clustering is incorporated although they mentioned that both algorithms can be improved by genetic algorithms. They also implemented a combination of associative classification and genetic algorithms in an effort to come up with the best system of heart disease prediction. A similar result by Akhil et al. [8] shows out that J48, a decision tree based algorithm, outperforms the Naive Bayes and neutral network on a different data set for heart disease prediction. Different data mining techniques for diagnosis have achieved varying probabilities for unique methods. In general under supervised classification methods like decision trees, Naive Bayes, and neural networks, decision trees seem to be better classifiers for heart disease diagnosis [9–11]. Regression trees, ID3, and CART can also be used on heart diseases prediction based on the important variables which are chest pain, exercises induced (exang), slope, and resting sugar level (restecg) where CART tend to be best classifier [12].

Although supervised classification seems to be effective, real life data normally comes as a complex mixture of both continuous and categorical data variables together with a lot of missing information. This also comes with difficulties in data manipulations since well-defined methods deal with either categorical or continuous data. Models involving mixed data generally experience variability and exhibit complex heterogeneous properties. Mixtures of categorical and continuous data are normally handled by generalised linear models (GLMs) of either Poisson or binomial maximum likelihood. However, in most cases these are normally not adequate enough to capture all the heterogeneity involved. GLMs in which the response is count (geometric, binomial/Poisson) are also sometimes referred to as count data problems. They appear in many real life settings, such as predicting the number items in economics, analysing infectious diseases in health and many modeling topics in the corporate world. Mixture models in a GLM context then become robust in capturing the heterogeneity where GLM cannot suffice. They allow the response for each observation y_{ij} to depend on x_j of the covariate vector where heterogeneity is exhibited and prior information is not available by grouping the likely homogeneous observations [13].

Mixture models are a form of unsupervised pattern recognition used to identify groups in the data based on a set of response variables or y-variable. Simultaneously they can be used to investigate relations of the groups themselves, or the within group expectations of the response variable given a set of covariates, or x variables. Count data mixture models often deal with extra dispersion problems especially in medical research where data is often count.

As such there has been increasing interest in modeling count data using the Poisson process and geometric and negative binomial [14, 15]. We attempt to present a detailed study on heart disease prediction using mixture model based clustering techniques on count data. Poisson mixture regression models will be here applied to heart disease data in order to determine how different predictor variables affect the heart rate diagnosis in different groups, unknown but to be identified. Unsupervised classification, unlike in the previous works, will be here applied to determine heart disease prediction efficiently.

Model based clustering will be here administered in a Poisson mixture regression model way. Although clustering has been incorporated in some classification work previously done, it was only incorporated as preprocessing step before classification.

The predictions will be accomplished via Poisson mixture regression models in a GLM context and identification of the most important attributes (risk factors) for each cluster will be done through the mixture model, thus a way to effectively help reduce the number of tests needed by cardiologist to diagnose heart disease. Linear regression mixtures have also been studied extensively, especially when no information about membership of the points assigned to each line was available [16].

However, we focus on mixture regression modeling on a count data problem using Poisson distributions. Application was done on Cleveland Clinic Foundation heart disease data under different Poisson mixture models and results analysed with a version of the Zero Inflated Poisson Mixture Regression model also included and analysed.

METHODS

Mixture Regression Model

In regression unobserved heterogeneity occurs if important covariates have been omitted either in the data collection or in some way such that important features are not accounted for in analysis, leading to biased parameter estimates. Mixture regression models seek to accommodate these different heterogeneity problems by grouping homogeneous observations into groups via a model based clustering. They can be standard, saturated, or with concomitant variables. The difference among these mixture regression models depends mainly on how the posterior probabilities are computed for the indicator variable. It has been shown that whilst standard variable models group and are classified based on the discriminant analysis rule and specify the joint distribution of the mixture distribution, the concomitant one is classified based on the logistic regression type rule and specifies the conditional distribution of response given the covariate [17]. In reality we tend to choose models with the best parameters compared to others.

Relaxing the normality and homogeneity assumptions in parameter distribution is quite possible by using mixture regression models especially using a GLM formulation as studied by Grun and Leish [18]. In a finite mixture regression model we consider a mixture of k components where each of the components follows a certain parametric distribution. Every component has a weight and prior probability for an observation to come from the given component and the mixture distribution is given by the weighted sum over the k components. Weights may depend on other variables not provided by data which are referred to as concomitant named here as concomitant variable mixture model. Traditionally, age and sex have been used to relate different segments of a population although they do not perform well in a prior segmented environment; thus the use of mixture models can be incorporated to improve classification in such segmented cases [19, 20]. Finite mixture of GLM format relaxes the assumption that regression coefficients and dispersion parameters are the same for all observations, giving a high degree of heterogeneity which can be incorporated into the clustering model.

Mixture Regression Classes

Two classes of mixture regression models were here considered: namely, the standard and concomitant variable mixture regression models. A component is here described by a GLM with different linear operator but the same error and link function resulting in a general standard mixture regression model of a GLM is given by

$$f(y|x, \Phi) = \sum_k \pi_k f_k(y|x; \beta_{0k}, \beta_k) \tag{1}$$

where y is the response variable assumed to follow a distribution of the exponential family, conditional upon component k, whilst the conditional expectation of the response variable is given as

$$E(y|x) = g^{-1}(\beta_{0k} + x'\beta_k) \tag{2}$$

with $g(\cdot)$ some link function. Here $\Phi = \{\beta_0, \beta, \pi\}$, $\beta_0 = (\beta_{0k})$, and $\beta = (\beta_k)$. Exponential family probability distributions are the commonly used and these include Poisson, binomial, normal, negative binomial, and geometric. We are going to concentrate on the Poisson due to the nature of the counts within the data to be considered. Often it is assumed that the component specific densities are from the same parametric family for each component; that is, $f_k = f$ for notational simplicity and that the link function is also the same for all components $g_k = g$. In a cluster-wise regression setting this will be an obvious model choice as no prior knowledge about differences in distributional families of the components is available.

Another popular extension is to have a so-called concomitant variable model for the prior class probabilities, such that the weight π_k also depends on a set of explanatory variables (e.g., using a multinomial logit model). In such models substantial changes in the parameter estimates with quite small changes in the mixing model are a possible way to conclude that the random effects do not have enough information in the data about the distribution (i.e., there are heterogeneity issues). A concomitant variable mixture regression model is based on regression model (1) but uses a parameterisation on the posterior probabilities which assume that the concomitant variables fall into two sets $x = (x_1, x_2)$ but the two sets may (partially or completely) overlap. The first set affects y whilst the second affects latent group indicator z variable assumed multinomial (i.e., $f(z|x; \pi) = \prod_k \pi_{k|x}$). Thus the marginal distribution of the response variable is given as

$$f(y \mid x_1, x_2; \Phi) = \sum_k \pi_{k|x_2} f_k(y \mid x_1; \beta_{0k}, \beta_k). \tag{3}$$

Here $\Phi = \{\beta_0, \beta, \gamma_0, \gamma\}$, $\beta_0 = (\beta_{0k})$, $\beta = (\beta_k)$, $\gamma_0 = (\gamma_{0k})$, and $\gamma = (\gamma_k)$. The γ_k's are derived from the component indicator function

$$\pi_{k|x} = \frac{\exp\left(\gamma_{0k} + x'\gamma_k\right)}{\sum_h \exp\left(\gamma_{0h} + x'\gamma_h\right)} \quad (4)$$

and are normally taken to be zero for identification. The component intercept parameter β_0 can sometimes be aliased with the model intercept resulting in big standard errors for the intercept.

The Poisson Mixture Regression Model

A Poisson regression model is an example of a GLM in which distribution of the response Y with covariate vector x has Poisson density function given as

$$f(y, \lambda) = \frac{e^{-\lambda}\lambda^{y}}{y!} I_A(y). \quad (5)$$

In (5), $E(Y) = \lambda$, the link function $g(\lambda) = \log(\lambda) = \beta^T x$, $A = \{0,1, 2,\ldots\}$ is the set of nonnegative integers, and $I_A(y)$ is the indicator function, which is one if y belongs to the set A and is zero otherwise. The variance var(Y) = $E(Y) = \lambda$ and there exist dispersion problems when either var(Y) > $E(Y)$ (overdispersion) or var(Y) < $E(Y)$ (underdispersion). In many problems overdispersion is more common than underdispersion. Analysis of count data that are overdispersed relative to the Poisson distribution has been studied extensively; it is also known that application of the standard Poisson model may result in misleading inferences. There is therefore need for alternative methods of analysis to deal with the extradispersion problem. Finite mixture models are appropriate when the extra variability comes from the unobserved heterogeneity of the population, which composes two or more subgroups mixed in various proportions [13, 18, 20, 21]. In particular, Poisson mixture regression can be fitted to such extradispersed count data in the presence of covariate information. A mixture Poisson regression model assumes the objectives are drawn from a finite mixture Poisson where these distributions differ in intercept and coefficient of the explanatory variable in the regression component or are suspected to be heterogeneous. Mixture GLM models assume the count outcomes are clustered or represent repeated measurements and random effects that can be incorporated within the regression model to account for the inherent correlation between observations. Poisson mixture regression models which include the Poisson mixture model, negative binomial (NB), hurdle, Zero Inflated Poisson (ZIP), and Zero Inflated Negative Binomial (ZINB) models cater for different dispersion parameters and heterogeneity aspects, [22]. In particular the ZIP and hurdle models account for both heterogeneity and the extra zeros in the data.

A finite mixture model based on Poisson distribution may be written as model (1) where

$$f_j(y_i,\theta_{ij}) = \frac{e^{-\lambda_{ij}}(\lambda_{ij})^{y_i}}{y_i!} I_A(y_i). \tag{6}$$

In (6) $\log\lambda_{ij} = \beta_j^T x_i$, $i = 1 : n, j = 1 : k$, and for the k component mixture

$$\begin{aligned} E(Y_i) &= \sum_{i=1}^{k} \pi_j \lambda_{ij}, \\ \operatorname{var}(Y_i) &= E(\operatorname{var}(Y_i \mid Z_i)) + \operatorname{var}(Y_i \mid Z_i) \\ &= E(Y_i) + v_{ij}, \end{aligned} \tag{7}$$

where $\lambda_{ij} = \exp(\beta_j^T x_i)$ is the mean of the ith response condition to its membership of the jth component of the mixture, $v_{ij} = \sum \pi_j \lambda_{ij}^2 - (\sum \pi_j \lambda_{ij})^2$, and Z_j is the component indicator vector of zeros and ones with $Z_{ij} = (Z_j)i$ which is one if it is from component i and zero otherwise. $v_{ij} = 0$, iff $\lambda_{1i} = \lambda_{2i} = \cdots \lambda_{ki}$, that is, if the means of all components are the same, an indication that mixture model copes better than the homogeneous model. Heterogeneity across individuals in a Poisson mixture regression model is tackled in two ways which includes formulation in such a way that mean event rate has a discrete mixture distribution (i.e., varying across a finite number of unobserved classes) and that the mean event rate within each class depends on the explanatory variables. The general Poisson mixture regression model then assumes the following:

(i) Unobserved mixing process can occupy any of the k components, where k is finite and unknown.

(ii) For each of the observed count y_i, there exists an unobserved random variable Z_j representing component at which y_i is generated and (Y_i, Z_i) are piecewise independent.

(iii) Z_j follows a discrete distribution with k points of support $P(Z_j = j) = \pi_j$, with $\sum \pi_j = 1$.

(iv) Conditional on the Z_i, Y_i follows a Poisson distribution.

Here y_i, $i = 1 : n$, represent the ith response variable such that y is Poisson(λ_{ij}) with probability $\pi_j(x)$, which are the mixing weights, where x and z are matrices and k is the number of components in the model. When discriminant operator using the normal log link function is used to separate observation we got standard Poisson mixture regression model.

Using logit and log-linear links to model $\pi_k(x)$ and λ_{ij}, which denote vector of the regression coefficients of the jth component, respectively, gives

$$\operatorname{logit}(\pi_j(x)) = \log\left(\frac{\pi_j(x)}{1-\pi_j(x)}\right) = x_i^t \beta_j \quad (8)$$

$$\log(\lambda_{ij}) = z_j^T \gamma_j \quad (9)$$

for β_j and γ_j unknown parameters. This is a presentation of the concomitant variable Poisson mixture model. Whilst the covariates of the Poisson rates z are restricted to be the same, covariates in the mixing proportions (x) are not restricted to being the same as those in the Poisson rates (z). In Poisson model, dispersion problems can be modified via a different link function, alternative frequency distribution, or both. However, there are times when overdispersion or heterogeneity problems do not seem to be eliminated and hence can be rescued by mixture models. These assume a mixture of distribution as stated before such that

(i) if $k = 1$ we obtain the Poisson regression model reducing to Poisson model if β_{11} is specified to null vector,

(ii) if β_{1i} is specified to zero vector we obtain a k component Poisson mixture model,

(iii) if $k = 2$ and $\lambda_{1k} = 0$, for all j, then we obtain a Poisson regression model with extra weight at $y_i = 0$.

Thus mean and variance can be determined depending on the applicable condition. Maximisation of this Poisson mixture model is done using expectation and maximisation algorithm famously known as the EM. Extensive work on the EM algorithm and how it works have been studied for long by many authors including McLachlan et al. [13, 21]. As such EM algorithm explanations will not be considered here.

Zero Inflated Poisson Mixture Model (ZIPMR)

ZIPMR serves when the data involved has so many zeros which originates from how the counts are being generated in the model. It serves as a special mixture regression model to cater for both heterogeneity and zeros. It has a mixture of binary distribution that is degenerate at zero and an ordinary count distribution which can be Poisson or negative binomial. It considers the zeros to be separate from the nonzero, although it can include some zeros in the analysis with nonzeros. Zero inflated models can be thought of as finite mixture models (i.e., where two data-generating mechanisms are supposed, one generating 0s and one generating the full range of counts) [15]. It has been applied in many problems including epidemiology, occupational health, and children growth development [12]. In our case the Zero Inflated Poisson

(ZIP) model was considered since there were structured zeros within the data considered indicating that they are true zeros from the counting process and zeros which came from dichotomising processes done during sampling. The ZIP model assumes that

$$y_i = \begin{cases} 0, & \text{if } s_i = 0 \\ \text{Poisson}\left(\lambda_i, \beta, \gamma\right), & \text{if } s_i = 1, \end{cases} \quad (10)$$

where s_i represents whether component i has heart disease or not and other parameters are as defined before.

APPLICATION

Data

The study used Cleveland Clinic Foundation heart disease data set available at http://archive.ics.uci.edu/ml/datasets/Heart+Disease. It consists of 76 raw attributes and 303 observations for each attribute of which only 14 attributes were used for this research as given by Detrano [23]. The response variable (given as num) is a count of the rate at which an observation is diagnosed with heart disease where zero indicates no disease whilst 1, 2, 3, and 4 are the different levels of the heart disease stage. Our focus, however, will be to predict heart diseases diagnosis rate via a model based clustering approach using Poisson mixture regression models. This is justified by the fact that our data choice has mixture of both categorical and continuous variables with the response variable being count which is heterogeneous piecewise with the response being count. In this context, the Poisson mixture model assumes an underlying partition of the population into k homogeneous components, where each component has a rate of heart failure indicated by number of diagnosis on heart disease (angiographic disease status) dependent on the 13 different covariates.

Why Model Based Mixture Models on Heart Disease?

Data exploration here not shown results indicated that the data has lots of unexplained abnormalities. Most continuous variables showed skewed distributions and varying levels of high peaks indicating multimodality of the variables. The response variable relationship with indicators was observed to be nonlinear in most cases. Since the data is Poisson count, a general linear Poisson model was also analysed and results showed that it was underdispersed with a couple of other unexplained heterogeneity properties. As underdispersion is hardly addressed by change of link function or transformations, we introduce Poisson mixture models as a holistic approach to cater for the unexplained heterogeneity and extradispersion problems. Single parametric distributions on their own can never suffice to describe the data in a robust manner no matter how we transform the data. Irregularities of models as well as unobserved heterogeneity within the same model can be dealt with simultaneously in a single mixture models.

This is because mixture models use a combination of both parametric density functions to represent each cluster and the overall population as mixture model thus capturing heterogeneity using several homogeneous groupings.

Since a mixture model can capture these by way of dividing the observations into grouped distributions, we seek to employ a finite mixture model to heart disease data with so many explanatory variables causing heart problems and try to capture the heterogeneity in individuals via discrete mixture components.

The source heterogeneity is here not known, so we have an incomplete data problem with assumed unknown number of homogeneous groups which will be determined via a finite mixture regression model. Our data has a count response variable and a mixture of both continuous and categorical predictor variables; thus we choose Poisson mixture regression model as appropriate for heterogeneity investigations as well as identifying the number of possible groupings.

RESULTS

Finite Poisson Mixture Regression (FMPR) Model without Covariates

A GLM Poisson mixture model without covariates is here introduced in order to predict the heart disease. This is a Poisson mixture model that models the response variable, given the component membership of the observations is mutually independent. Flex mix package in R was used to determine different number of components (i.e., $\beta_{1i} = 0$). The summary results of the Poisson mixture model without covariates are shown in Table 1.

Table 1: Summary for FMPR models without covariates.

K	1	2	3	4	5
AIC	865	**797**	801	805	809
BIC	868	**808**	819	831	842

Table 1 shows the AIC and BIC values rounded to 3 significant figures for the Poisson mixture model without covariates under varying components, *K*. Based on both the AIC and BIC criteria a two-component Poisson mixture model best describes the data. We can therefore describe the heart disease diagnosis by a two-component mixture Poisson model.

Two-Component FMPR Model without Covariates Properties

The properties of the chosen two-component Poisson mixture model without covariates are here summarised as Figure 1. The components from a Poisson mixture regression model without covariate are well separated as indicated by the rootograms in Figure 1 showing the posterior probabilities of the observation per given components. The prior probability of an observation getting to components one and two is 0.549 and 0.451, respectively. The mean component rate for diagnosis shows that component one has a high mean rate compared to component two.

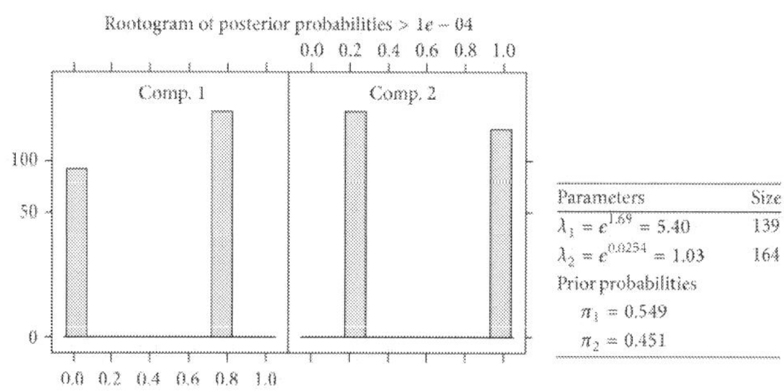

Figure 1: Information criteria graphs.

FMPR with Covariates

A mixture Poisson regression model with the covariates is here analysed and applied to the heart disease data in order to predict the heart disease diagnosis using the 13 explanatory variables. Analysis was made using both the standard and a multinomial concomitant variable FMPR models and models compared with the GLM Poisson regression model (single component).

The standard FMPR model has the single component Poisson mixture as the best model which correspond to the GLM Poisson model. It had a BIC value of 680 whilst all other models where above 700. The concomitant variable model supposes that if we let y_i denote the ith response, then the heart disease Poisson mixture regression model with 13 covariates assumes that each diagnosis count y_i is associated with at least one of the covariates given; diagnosis counts are independent and follow a mixture Poisson regression model with conditional means equal to the number of diagnoses. Finally, it also assumes that mean rate for the diagnosis is given by the exponential link function as a multiplicative factor $\lambda_{ij} = e^{\beta_{0j} + \beta_{1j} x_{1j} + \cdots + \beta_{kj} x_{kj}}$, where x_{ij} represent the ith observation in the jth component of predictor x.

We also determined the number of components required for the concomitant FMPR model and results of the models considered and analysed are shown in Figure 2: both AIC and BIC model criteria seem to favour a 2-component FMPR model as shown by Figure 2 against even a single component (GLM Poisson). We therefore chose a two-component FMPR model as best and simple model due to its lowest BIC of 647 compared to all other mixture models including the standard FMPR and GLM Poisson models which has BIC value of 680.

	Iter	Converged	k	k0	log Lik	AIC	BIC	ICL
1	2	TRUE	1	1	−300.30	629	681	681
2	14	TRUE	2	2	−200.67	487	647	656
3	20	TRUE	3	3	−192.77	530	797	854
4	17	TRUE	4	4	−189.31	581	956	1082
5	16	TRUE	5	5	−186.58	633	1116	1310
6	16	TRUE	6	6	−187.22	692	1283	1482
7	21	TRUE	7	7	−186.50	749	1447	1626
8	19	TRUE	8	8	−185.99	806	1612	1953

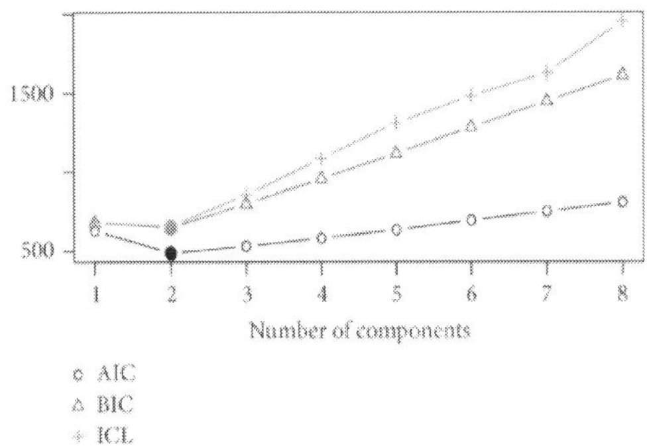

Figure 2: FMPR model selection: the Information criteria values and graph.

Two-Component Poisson Mixture Regression Results

Since the whole aim is to be able to predict heart disease, we analysed a Poisson mixture regression model with the already established 2 components for the data. A summary with parameter values and properties of the two-component Poisson mixture regression (FMPR) model is given as Figure 3.

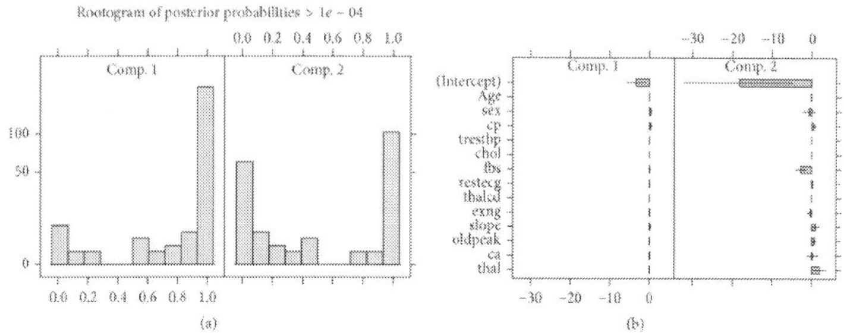

Figure 3: Poisson mixture regression plots: (a) concomitant FMPR model rootogram and (b) concomitant FMPR parameter CI.

Figure 3(a) shows the component distributions for the 2-component FMPR via the rootograms which shows that components are well separated. The parameter estimates confidence intervals for each component are shown as Figure 3(b). There seem to have wide confidence intervals for parameters in component 2 than they are in component 1. The model properties are summarised in Table 2 which shows that the 2-component FMPR model had 198 observations classified under cluster 1 at a probability of 0.642 whilst 105 are in cluster 2 with a prior probability of 0.358. These groups are mostly likely to be those who will develop the disease and those who will not.

Table 2: Summary of the FMPR model.

Model	DF	AIC	BIC	Prio.Prob.	Size
FMPR	43	490	650	$\varpi_1 = 0.642$	198
				$\varpi_2 = 0.358$	105

Parameter Estimates

The parameter estimates for the 2-concomitant variable FMPR model describe the average count rate for diagnosis for each component is given multiplicative exponential of the significant covariates within the component. Exponents of the parameters are shown in Table 3: the two-component FMPR models generally show that there is a general no change on the heart diagnosis rate with increases in age, cholesterol level (chol), maximum heart rate (thalacd), and resting blood pressure (trestbp) levels in component 1 whilst component 2 is not affected by age, trestbp, thalcd, and slope with a slight change due to cholesterol and restecg. The major risk factors of heart disease for component 2 include chest pain (cp), lack of exercises (exang), high fasting blood sugar (fbs), sex (males are at risk than woman), number of major vessels coloured by flourscopy (ca), oldpeak, and defect type (thal) whilst component 1 has main risk factors as resting electrographic results (restecg) and slope of peak exercise (slope). Analysis on the heterogeneity within a variable can also be inferred; for example, change in one unit increase in heart disease diagnosis has 56% increase more for males in component 2 than 1 compared to females. Therefore we are able to infer the differences not only between components but also within heart disease risks. Heart disease diagnosis can there be described differently for those in different components varying coefficients as well as risks that determine the rate of diagnosis.

Table 3: Parameter estimates values.

FMPR model with concomitant		
	Comp. 1	**Comp. 2**
coef.(intercept)	0.867	$7.33e-07$
coef.Age	1.00	$9.93e-01$
coef.sex	1.07	1.64
coef.cp	1.10	8.45
coef.trestbp	0.998	1.00
coef.chol	0.10	1.01
coef.fbs	1.03	1.50
coef.restecg	1.09	1.05
coef.thalcd	0.998	$9.97e-01$
coef.exng	1.19	3.33
coef.oldpeak	1.07	1.39
coef.slope	1.06	$9.18e-01$
coef.ca	1.13	1.40
coef.thal	1.07	1.25

Zero Inflated Poisson Mixture Regression (ZIPMR) Model Results

The existence of so many zeros in the counts from the dichotomising process necessitated the use of ZIPMR model in the analysis. Results from the ZIPMR model under different components are given as Figure 4.

	Iter	Converged	k	k0	logLik	AIC	BIC	ICL
1	2	TRUE	1	1	−10365.3	20700	20700	20700
2	9	TRUE	2	2	−201.7	461	569	570
3	13	TRUE	3	3	−186.2	488	704	738
4	14	TRUE	4	4	−186.4	547	870	949
5	15	TRUE	5	5	−187.0	605	1040	1140

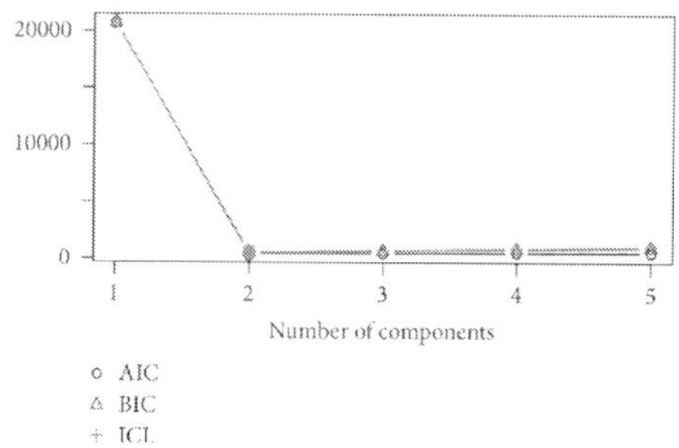

Figure 4: ZIPMR model selection.

The best model criteria favour a two-component ZIPMR as best model as indicated by low AIC, BIC, and ICL values in Figure 4 compared to any other model including the GLM ZIP Poisson model which has the highest BIC value. ZIPMR model has one component which has no disease and the second component which responds to the heart disease diagnosis based on the multinomial covariates; thus a 2-component model best describes the heart disease prediction as those who are likely to get disease and those who will rarely develop the disease. The cluster and model properties of ZIPMR model are given in Table 4.

Table 4: Summary of the ZIPMR models.

Model	AIC	BIC	Prior prob.	size
$ZIPR_1$	462	569	$\varpi_1 = 0.54$	164
			$\varpi_2 = 0.46$	139

Table 4 shows the AIC and BIC values for the ZIPMR model as well as component distribution where 164 observations are less likely to be diagnosed with heart disease at 54% rate in component 1 whilst there are

46% chances for 139 observations to get heart disease as described by component 2. The ZIPMR model seems to describe the heterogeneity and fit the data simpler and better than the normal 2-Poisson-mixture-component model due to its low BIC. The model also has highly separated components as figured by Figure 5(a) an indication of better separation among clusters. As component 2 describes those most likely to develop the disease, most parameters in the same component are also highly significant as indicated by the confidence intervals in Figure 5(b). Rating from the rootogram, the concomitant ZIPMR model has highly separated components as indicated by Figure 5(a).

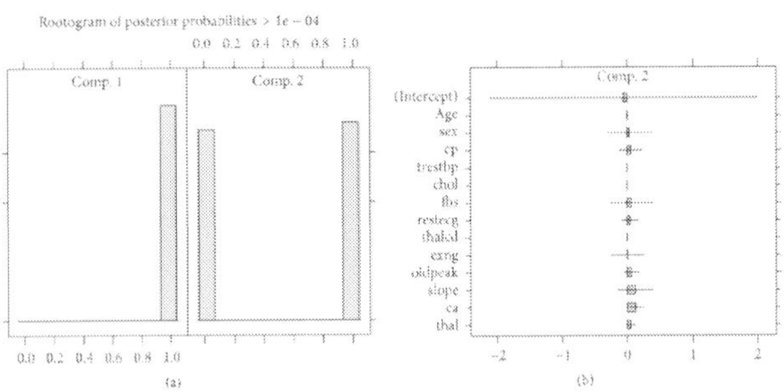

Figure 5: ZIP model results graphs.

Parameter Values of the ZIPMR Models

To determine which predictor variables predict heart disease the most, we find the exponential model parameters as shown by the parameter values given in Table 5. The ZIPMR shows that the first component does not change with any variations of predictors as expected. It represents those without the heart disease, or who are not at risk with any of the variables given. However, most second component parameter values change with a change in the heart rate diagnosis. These outstanding risk factors include chest pains, lack of exercise, high fasting blood sugar levels, exercise slope, number of vessels, and defects state. Thus we are able to predict the heart disease diagnosis rate using a 2-component Poisson mixture model with a degenerate at zero. We can safely conclude that a change in fasting blood sugars levels or chest pain or exercise slope can contribute to rate at which one develops heart disease. Nine of the 13 explanatory variables which include sex, cp, fbs, restecg, exercise, slope, oldpeak, ca, and thal seem to give more risk in heart prediction than cholesterol, age, and maximum heart rate.

Table 5: ZIPR models estimated parameter values.

ZIPR with concomitant		
	Comp. 1	**Comp. 2**
coef.(intercept)	0	0.120
coef.Age	1	1.00
coef.sex	1	1.28
coef.cp	1	1.24
coef.trestbp	1	1.00
coef.chol	1	1.00
coef.fbs	1	1.04
coef.restecg	1	1.08
coef.thalcd	1	0.997
coef.exng	1	1.12
coef.slope	1	1.23
coef.oldpeak	1	1.10
coef.ca	1	1.23
coef.thal	1	1.12

DISCUSSION

Use of Finite Mixture Regression Models

The use of GLM mixture Poisson regression models was reviewed under different classes which include standard and concomitant variable models. The general use of the mixture regression models was to capture the heterogeneity which cannot be catered for via normal general Poisson regression models, by clustering homogeneous observations in components. This seemed to add a lot of value on the concomitant model than it does on the standard mixture model. Inclusion of mixture regression models greatly improves the model fit on the data as indicated by very low BIC values compared to GLM models. Thus improving on the efficiency at which we can predict heart disease since different components has different risk factors or predictors individuals can be diagnosed into components depending on whether they are major risk or minor risk factors. Mixture models also capture the extra variability via a concomitant variable Poisson mixture regression model, which tend to capture more heterogeneity and hence a better model than both the standard Poisson mixture regression model and the single component (general Poisson regression model here not shown). The presence of so many zeros within the count data also triggered us to incorporate a Zero Inflated mixture Poisson regression model to handle the many zeros for Poisson models. The idea of handling the zeros in a special manner as expected improves the models as it not only does fit the data properly but also made a reality assumption of putting those who are classified not to have the disease into a single cluster whilst those with disease in another component. The assumption is that those with the disease are modeled by Poisson mixture regression to predict the diagnosis allowing variability on the covariates. We noted that the Zero Inflated Poisson mixture model with a concomitant variable is even better than just a two-Poisson-mixture regression model as it has better and lower BIC and AIC values. Thus mixture models give more insight and better describe the data than supervised classification method does above its ability to efficiently predict heart disease and identify major risk factors.

Risk Factors

Although most heart disease prediction models assume that cholesterol, lack of exercise, and blood pressure are the main risk factors, we deduced that Poisson mixture regression models explain also the different levels at which these risk factors affect heart disease together with other correlated risk factors within the 2 heterogeneous groups involved in heart disease. On the contrary when Poisson mixture regression models are used, it showed that males are at more risk than females for heart disease whilst cholesterol does not seem to be a dangerous risk factor on heart diseases. Chest pain is also indicated among the major risk factors with nonanginal pain associated with heart disease more than angina pains, lack of exercise as before attributed to heart disease, those with abnormal resting electrocardiographic level who are at higher risk, the type of defects on the chambers, and number of major vessels an individual has. Therefore mixture models do not only fit the data better but also give more information on the heterogeneous characteristics of different individuals in the prediction of heart diseases. We can safely conclude that heart disease can be efficiently predicted by a two-component Poisson mixture regression model, with each component describing attributes differently. Thus the different components cater and captures the different heterogeneity properties exhibited by the general Poisson regression model hence the conclusion that model based clustering adds value to the prediction of heart diseases. The mean heart diagnosis is here given as a multiplicative factor of the parameter estimates and is differently predicted by two-component concomitant Poisson mixture regression model, with the best model catering for the zeros in ZIP mixture regression model.

CONCLUSION

Mixture model helps inference on the different groupings not only within the sample but also within each of the groupings. Heart disease can therefore be effectively diagnosed in an individual depending on the sex, whether they do exercise or not, and level of chest pain with a two-component Poisson finite mixture regression model. Hence heterogeneity across components is explained via different regression parameters and mixing probabilities. The risk factors can also be efficiently identified within each group (component) and individuals depending on the component.

Acknowledgments

The authors are grateful firstly to the Turkish Government scholarship who made it possible for them to pursue the study and all their colleagues who were of support and help.

Disclosure

The Turkish Government scholarship has been responsible for paying the corresponding author's tuition fees for the past 3 years; however there are no other influences on research work from the scholarship authorities. Data here used was taken from the UCI database on the Cleveland heart disease.

Competing Interests

The authors declare that they have no competing interests.

REFERENCES

1. Go A. S., Mozaffarian D., Roger V. L., et al. Heart disease and stroke statistics—2014 update: a report from the american heart association. *Circulation.* 2014;2014 doi: 10.1161/01.cir.0000441139.02102.80.
2. Bagirov A. M., Rubinov A. M., Soukhoroukova N. V. Unsupervised and supervised data classification via nonsmooth and global optimization. *Top.* 2003;11(1):1–75. doi: 10.1007/BF02578945.
3. Masethe H. D., Masethe M. A. Prediction of heart disease using classification algorithms. Proceedings of the World Congress on Engineering and Computer Science (WCECS '14); October 2014; San Francisco, Calif, USA.
4. Anooj P. Clinical decision support system: risk level prediction of heart disease using weighted fuzzy rules. *Journal of King Saud University—Computer and Information Sciences.* 2012;24(1):27–40. doi: 10.1016/j.jksuci.2011.09.002.
5. Taneja A. Heart disease prediction system using data mining techniques. *Oriental Journal of Computer Science and Technology.* 2013;6(4):457–466.
6. Kaur B., Singh W. Review on heart disease prediction system using data mining techniques. *International Journal on Recent and Innovation Trends in Computing and Communication.* 2014;2(10):3003–3008.
7. Soni J., Ansari U., Sharma D., Soni S. Predictive data mining for medical diagnosis: an overview of heart disease prediction. *International Journal of Computer Applications.* 2011;17(8):43–48. doi: 10.5120/2237-2860.
8. Akhil M., Chandra P., Deekshatulu B. L. Heart disease prediction system using associative classification and genetic algorithm. *Proceedings of the International Conference on Emerging Trends in Electrical, Electronics and Communication Technologies (ICECIT '12)*, 2012.
9. Dangare C. S., Apte S. S. Improved study of heart disease prediction system using data mining classification techniques. *International Journal of Computer Applications.* 2012;47(10):44–48. doi: 10.5120/7228-0076.
10. Bhatla N., Jyoti K. An analysis of heart disease prediction using different data mining techniques. *International Journal of Engineering Research & Technology.* 2012;1(8):14.
11. Chaurasia V., Pal S. Early prediction of heart diseases using data mining

techniques. *Caribbean Journal of Science and Technology*. 2013;1:208–217.

12. Fang R. *Zero Inflated Negative Binomial for Overdispersed Count Data with Excess Zeros and Repeated Measures, An Application to Humana Microbiota Sequential Data*. NCCMM College; 2008.
13. McLachlan G., Peel J. *Finite Mixture Models*. John Wiley & Sons; 2000.
14. Greene W. Functional form and heterogeneity in models for count data. *Foundations and Trends in Econometrics*. 2007;1(2):113–218. doi: 10.1561/0800000008.
15. Hilbe J. M. *Modelling Count Data*. Cambridge University Press; 2014.
16. Wang K., Yau K. K. W., Lee A. H., McLachlan G. J. Two-component Poisson mixture regression modelling of count data with bivariate random effects. *Mathematical and Computer Modelling*. 2007;46(11-12):1468–1476. doi: 10.1016/j.mcm.2007.02.003.
17. Wedel M. Concomitant variables in finite mixture models. *Statistica Neerlandica*. 2002;56(3):362–375. doi: 10.1111/1467-9574.t01-1-00072.
18. Grun B., Leish F. *FlexMix: Finite Mixtures of Generalised Linear Regression Models*. LMU; 2007.
19. Grun B., Leisch F. FlexMix version 2: finite mixtures with concomitant variables and varying and constant parameters. *Journal of Statistical Software*. 1988;28(4):1–35.
20. Dalrymple M. L., Hudson I. L., Ford R. P. K. Finite mixture, zero-inflated poisson and hurdle models with application to SIDS. *Computational Statistics and Data Analysis*. 2003;41(3-4):491–504. doi: 10.1016/s0167-9473(02)00187-1.
21. McLachlan G. F. On the EM algorithm for overdispersed data. *Statistical Methods in Medical Research*. 1997;6:76–98. doi: 10.1191/096228097666670466.
22. Zhou M., Carin L. Negative binomial process count and mixture modeling. *IEEE Transactions on Pattern Analysis and Machine Intelligence*. 2015;37(2):307–320. doi: 10.1109/TPAMI.2013.211.
23. Detrano R. *Long Beach and Cleveland Clinic Foundation*. V. A. Medical Center; http://archive.ics.uci.edu/ml/datasets/Heart+Disease.

CHAPTER

11

A Hierarchical Gamma Mixture Model-Based Method for Classification of High-Dimensional Data

Muhammad Azhar, Mark Junjie Li and Joshua Zhexue Huang

College of Computer Science and Software Engineering, Shenzhen University, Shenzhen 518060, China

ABSTRACT

Data classification is an important research topic in the field of data mining. With the rapid development in social media sites and IoT devices, data have grown tremendously in volume and complexity, which has resulted in a lot of large and complex high-dimensional data. Classifying such high-dimensional complex data with a large number of classes has been a great challenge for current state-of-the-art methods. This paper presents

Citation: (APA): Azhar, M., Li, M. J., & Zhexue Huang, J. (2019). A Hierarchical Gamma Mixture Model-Based Method for Classification of High-Dimensional Data. Entropy, 21(9), 906.(21 pages) DOI: https://doi.org/10.3390/e21090906 URL: https://www.mdpi.com/1099-4300/21/9/906/htm

a novel, hierarchical, gamma mixture model-based unsupervised method for classifying high-dimensional data with a large number of classes. In this method, we first partition the features of the dataset into feature strata by using k-means. Then, a set of subspace data sets is generated from the feature strata by using the stratified subspace sampling method. After that, the GMM Tree algorithm is used to identify the number of clusters and initial clusters in each subspace dataset and passing these initial cluster centers to k-means to generate base subspace clustering results. Then, the subspace clustering result is integrated into an object cluster association (OCA) matrix by using the link-based method. The ensemble clustering result is generated from the OCA matrix by the k-means algorithm with the number of clusters identified by the GMM Tree algorithm. After producing the ensemble clustering result, the dominant class label is assigned to each cluster after computing the purity. A classification is made on the object by computing the distance between the new object and the center of each cluster in the classifier, and the class label of the cluster is assigned to the new object which has the shortest distance. A series of experiments were conducted on twelve synthetic and eight real-world data sets, with different numbers of classes, features, and objects. The experimental results have shown that the new method outperforms other state-of-the-art techniques to classify data in most of the data sets.

Keywords: data mining, unsupervised classification, decision cluster, gamma mixture model, expectation maximization, high-dimensional data, curse of dimensionality

INTRODUCTION

The classification of data is an important research topic in the field of data mining [1,2,3,4,5,6]. Intrusion detection, weather forecasting, face recognition, product recommendations, and so forth are some important applications of classification. Currently, there are several methods for classification in which hierarchical techniques are the most famous due to their clear understanding and classification accuracy. The decision tree [7,8,9] and its ensemble methods, like Bagging [10], Boosting [11], and Random Forests [12] are the most important hierarchical techniques.

In addition to pure classification techniques, clustering methods have also been used for classification purposes [13,14,15]. An early example of such methods was proposed in 1980 [16] that uses the k-means clustering method to build a cluster tree classification model. In this method, a binary cluster tree was built by interactively executing the k-means clustering algorithm. At each node, a further partition was determined by the percentage of the dominant class in the cluster node. However, only small numeric data could be classified, and every time, only two subclusters could be formed. In 2000, Huang et al. [17] proposed a new interactive approach to build a decision cluster classification model. In this approach, the k-prototypes clustering algorithm was used to partition the training data, and a visual cluster validation method [18] was adopted to verify the partitioning result at each cluster node.

With the expeditious advancement in IoT devices and social network sites [19,20], data have grown exponentially in volume and complexity, resulting in a lot of large and complex high-dimensional data. The large number of features in high-dimensional data have created the issue of the curse of dimensionality [21,22] which is characterized by correlated, sparse, noisy, and uninformative features [23]. Due to such types of complex data with a large number of classes, the existing classification techniques face a lot of challenges. Thus, the traditional decision tree performs well on a low-dimensional dataset, but in the case of high-dimensional data with multiple classes, this technique is unable to get high accuracy [24,25,26]. The main reason behind this shortcoming is how the build tree only uses a small number of features from high-dimensional data. This is because each partitioning step in building a decision tree model only considers one feature, while the information is usually stored among many features. When there is a large number of classes, overfitting occurs due to the generation of a large number of leaves [27,28,29].

To classify such high-dimensional complex data with a large number of classes, in recent years researchers have proposed several techniques. Some of these are pure classification techniques [10,11,12], while others use clustering algorithms to classify data [30,31,32,33,34]. The major issue with these techniques is the poor performance of classifying high-dimensional data with a large number of classes in terms of classification accuracy and computation cost.

To solve this key issue of classifying high-dimensional data with a large number of classes, we propose a new Hierarchical Gamma Mixture Model-based Unsupervised Method in this paper. In this hierarchical method, we apply a subspace ensemble approach to deal with this challenging problem by integrating multiple techniques in an innovative solution, named as the Stratified Subspace Sampling GMM-based Supervised method, or SSS-

GMM for short. In this method, we first use the GMM Tree [35] to estimate the number of feature groups/strata given to the k-means algorithm as an input parameter. The k-means algorithm partitions the set of features of the high-dimensional dataset into a set of feature strata. After that, a stratified subspace sampling method is used to sample a set of subsets of features on feature strata by using sampling without replacement. A set of subspace data sets are generated from the set of feature subsets. These subspace data sets, produced by a stratified sampling method, are good representations of the high-dimensional data. After generating the subspace data sets, the GMM Tree algorithm is used again on each individual subspace dataset to estimate the number of clusters and initial cluster centers which are given as input parameters to the k-means algorithm to cluster the subspace dataset. Then, the link-based method [36] is used to integrate the clustering results generated from each subspace dataset into an object cluster association (OCA) matrix, on which the k-means algorithm is used to produce the ensemble clustering result with the number of clusters identified by the GMM Tree algorithm.

After producing the ensemble clustering result, the class labels are assigned to the clustering result as follows: (1) The purity of each cluster in the ensemble clustering result are computed based on the percentage of each class. (2) The dominant class label is assigned to each cluster after computing the purity and the clusters with less purity where α is discarded, and where α is the threshold set for the purity of a cluster. After assigning the class labels to the clusters, the center of each cluster is computed by finding the mean object. This clustering result with an assignment of a dominant class is used as a classifier to classify new objects as follows: (1) The distances between a new object and the center of each cluster in the classifier are computed, and the class label of the cluster is assigned to the new object which has the shortest distance.

We have conducted experiments on 12 synthetic data sets and eight real-world data sets, and compared the results of SSS-GMM with the results of k-NN, Bagging, C4.5, Random Forest, and Adaboost in terms of classification accuracy. The results show that SSS-GMM significantly outperforms all other algorithms in classifying the high-dimensional complex data with a large number of classes.

The two key contributions of this paper are:

- We integrate the multiple techniques of stratified sampling, subspace clustering, GMM Tree, k-means, and the link-based approach in an innovative algorithm to solve the challenging problem of classifying the high-dimensional complex data with the curse of dimensionality characteristics and a large number of classes.

- We report the classification accuracy by the new method on a high-dimensional complex dataset with a large number of classes and demonstrate that the new method is capable of dealing with such high-dimensional data sets. To our best knowledge, similar results were rarely reported in existing publications.

The rest of this paper is organized as follows. Related work is presented in Section 2. Section 3 reviews the GMM Tree method. Section 4 presents the SSS-GMM. Section 5 shows the experimental results on both synthetic and real-world data sets. Conclusions and future work are given in Section 6.

RELATED WORK

In the last few decades, researchers have proposed several clustering-based classification methods to solve the supervised classification problems [13,14,16,37,38]. According to the research provided, it has been proved that classification can be considered as a high-level model in which one class can be mapped to one, or more than one cluster. Thus, classification can be considered as a clustering problem that can also be solved by the clustering techniques.

An early example of such types of methods is the interactive approach proposed by Mui et al. [16]. In this method, the k-means [39] algorithm was used to find k clusters by partitioning the training dataset without using the class label information. Then, the dominant class was found in each cluster by computing the percentage of each class by using the class label information. In the case of purity greater than 90 percent, the dominant class label was assigned to this cluster as its class. Otherwise, the cluster needed further partitioning with the k-means algorithm. After finding all clusters, the centers of each cluster was computed and a k-NN like classification model formed, in which each cluster has a dominant class and its center.

To classify the objects in the test set, the distances between the objects of the test set and the cluster centers were computed, and the class label of the cluster assigned to the new object which had the shortest distance to the new object. Since the number of clusters was much less than the number of training samples in a k-NN model, this cluster center-based classification model performed better than the k-NN model.

Zhang et al. [13] proposed a cluster-based tree algorithm to improve k-NN classification, consisting of tree construction and classification steps. This cluster-based tree algorithm performs better than the standard k-NN. Kyriakopoulou et al. [15] made another contribution to test clustering methods for classification. In their work, a new classification algorithm was proposed in which both training and testing data sets are clustered. In the next step, the dataset was augmented with meta-features, and finally, a classifier was trained on the expanded dataset. This method follows a probability mixture model in which clusters are explored based on the distribution of objects of one class in the multidimensional data space. The benefit of this method is that a classifier learns from a small training dataset by combining unsupervised learning methods with classification methods. However, a major shortcoming is that it does not perform well on high-dimensional data sets.

In 2000, Huang et al. proposed a decision clusters classifier model [17] that used the k-prototype clustering method to construct a decision clusters tree by using an interactive Fast Map algorithm. To perform classification, the k-NN-like algorithm was used. A major limitation of this method was the manual procedure of the technique, because both the building model and validation of the clusters were based on human judgment. Compared to this manual approach, the Automatic Decision Cluster Classifier (ADCC), proposed by Yan Li et al. [40], was an automatic hierarchical clustering method which used clustering algorithms in parallel to class label information during the tree construction to generate a classifier.

To improve classification accuracy, researchers havc proposed several methods which generate and integrate multiple classifiers. The early work in this field has been done by Quinlan [41] and [42]. After that, ensemble classifier methods have been proposed, in which the final result could be obtained by combining the predictions of multiple classifiers. Bagging [10], Boosting [11], and random forests [12] are prominent examples of such ensemble methods which generate multiple diverse classifiers from the training dataset and use these classifiers for final classification results. Boosting uses all objects of the training data in each iteration and assigns a weight for each object based on the importance which is calculated according to the incorrectly classified objects. Thus, incorrectly classified objects will get more importance in the next iterations.

Bagging generates multiple data sets by sampling the objects with replacement from the training data. Bagging constructs classifiers from each bagged dataset.

In both bagging and boosting methods, the multiple classifiers make an ensemble and are used for voting to assign the class label of the new objects. In boosting, classifiers have different vote importance, while bagging assigns the same vote for every component classifier.

As we used the clustering method to do classification, finding the number of clusters was a classical issue to do clustering. Thus, we used our GMM Tree method to find the number of clusters and the initial cluster centers, which were used to explore decision clusters. An overview of a GMM Tree is given below.

OVERVIEW OF GMM TREE

This section presents an overview of the GMM Tree method [**35**], which plays a vital role in our proposed SSS-GMM. The GMM Tree is a hierarchical method that estimates the number of clusters and the initial cluster centers in a dataset by generating multiple levels of Gamma Mixture Models (GMMs).

The estimated number of cluster and initial cluster centers are used as input parameters to the clustering algorithm to find the decision clusters in the training dataset, and these decision clusters are then used to classify the objects of the test dataset. The important steps of the GMM Tree method is discussed below.

Given a dataset X with N objects in which each object is a m-dimensional vector, the GMM Tree method allocates an observation point p in the data space, and the distances between the N objects and the allocated observation point p are calculated using the Euclidean distance function.

This process is described in Figure 1 in which Figure 1a shows an example of dataset X in a two-dimensional space. The black dot (Obs. Point) is assigned as an observation point to the data space.

The distribution of distances between the objects in the dataset and the observation point is shown in Figure 1b in which the peaks of the distance distribution shows the dense regions in the dataset of Figure 1a, that is, the clusters.

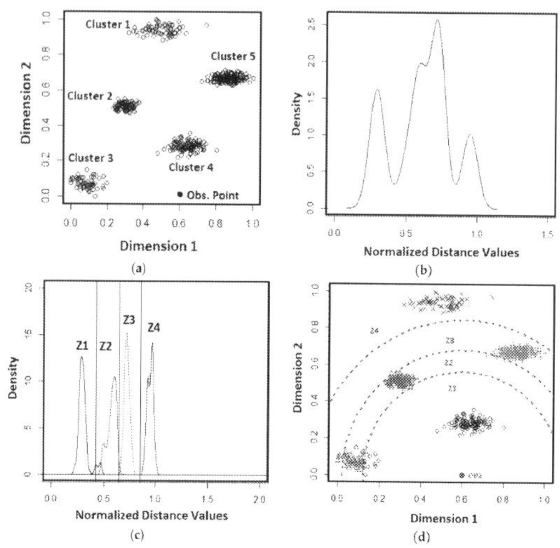

Figure 1: Mappings of five clusters to a distance distribution with respect to the observation point. (a) A sample dataset *X* in two dimensional space; (b) The distance distribution generated by computing the distances between the objects in the dataset *X* and the observation point; (c) The GMM components generated from the distance distribution; (d) Partition of the dataset *X* based on the GMM components generated.

Let $Xd=\{x_1,x_2,\ldots,x_N\}$ be a set of distance values, the distance distribution is modeled as a GMM which is defined as:

$$P\left(x|\theta\right) = \sum_{\mu=1}^{M} \pi_\mu g\left(x|\theta_\mu\right), x \geq 0 \tag{1}$$

where M is the number of Gamma components, θ_μ are the parameters of Gamma component μ, including shape parameter α_μ and scale parameter β_μ, and π_μ is the mixing proportion of component μ. The condition $\sum_{\mu=1}^{M} \pi_\mu = 1$ must hold to confirm that $P(x|\theta)$ is a well defined probability distribution. The probability density function of Gamma component μ is

$$g\left(x|\alpha_\mu, \beta_\mu\right) = \frac{x^{\alpha_\mu - 1}}{\Gamma\left(\alpha_\mu\right) \beta_\mu^{\alpha_\mu}} e^{\frac{(-x)}{\beta_\mu}}, \alpha_\mu > 0, \beta_\mu > 0, \tag{2}$$

where $\Gamma(x)$ is a Gamma function. This Gamma function is defined as $\Gamma(x) = \int_0^\infty f^{x-1} e^{-f} df$, which is a definite integral for $\Re$ [x].

The parameters of the Gamma Mixture Model, defined in Equation (1), are solved by maximizing the log likelihood function, which is defined as

$$\zeta\left(\theta|X_d\right) = \sum_{i=1}^{N} log\left(\sum_{\mu=1}^{M} \pi_\mu g\left(x_i|\alpha_\mu,\beta_\mu\right)\right), \tag{3}$$

This Equation (3) is solved using the Expectation Maximization algorithm [43].

The GMMs generated with different numbers of components and solved by Expectation Maximization algorithm, the best-fitted GMM is selected by using second-order variant of AICc (Akaike Information Criterion) [44]. AICc is defined as

$$AICc = -2\log\left(\zeta\left(\theta^*\right)\right) + 2\text{Y}\left(\frac{N}{\text{N}-\text{Y}-1}\right), \tag{4}$$

where N is the total number of distances in Xd, Y is the number of parameters, and ζ(θ∗) is the maximum log-likelihood value of the Gamma Mixture Model. The best fitted GMM is selected based on the the smallest value of AICc.

The dataset X is partitioned into M subsets based on the components of the best-fitted GMM, in which each subset is identified by one GMM component. As discussed in example, Figure 1c shows the vertical lines which partition the components, generated from the distance distribution, into four zones. The partitions of the four zones on the original data space is shown in Figure 1d. It is clearly seen from the Figure 1d that some component models identify more than one cluster, such as the model Z2 while some other component models identify one cluster, such as the models of Z1, Z3, and Z4.

The criteria to decide whether the subset of points needs further partitioning or not, are as follows:

- If the number of objects in the subset is greater than the given threshold of minimum points, a new observation point is assigned to the data space, and the distance distribution vector is computed with respect to the observation point. If the distance distribution contains only one peak, the subset is made as a lead node. Otherwise, a new GMM is built from the distance distribution to further partition the subset.
- In case of the number of objects in the subset is smaller than the

threshold specified for the minimum number of points, the subset is made a leaf node without further partitioning.

The process continues until a GMM tree is built by generating all leaf nodes. An example of GMM-tree is shown in Figure 2.

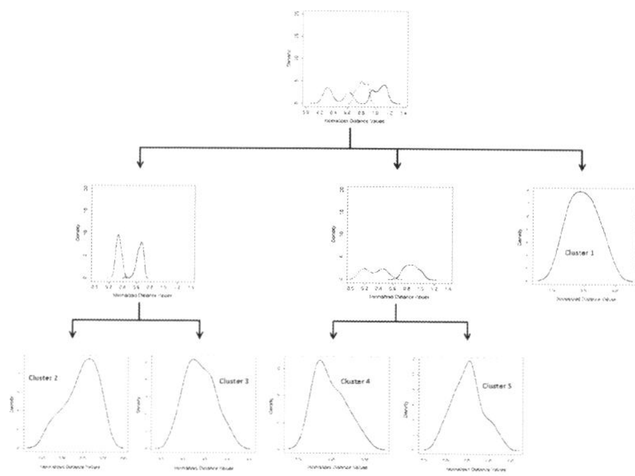

Figure 2: A GMM tree for identifying the number of clusters in a dataset.

Each leaf node of the GMM tree represents a cluster in the original dataset. A postprocessing step using the *k* nearest neighbor method is carried out on the set of objects in each leaf node due to the complexity of the dataset. The postprocessing step are discussed in detail in [**35**]. Our purpose is to illustrate the basic process of generating the GMM Tree. If a leaf node of the GMM Tree does not have high density due to less number of objects, then the objects are considered as outliers and that leaf node is ignored. In case of more than one dense region in a leaf node, the centroid objects of each dense region are taken as initial cluster centers. The number of initial cluster centers is considered as the number of cluster candidates which is the output of the GMM Tree algorithm. The number of clusters and the initial cluster centers found by the GMM Tree algorithm was used as an input parameter to the clustering algorithm used in our proposed method to find decision clusters which were used to classify the dataset.

STRATIFIED SUBSPACE SAMPLING GMM-BASED METHOD (SSS-GMM)

This section presents a novel ensemble method to classify high-dimensional complex data with a large number of classes. In this method, we first cluster the data by using the ensemble clustering approach, and then we train the ensemble clustering result to make decision clusters by assigning the class labels to clusters based on the dominant class. These decision clusters are used to classify the new objects by computing the distances between the new objects and the centers of the decision clusters. The class label of the decision cluster is assigned to the new objects, which has a minimum distance from the center of the decision cluster. This ensemble method is named the stratified subspace sampling GMM-based method (SSS-GMM). In the following subsections, we present the key steps of SSS-GMM in detail.

Generation of Feature Strata from the Training Dataset Dtrain

Given a high-dimensional training dataset D_{train}, we first partitioned the features of D_{train} into *L* feature clusters/strata in such a way that the features belonging to the same stratum were highly correlated. To find these closely related feature strata, we used k-means. As k-means requires the number of feature strata *L* as an input parameter, the GMM Tree method was used on the features of D_{train} to discover the number of feature clusters *L* (or feature strata) and initial feature cluster centers. This *L* number of feature cluster centers were given to k-means as an input parameter, and the k-means algorithm generated feature clusters by grouping the features of D_{train} into *L* feature strata $\{F_1,F_2,\ldots,F_L\}$ based on correlation. This process is discussed in detail in [45].

In the case of a large number of objects in D_{train}, a percentage of objects ϑ was selected from D_{train} and the GMM Tree algorithm was used to estimate the number of feature strata ι_i. This process of selecting ϑ percent of objects and finding the number of clusters was repeated v times. The number of feature strata *L* was computed by $Max(\iota_1,\iota_2, \ldots, \iota_v)$, which was given to the k-means algorithm to partition the features of D_{train} into *L* feature strata {F1,F2,…,FL}.

Generation of Subspace Data Sets from the Training Dataset Dtrain

After generation of the feature strata $\{F_1,F_2,\ldots,F_L\}$, subspace data sets were generated by stratified sampling without replacement, as follows. To generate *T* subspace data sets $\{D_1,D_2,\ldots,D_T\}$ from D_{train}, $d_\tau \times (p_l/p)$ features were randomly selected without replacement from each feature stratum F_l, where d_τ is the number of features in subspace dataset D_τ for $(1 \le \tau \le T)$, and p_l/p is the proportion of features in Fl. These selected features from each stratum were combined into a single subset of d_τ features. This process was repeated T=p/d times, and *T* subsets of features were generated. After that, *T* subspace data sets $\{D_1,D_2,\ldots,D_T\}$ were generated by extracting the the corresponding data from D_{train} for each feature's subset in such a way that $D_\tau \cap D_t = \emptyset$ for ($\tau \neq t$) and $\cup_{\tau=1}^{T} D_\tau = D$.

Generation of Clustering Results from Subspace Data Sets

After generating the subspace data sets $\{D_1,D_2,\ldots,D_T\}$, the next step was to partition each subspace dataset into clusters. For this purpose, the GMM Tree algorithm was used first to estimate the number of clusters and initial cluster centers in each subspace dataset, and then the k-means algorithm was used to partition each subspace dataset to generate the clustering result.

After producing the clustering results from each subspace dataset D_τ, we needed to ensemble these base clustering results of subspace data sets into the global clustering results of D_{train}.

Generation of Ensemble Clusters from Individual Subspace Clustering Results

To generate an ensemble clustering result from the base clustering result $\Psi_1,\Psi_2,\ldots,\Psi_T$, we used the link-based approach in [36] to ensemble the Ψ_τ into an object cluster association (OCA) matrix, where each base clustering result $\Psi_\tau = \{C_1^\tau, C_2^\tau, \ldots, C_{k_\tau}^\tau\}$ represented the clustering result of the τth subspace dataset. Each cluster C_i^τ contained a subset of objects in D_{train} where $C_i \neq \emptyset$ for ($1 \leq i \leq k_\tau$), $\cup_{\tau=1}^{k_\tau} C_i = D_{train}$ and $C_i \cap C_j = \emptyset$ for $(i \neq j)$.

For any two base clusterings Ψ_i and Ψ_j, the link $e_{x,y}$ between cluster $C_x \in \Psi_i$ and cluster $C_y \in \Psi_j$ was computed as

$$e_{x,y} = \frac{|\, C_x \cap C_y \,|}{|\, C_x \cup C_y \,|}, \tag{5}$$

where $1 \leq x \leq k_i$ and $1 \leq y \leq k_j$.

Two clusters $C_x \in \Psi_i$ and $C_y \in \Psi_j$ were linked if $e_{x,y} > 0$. By computing these links between clusters in different base clustering results, the similarity was computed between two clusters of the same clustering result that is, $C_x, C_y \in \Psi_\tau$ as

$$SIM(C_x, C_y) = \frac{WTQ_{x,y}}{WTQ_{max}} * \beta, \tag{6}$$

where $\beta \in [0,1]$ is a threshold for the confidence level of considering the two clusters as being similar, and $WTQ_{x,y}$ is called the Weighted Triple Quality that was used to measure the strength of the indirect connection between C_x and C_y through clusters in other clustering results which form triples with C_x and C_y through links. $WTQ_{x,y}$ is calculated as follows:

- Find all clusters in other clustering results which are directly linked to C_x and C_y to form triples. Assume the set of *k* clusters directly linked to C_x and C_y is N_k.
- For each cluster C_t in N_k, let $W_t = e_{t,x} + e_{t,y}$ and $WTQ^t_{x,y} = \frac{1}{W_t}$.
- For all clusters in N_k, $WTQ_{x,y}$ is calculated as

$$WTQ_{x,y} = \sum_{t=1}^{k} WTQ^t_{x,y}. \tag{7}$$

WTQ_{max} is the maximum value of all $WTQ_{x,y}$.

Given a clustering result Ψτ, the association of each object xi in *D* to the clusters {Cτ1,Cτ2,…,Cτkτ} is calculated as

$$OCA(x_i, C_t) = \begin{cases} 1 & \text{if } C_t = C^*(x_i) \\ \text{SIM}(C_t, C^*(x_i)) & \text{otherwise} \end{cases} \tag{8}$$

where $C^*(x_i)$ is the cluster to which x_i belongs, and $SIM(C_t, C^*(x_i)) \in [0,1]$ is calculated in Equation (6).

By computing the cluster associations of all clusters in $\Psi_1, \Psi_2, \ldots, \Psi_\tau$ and all objects in D_{train} with Equation (8), a new object cluster association matrix (OCA matrix for short) was obtained as a new representation of D_{train}. In the OCA matrix, each row indicated an object in D_{train}, while each column was a cluster in one of the τ base clustering results $\Psi_1, \Psi_2, \ldots, \Psi_\tau$. Thus, each element in the OCA matrix indicates the association level of the row object to the column cluster.

This computed OCA matrix is the new representation of dataset D_{train} which was used to compute the ensemble clustering result. For this purpose, we first ran the GMM Tree algorithm again on an OCA matrix to find the number of clusters and initial cluster centers, and then used them as inputs to k-means to cluster the OCA and produce the final ensemble clustering result.

Assignment of Class Labels to the Clusters in Ensemble Clustering Result

After producing the ensemble clustering result, the class labels were assigned to the clusters as follows:

- The purity of each cluster in the ensemble clustering result, produced by k-means in the previous step, were computed based on the percentage of each class.
- The dominant class label was assigned to each cluster after computing the purity.
- As we only needed the clusters which were pure, we kept the clusters in which objects belonging to the dominant class were greater than threshold α percent. The clusters in which dominant class objects were less than the threshold α were discarded.

Classification of the Objects in the Testing Dataset D_{test} *Based on Generated Classifier*

After assigning the class labels to the clusters in the previous step, the center of each cluster was computed by finding the mean object. This clustering result with an assignment of the dominant class was used as a classifier to classify new objects as follows:

- The distances between objects of the testing dataset D_{test} and the center of each cluster in the classifier were computed.
- The class label of the cluster to the objects of D_{test} based on the shortest distance was assigned.

The steps of the SSS-GMM are summarized in Figure 3.

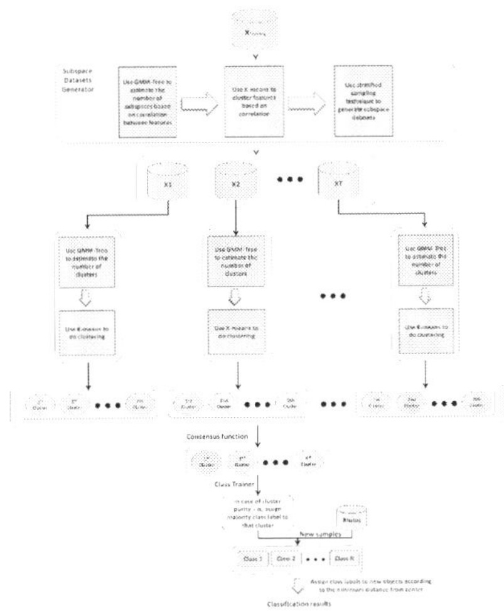

Figure 3: The process of SSS-GMM to classify a high-dimensional dataset.

The pseudo code was given in Algorithm 1, which took a high-dimensional training dataset $D_{train}=(x_1,x_2,\ldots,x_n)$, test dataset D_{test}, number of iterations v to find feature strata, percentage of objects μ to estimate the number of feature strata *L*, and the number of subspace data sets T as inputs, whereas its output was the classification accuracy on the test set D_{test}.

Algorithm 1: SSS-GMM algorithm

Input: A high-dimensional training dataset $D_{train} = (x_1, x_2, \ldots, x_n)$, test dataset D_{test}, number of iterations v to find feature strata, percentage of objects ϑ to estimate the number of feature strata L, and number of subspace data sets, T.

Output: Classification accuracy

for $i := 1$ *to* v **do**
  $D_{train}(i) \Leftarrow$ Randomly extract ϑ percent objects from training dataset D_{train};
  $t_i \Leftarrow$ Compute estimated number of feature strata by GMM Tree(Transpose($D_{train}(i)$));
$L \Leftarrow \mathrm{Max}(t_1, t_2, \ldots, t_v)$;
Split training dataset D_{train} into L feature strata $\{F_1, F_2, \ldots, F_L\}$ by using k-means by giving L initial feature cluster centers as input;
for $\tau := 1$ *to* T **do**
  Randomly pick $d_\tau \times (p_l/p)$ features from each feature stratum F_l, where $\sum_{l=1}^{L} p_l = p$, $\sum_{l=1}^{L} p_l/p = 1$;
  Generate subspace dataset D_τ by merging the randomly picked $d_\tau \times (p_l/p)$ features;
  Generate a set of k_τ cluster centers $\Omega_\tau = \{c_1^\tau, c_2^\tau, \ldots, c_{k_\tau}^\tau\}$ from GMM Tree(D_τ);
  Use k-means(D_τ, Ω_τ) to generate a set of k_τ clusters $\Psi_\tau = \{C_1^\tau, C_2^\tau, \ldots, C_{k_\tau}^\tau\}$;
Generate Ψ from $\Psi_1, \Psi_2, \ldots, \Psi_T$ where $\Psi_\tau = \{C_1^\tau, C_2^\tau, \ldots, C_{k_\tau}^\tau\}$;
Generate OCA matrix according to the Equations (5)–(8) from Ψ;
Determine the number of clusters K and initial cluster centers in OCA matrix by GMM Tree;
Use k-means method to cluster the data;
Compute the purity of each cluster in the ensemble clustering result based on the percentage of each class;
Assign the dominant class label to each cluster;
Discard the clusters with purity less that α;
Compute the center of each cluster by finding the mean object;
Compute the distances between object of test dataset D_{test} and the center of each cluster in the classifier;
Assign class label of the cluster to the new object which has shortest distance;
Output the classification accuracy ;

EXPERIMENTS

In this section, we present the performance analysis of SSS-GMM in comparison with k-NN, Bagging, C4.5, Random Forest, and Adaboost in terms of accuracy and computational time. We have used both synthetic and the real-world data sets in the experiments to demonstrate the performance of our proposed method. These synthetic and real-world data sets are diverse in terms of the number of features (p), objects (N), and classes (C) which are helpful to analyze the performance of our proposed SSS-GMM. First, the characteristics of the synthetic and real-world data sets are described. Then, the experimental settings are discussed. Finally, the experimental results of SSS-GMM are discussed in comparison with the other state-of-the-art methods in terms of accuracy. Since our proposed SSS-GMM relies on the construction of a decision cluster tree, which is, to some extent, like a decision tree, it would be more convincing to compare our method with state-of-the-art decision tree-like algorithms, including Bagging, C4.5, Random Forest, and Adaboost. In addition to this, we have compared our proposed method with k-NN because we have also used k-NN in building our proposed classifier.

Data Sets

The experiments were conducted on 12 synthetic data sets and eight real-world data sets. R package clusterGeneration [46] was used to generate the 12 synthetic data sets. The characteristics of these synthetic data sets are listed in Table 1. The first column from the left shows the dataset name. The second column describes the number of objects N in each dataset. The third and the fourth columns show the number of features p and the number of clusters K given in each dataset, respectively. The same value of cluster separation index = 0.5 was used for each synthetic dataset which indicated the degree of separation between a cluster and its nearest neighboring cluster. We used this constant value for each synthetic dataset to see the trend change in the accuracy based on the change in the number of features and the number of classes. Each cluster contained 100 objects following a normal distribution. Thus, to generate the synthetic data sets with the settings according to Table 1, we set parameters of clusterGeneration as sepVal (cluster separation index) = 0.5, numClust (Number of clusters) according to Table 1: column 4 (Classes), clustSizeEq (cluster objects size) = 100, numNonNoisy (Number of features) according to Table 1: column 3 (features).

Table 1: Characteristics of the synthetic data sets

Data Sets	Objects (N)	Features (p)	Classes (C)
DS1	15,000	200	150
DS2	20,000	200	200
DS3	25,000	200	250
DS4	30,000	200	300
DS5	35,000	200	350
DS6	40,000	200	400
DS7	15,000	1000	150
DS8	20,000	1000	200
DS9	25,000	1000	250
DS10	30,000	1000	300
DS11	35,000	1000	350
DS12	40,000	1000	400

To evaluate the performance of our proposed method on real data, eight real-world data sets were selected. The details of the real-world data sets have been summarized in Table 2. These data sets were high-dimensional, and the features ranged from 64 to 8460, taken from the UCI [47] and KEEL repository [48]. All data sets were multiclass, with a range of 10 to 983. Optdigits was the Optical Recognition of Handwritten Digits Data Set. The fbis dataset was from the Foreign Broadcast Information Service data of TREC-5. Wap is a skew dataset with 20 classes. The remaining data sets were selected from the KEEL dataset repository, while the Aloi(short) dataset was taken from Amsterdam Library of Object Images dataset repository [49]. The Aloi(short) dataset contained 10,800 objects representing 100 objects (clusters). These data sets had diverse representations in application domains and the numbers of objects, features, and classes.

Table 2: Characteristics of the real-world data sets

Number	Data Sets	Objects	Features	Classes
1	Optdigits	5620	64	10
2	fbis	2463	2000	17
3	wap	1560	8460	20
4	Isolet	6238	617	26
5	Aloi(short)	10,800	128	100
6	Corel5k	5000	499	374
7	Mediamill	43,907	120	101
8	Delicious	16,150	500	983

The class labels were used to train our proposed method during the assignment of class labels to generated clusters. These class labels were also used as ground truth to compare performance accuracy on the testing set. The objects in some of these data sets belonged to more than one class. Thus, classifying these kinds of data sets was more difficult due to high overlapping.

Experimental Settings

Here, we show the comparison results of SSS-GMM and other five classification methods: decision tree (C4.5), Bagging, Original k-NN, Random Forest, and Adaboost on eight real data sets. In our experiments, we used default parameters for C4.5, Random Forest, and k-NN in R. For C4.5, the minimum number of instances per leaf was set as 2. For Random Forest, the random number seed was set to 1. For k-NN, the number of neighbors, k, equaled to 1. The base classifier to be used in Bagging and AdaBoost was C4.5. The default values for the GMM Tree algorithm (used in our proposed method) were set according to [35]. The package mixtools [50] was used for GMM and EM. For SSS-GMM, we set μ as 20 percent and the number of iterations ν as 10. In this way, 20 percent of objects were randomly extracted 10 times to find the number of feature strata, and then the maximum of these 10 iterations was used as the initial estimation of the number of feature strata in the dataset *D*. The threshold parameter α was set as 90 percent to set the purity level of each cluster. We used the 10-fold method to get classification accuracy. The number of subspace data sets M was set as 10, which can be changed by the user but should not be very high because it will generate more GMM Trees and increase the computation cost. The value of β used in Equation (6) was set to 0.8 as the default value [36]. The minimum number of objects was set to be 10 as a termination threshold for a leaf node. If a subset has 10 objects or less, this subset is either a small cluster or a set of outliers. There is no need to partition it further.

The experiments were conducted on machines with 128 GB memory running on Windows Server 2012 R2 Standard and 3.30 GHz Intel(R) Core(TM)i5-4590 CPU with 12 GB of memory running Windows 7 of 64-bit.

Experimental Results

In this section, we discuss the results on the comparison of our proposed algorithm with other methods in classification accuracy. The comparative results of SSS-GMM and other classification methods on synthetic data sets are given in Table 3. We can see that on comparatively low-dimensional datasets, that is, from DS1-DS6, our proposed SSS-GMM performed better than C4.5, Bagging, Original k-NN, and Random Forest, except Adaboost. While on high-dimensional data sets, that is, from DS7-DS12, our method outperformed all other methods. As a whole, our proposed SSS-GMM achieved the best performance in eight data sets out of 12 synthetic data sets, and was only slightly worse than the best method in other four data sets. In general, our proposed SSS-GMM was the best-performing method, followed by Adaboost. The accuracy performance is shown in Figure 4 where Figure 4a shows the accuracy of all methods on data sets DS1-DS6 and Figure 4b shows the accuracy on data sets DS7-DS12.

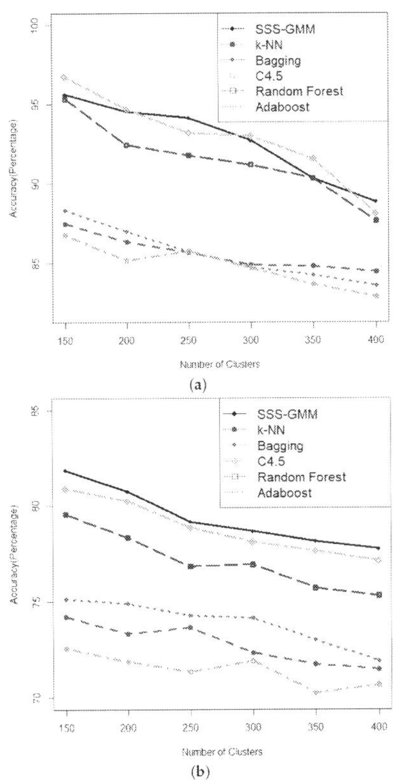

Figure 4: Accuracy comparison on synthetic data sets (SSS-GMM, k-NN, Bagging, C4.5, Random Forest, Adaboost). (a) Synthetic data sets (DS1-DS6) Features = 200 and Objects (15,000–40,000). (b) Synthetic data sets (DS7-DS12) Features = 1000 and Objects (15,000–40,000).

Table 3: Synthetic data sets: Comparison of SSS-GMM with other five state-of-the-art methods in terms of accuracy

Data Sets	SSS-GMM	k-NN	Bagging	C4.5	Random Forest	Adaboost
DS1	95.61	87.45	88.34	86.76	95.34	**96.73**
DS2	94.54	86.34	86.98	85.15	92.45	**94.67**
DS3	**94.15**	85.67	85.71	85.74	91.77	93.18
DS4	92.72	84.86	84.73	84.67	91.19	**92.99**
DS5	90.34	84.78	84.23	83.66	90.32	**91.56**
DS6	**88.85**	84.45	83.56	82.87	87.67	88.11
DS7	**81.85**	74.23	75.14	72.56	79.56	80.89
DS8	**80.76**	73.34	74.92	71.87	78.34	80.23
DS9	**79.17**	73.67	74.29	71.34	76.84	78.86
DS10	**78.67**	72.34	74.15	71.89	76.93	78.11
DS11	**78.15**	71.74	73.02	70.23	75.72	77.67
DS12	**77.75**	71.47	71.89	70.67	75.32	77.13

Table 4 shows the classification accuracy performance of SSS-GMM and other classification methods on real data sets. A confusion matrix of Optdigits(test set) generated by SSS-GMM is shown in Figure 5 which was used to find the classification accuracy on the test set of Optdigits dataset. We can see that, except for two data sets (Optdigits and Corel5k), our proposed SSS-GMM outperformed all methods. As a whole, our proposed SSS-GMM achieved the best performance in six real-world data sets out of eight data sets. Thus, our proposed SSS-GMM was also the best-performing method, followed by Adaboost on real-world data sets. Adaboost is a state-of-the-art method which can handle low-dimensional data without noise. However, in the case of high-dimensional noisy data and outliers, its performance deteriorates. Thus, in the case of most of synthetic data sets, the performance of Adaboost and SSS-GMM is similar, but in the case of real data sets, our proposed SSS-GMM performed better. The accuracy performance on various data sets is shown in Figure 6 and Figure 7. Figure 6 shows the classification accuracy on the data sets which had a relatively less number of classes (i.e., 10–26 classes) as compared to the data sets used in Figure 7 where all data sets have more than 100 classes.

Ground Truth

Predicted	0	1	2	3	4	5	6	7	8	9
0	167	0	1	0	0	3	0	2	0	5
1	0	174	3	0	2	0	1	0	0	2
2	1	0	170	0	0	2	0	4	0	0
3	0	0	1	169	0	9	0	2	0	2
4	0	0	2	0	172	1	0	1	2	3
5	0	0	1	0	4	178	0	0	0	0
6	0	3	0	2	1	2	168	5	0	0
7	1	0	0	0	1	0	3	174	0	0
8	0	2	0	0	0	3	0	0	169	0
9	1	3	0	0	0	1	0	1	0	174

Figure 5: Confusion matrix of Optdigits (test set) generated by SSS-GMM.

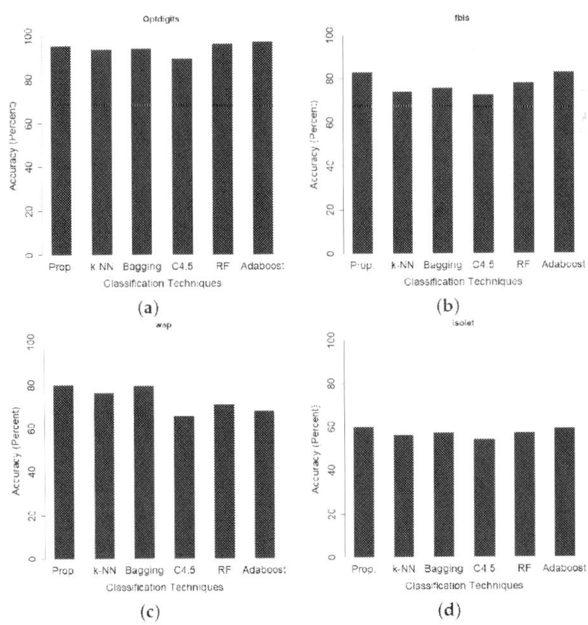

Figure 6: Accuracy comparison (Prop = SSS-GMM, k-NN, Bagging, C4.5, RF = Random Forest, Adaboost) data sets: (a) Optdigits, (b) fbis, (c) wap, (d) Isolet.

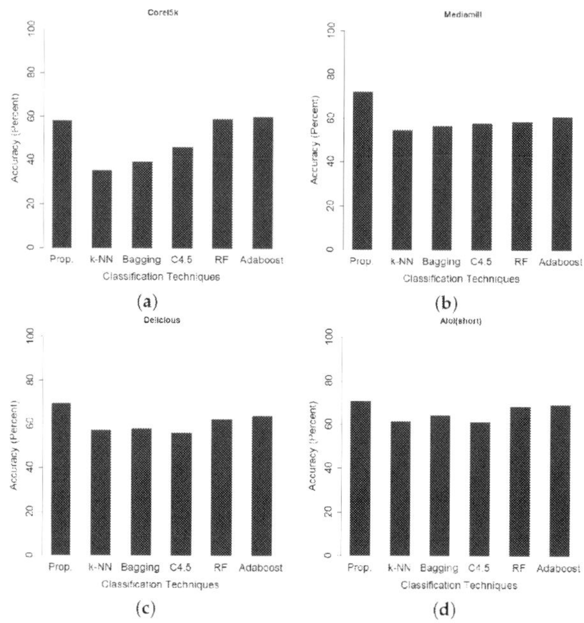

Figure 7: Accuracy comparison (Prop = SSS-GMM, k-NN, Bagging, C4.5, RF = Random Forest, Adaboost) data sets: (a) Corel5k, (b)Mediamill, (c) Delicious, (d) Aloi(short).

Table 4: Real data sets: Comparison of SSS-GMM with other five state-of-the-art methods in terms of accuracy

Data Sets	SSS-GMM	k-NN	Bagging	C4.5	Random Forest	Adaboost
Optdigits	95.43	93.72	94.11	89.72	96.42	**97.20**
fbis	**82.99**	74.23	75.82	72.84	78.03	82.92
wap	**80.23**	76.41	79.63	65.64	71.15	67.95
Isolet	**60.12**	56.21	57.21	54.21	57.21	59.21
Aloi(short)	**70.51**	61.43	64.32	61.22	68.34	69.13
Corel5k	58.23	35.45	39.25	46.21	59.21	**59.99**
Mediamill	**72.14**	54.66	56.65	57.81	58.45	61.13
Delicious	**69.32**	57.34	57.99	56.13	62.36	63.83

Table 5 shows the time comparison of SSS-GMM with five other state-of-the-art methods on DS7-DS12. We selected these data sets due to their high dimensions and large number of classes. From the results, it is clear that the testing time of SSS-GMM is at a minimum as compared to other state-of-the art methods, which can compensate for the shortcoming of SSS-GMM of taking time during training. The testing time of SSS-GMM was shortest because it only needs to find the distances between centers of the clusters and the objects of the testing data sets. Thus, overall, our proposed SSS-GMM outperformed other methods in terms of accuracy and testing time.

Table 5: Synthetic data sets: Training/testing time comparison of SSS-GMM with five other state-of-the-art methods (time is in Minutes:Seconds format)

Data Sets	**SSS-GMM**	**K-NN**	**C4.5**	**Bagging**	**Random Forest**	**Adaboost**
DS7	05:24/00:16	00:00/02:13	02:02/00:33	05:33/00:33	04:12/01:02	02:51/00:21
DS8	06:01/00:19	00:00/02:32	02:09/00:48	06:13/00:39	04:53/01:14	03:13/00:29
DS9	06:42/00:21	00:00/02:51	02:22/00:59	07:03/00:44	05:24/01:23	03:36/00:35
DS10	07:33/00:24	00:00/03:11	02:29/01:13	07:41/00:49	05:56/01:29	03:54/00:43
DS11	07:51/00:29	00:00/03:23	02:36/01:41	08:32/00:54	06:11/01:36	04:23/00:56
DS12	08:21/00:37	00:00/03:49	02:48/01:49	08:58/00:59	06:42/01:45	04:52/01:11

To validate the clustering result, we investigated the quality of the clustering results. To measure the clusters' quality, we used ARI [51] and Purity [52]. The Purity was defined as:

$$Purity = \frac{1}{N}\sum_{k=1}^{K} max_{1\leq q\leq \bar{K}}|C_k \cap Y_q| \quad (9)$$

where K is the number of true clusters, C_k is the set of objects in true cluster k, $\bar{K}$ is the number of clusters by k-means, N is the total number of objects in the dataset, and Y_q is the set of objects in cluster q by k-means. $\max 1\leq q\leq \bar{K}$ denotes that for $\bar{K}$ clusters by k-means, only the cluster q whose intersection with C_k has the largest number of objects is considered. $|.|$ indicates the number of objects in the intersection of C_k and Y_q. The value of Purity is between 0 to 1.

ARI[51]wasalsousedtomeasuretheclusteringperformance.Itisameasure of agreement between two partitions: one given by the clustering process, and the other defined by the ground truth. Given a dataset D of N objects, and two partitions of these N objects, namely, $C=C_1,C_2,\ldots,C_K$ being a partition of N objects into K clusters and $Y=Y_1,Y_2,\ldots,Y_P$ being a partition of N objects into P classes (the ground truth), ARI is defined as:

$$ARI = \frac{\sum_{jk}\binom{N_{jk}}{2} - \left[\sum_j \binom{\alpha_j}{2} \sum_k \binom{\beta_k}{2}\right] / \binom{N}{2}}{1/2\left[\sum_j \binom{\alpha_j}{2} \sum_k \binom{\beta_k}{2}\right] - \left[\sum_j \binom{\alpha_j}{2} \sum_k \binom{\beta_k}{2}\right] / \binom{N}{2}} \tag{10}$$

where N_{jk} is the number of objects in cluster C_j and the second partition, Y_k, α_j is the number of objects in the first partition, C_j, and β_k is the number of objects in the second partition, Y_k.

The Purity and ARI results on synthetic data sets are shown in Table 6, while the Purity and ARI values measured by SSS-GMM on real data sets are shown in Table 7. From the results, it is clearly seen that SSS-GMM performed very well in finding clusters on synthetic data sets, while in the case of real data sets, it also performed well.

Table 6: Purity and ARI measured by SSS-GMM on synthetic data sets

Data Sets	Classes	Purity	ARI
DS1	150	0.896	0.866
DS2	200	0.887	0.857
DS3	250	0.876	0.856
DS4	300	0.868	0.845
DS5	350	0.865	0.849
DS6	400	0.854	0.838
DS7	150	0.797	0.768
DS8	200	0.794	0.762
DS9	250	0.783	0.744
DS10	300	0.787	0.739
DS11	350	0.778	0.728
DS12	400	0.776	0.734

Table 7: Purity and ARI measured by SSS-GMM on real data sets

Data Sets	Classes	Purity	ARI
ISOLET	26	0.601	0.614
Aloi(short)	100	0.718	0.696
Corel5k	374	0.636	0.612
Mediamill	101	0.616	0.592
Delicious	983	0.706	0.681
Bibtex	159	0.528	0.532
Bookmarks	208	0.514	0.547
Topics	101	0.583	0.594

Industries	313	0.534	0.511
Regions	228	0.519	0.513

Finally, we have also shown the number of clusters found during construction of the SSS-GMM for validation purposes. As shown in Table 8 and Table 9, we found that our proposed SSS-GMM successfully discovered most of the classes in both synthetic and real-world data sets.

Table 8: Clusters found in the synthetic data sets by SSS-GMM

Dataset	**Actual Clusters**	**Total Clusters Found**	**Unique Clusters Found**
DS1	150	148	148
DS2	200	194	194
DS3	250	244	242
DS4	300	294	291
DS5	350	345	341
DS6	400	395	391
DS7	150	147	147
DS8	200	194	193
DS9	250	243	241
DS10	300	294	291
DS11	350	344	342
DS12	400	394	392

Table 9: Clusters found in the real-world data sets by SSS-GMM

Number	**Dataset**	**Classes**	**Total Clusters Found**	**Unique Clusters Found**
1	Optdigits	10	14	10
2	fbis	17	20	17
3	wap	20	24	20
4	Isolet	26	39	26
5	Aloi(short)	100	128	97
6	Corel5k	374	423	368
7	Mediamill	101	121	97
8	Delicious	983	1012	977

From the above experimental results, we can conclude that our algorithm performed better than all other methods on most of the data sets which were diverse in nature in terms of the number of features and the number of classes.

CONCLUSIONS AND FUTURE WORK

In this paper, we proposed a new hierarchical Gamma Mixture Model-based method (named SSS-GMM) for classifying high-dimensional data with a large number of classes which used a subspace ensemble approach to deal with this challenging problem by integrating multiple techniques in an innovative way. For this purpose, we first used the GMM Tree to find the number of feature strata, and then a k-means algorithm to divide the set of features of the dataset into feature strata. Then, the stratified subspace sampling method was used to sample subspace features from the feature strata and generate a set of subspace data sets from the high-dimensional dataset. After that, the GMM Tree algorithm was used again to identify the number of clusters and initial clusters in each subspace dataset for the k-means algorithm to cluster the subspace dataset. Then, the link-based method was used to integrate the subspace clustering results into an object cluster association matrix, from which the ensemble clustering result was generated by the k-means algorithm with the number of clusters identified by the GMM Tree algorithm. After producing the ensemble clustering result, the dominant class label was assigned to each cluster after computing the purity. A classification was made on the object by computing the distances between the new object and the center of each cluster in the classifier, and the class label of the cluster was assigned to the new object which had the shortest distance. A series of experiments were conducted on twelve synthetic and eight real-world data sets with different numbers of classes, features, and objects. The experimental results have shown that the new method performs better in classifying data in all data sets as compared to the other state-of-the-art techniques.

Our future work consists of analysing SSS-GMM on noisy data with thousands of features and classes. We will also analyse the performance of this method on high-dimensional regression data sets.

AUTHOR CONTRIBUTIONS

Conceptualization, M.A., and J.Z.H.; Methodology, M.A., and J.Z.H.; Software, M.A.; Validation, M.J.L. and M.A.; Formal Analysis, M.A., and J.Z.H.; Investigation, M.A., and M.J.L.; Writing, M.A., and and J.Z.H.

ACKNOWLEDGMENTS

We would like to thank Shenzhen University for all technical support and equipment.

REFERENCES

1. Valencia, F.; Gómez-Espinosa, A.; Valdés-Aguirre, B. Price Movement Prediction of Cryptocurrencies Using Sentiment Analysis and Machine Learning. *Entropy* 2019, *21*, 589.
2. Khan, K.; Attique, M.; Syed, I.; Sarwar, G.; Irfan, M.A.; Khan, R.U. A Unified Framework for Head Pose, Age and Gender Classification through End-to-End Face Segmentation. *Entropy* 2019, *21*, 647.
3. Zhang, W.; Zhou, J. A Comprehensive Fault Diagnosis Method for Rolling Bearings Based on Refined Composite Multiscale Dispersion Entropy and Fast Ensemble Empirical Mode Decomposition. *Entropy* 2019, *21*, 680.
4. Xue, N.; Luo, X.; Gao, Y.; Wang, W.; Wang, L.; Huang, C.; Zhao, W. Kernel Mixture Correntropy Conjugate Gradient Algorithm for Time Series Prediction. *Entropy* 2019, *21*, 785.
5. Wei, Y.; Fang, S.; Wang, X. Automatic Modulation Classification of Digital Communication Signals Using SVM Based on Hybrid Features, Cyclostationary, and Information Entropy. *Entropy* 2019, *21*, 745.
6. Sagheer, A.; Zidan, M.; Abdelsamea, M.M. A Novel Autonomous Perceptron Model for Pattern Classification Applications. *Entropy* 2019, *21*, 763.
7. Quinlan, J.R. Induction of decision trees. *Mach. Learn.* 1986, *1*, 81–106.
8. Quinlan, J.R. *C4.5: Programs for Machine Learning*; Morgan Kaufmann Publishers Inc.: San Francisco, CA, USA, 1993.
9. Rastogi, R.; Shim, K. PUBLIC: A Decision Tree Classifier that Integrates Building and Pruning. *Data Min. Knowl. Discov.* 2000, *4*, 315–344.
10. L, B. Bagging predictors. *Mach. Learn.* 1996, *24*, 123–140.
11. Freund, Y.; Schapire, R.E. Experiments with a New Boosting Algorithm. Available online: http://citeseerx.ist.psu.edu/viewdoc/download?doi=10.1.1.51.6252&rep=rep1&type=pdf (accessed on 10 September 2019).
12. Breiman, L. Random forests. *Mach. Learn.* 2001, *45*, 5–32.
13. Zhang, B.; Srihari, S.N. Fast k-nearest neighbor classification using cluster-based trees. *IEEE Trans. Pattern Anal. Mach. Intell.* 2004, *26*, 525–528.

14. Zeng, H.J.; Wang, X.H.; Chen, Z.; Lu, H.; Ma, W.Y. CBC: Clustering based text classification requiring minimal labeled data. In Proceedings of the Third IEEE International Conference on Data Mining, Melbourne, FL, USA, 22–22 November 2003; pp. 443–450.
15. Kyriakopoulou, A.; Kalamboukis, T. Combining clustering with classification for spam detection in social bookmarking systems. In Proceedings of the European Conference on Machine Learning and Principles and Practice of Knowledge Discovery in Databases, Antwerp, Belgium, 15–19 September 2008.
16. Mui, J.K.; Fu, K.S. Automated classification of nucleated blood cells using a binary tree classifier. *IEEE Trans. Pattern Anal. Mach. Intell.* 1980, *PAMI-2*, 429–443.
17. Huang, Z.; Ng, M.K.; Lin, T.; Cheung, D. An interactive approach to building classiffication models by clustering and cluster validation. In Proceedings of the International Conference on Intelligent Data Engineering and Automated Learning, Hong Kong, China, 13–15 December 2000; pp. 23–28.
18. Huang, Z.; Lin, T. A visual method of cluster validation with Fastmap. In Proceedings of the Pacific-Asia Conference on Knowledge Discovery and Data Mining, Kyoto, Japan, 18–20 April 2000; pp. 153–164.
19. Gubbi, J.; Buyya, R.; Marusic, S.; Palaniswami, M. Internet of Things (IoT): A vision, architectural elements, and future directions. *Future Gener. Comput. Syst.* 2013, *29*, 1645–1660.
20. Rathore, P.; Rao, A.S.; Rajasegarar, S.; Vanz, E.; Gubbi, J.; Palaniswami, M. Real-Time Urban Microclimate Analysis Using Internet of Things. *IEEE Internet Things J.* 2018, *5*, 500–511.
21. Bishop, C.M. *Pattern Recognition and Machine Learning (Information Science and Statistics)*; Springer: Berlin, Germany, 2006.
22. Bellman, R.E. *Adaptive Control Processes: A Guided Tour*; Princeton University Press: Princeton, NJ, USA, 2015; Volume 2045.
23. Kriegel, H.P.; Kröger, P.; Zimek, A. Clustering high-dimensional data. *ACM Trans. Knowl. Discov. Data* 2009, *3*, 1–58.
24. Piatetsky-Shapiro, G.; Djeraba, C.; Getoor, L.; Grossman, R.; Feldman, R.; Zaki, M. What are the grand challenges for data mining?: KDD-2006 panel report. *ACM SIGKDD Explor. Newsl.* 2006, *8*, 70–77.
25. Wu, Q.; Ye, Y.; Liu, Y.; Ng, M.K. SNP Selection and Classification of Genome-Wide SNP Data Using Stratified Sampling Random

Forests. *IEEE Trans. NanoBiosci.* 2012, *11*, 216–227.

26. Ye, Y.; Wu, Q.; Zhexue Huang, J.; Ng, M.K.; Li, X. Stratified sampling for feature subspace selection in random forests for high-dimensional data. *Pattern Recognit.* 2013, *46*, 769–787.
27. Sánchez-Rodríguez, D.; Hernández-Morera, P.; Quinteiro, J.; Alonso-González, I. A Low Complexity System Based on Multiple Weighted Decision Trees for Indoor Localization. *Sensors* 2015, *15*, 14809–14829.
28. Cai, L.; Gu, J.; Ma, J.; Jin, Z. Probabilistic Wind Power Forecasting Approach via Instance-Based Transfer Learning Embedded Gradient Boosting Decision Trees. *Energies* 2019, *12*, 159.
29. Mao, Q.; Ma, H.; Zhang, X.; Zhang, G. An Improved Skewness Decision Tree SVM Algorithm for the Classification of Steel Cord Conveyor Belt Defects. *Appl. Sci.* 2018, *8*, 2574.
30. Ahn, E.; Kumar, A.; Feng, D.; Fulham, M.; Kim, J. Unsupervised Feature Learning with K-means and An Ensemble of Deep Convolutional Neural Networks for Medical Image Classification. *arXiv* 2019, arXiv:1906.03359.
31. Richardson, A.; Goodenough, D.G.; Chen, H. Hierarchical unsupervised nonparametric classification of polarimetric SAR time series data. In Proceedings of the 2014 IEEE Geoscience and Remote Sensing Symposium, Quebec City, QC, Canada, 13–18 July 2014; pp. 4730–4733.
32. Marapareddy, R.; Aanstoos, J.V.; Younan, N.H. Unsupervised classification of SAR imagery using polarimetric decomposition to preserve scattering characteristics. In Proceedings of the 2015 IEEE Applied Imagery Pattern Recognition Workshop (AIPR), Washington, DC, USA, 13–15 October 2015; pp. 1–5.
33. Yumus, D.; Ozkazanc, Y. Land Cover Classification for Synthetic Aperture Radar Imagery by Using Unsupervised Methods. In Proceedings of the 2019 9th International Conference on Recent Advances in Space Technologies (RAST), Istanbul, Turkey, 11–14 June 2019; pp. 435–440.
34. Yin, X.; Song, H.; Yang, W.; He, C.; Xu, X. Unsupervised PolSAR image classification based on ensemble partitioning. In Proceedings of the 2013 IEEE International Geoscience and Remote Sensing Symposium- IGARSS, Melbourne, VIC, Australia, 21–26 July 2013;

pp. 3183–3186.

35. Azhar, M.; Huang, J.Z.; Masud, M.A.; Li, M.J.; Cui, L.Z. Identify the Number of Clusters from Complex Data with Many Clusters using GMM Trees and Forests. *Appl. Soft Comput.* 2019. under review.
36. Iam-On, N.; Boongeon, T.; Garrett, S.; Price, C. A link-based cluster ensemble approach for categorical data clustering. *IEEE Trans. Knowl. Data Eng.* 2010, *24*, 413–425.
37. Kyriakopoulou, A.; Kalamboukis, T. Text classification using clustering. In Proceedings of the Discovery Challenge Workshop at ECML/PKDD 2006, Berlin, Germany, 22 September 2006; pp. 28–38.
38. Ye, N.; Li, X. A machine learning algorithm based on supervised clustering and classification. In Proceedings of the International Computer Science Conference on Active Media Technology, Hong Kong, China, 18–20 December 2001; pp. 327–334.
39. Macqueen, J. Some methods for classification and analysis of multivariate observations. In Proceedings of the 5th Berkeley Symposium on Mathematical Statistics and Probability, Berkeley, CA, USA, 21 June–18 July 1965, 27 December 1965–7 January 1966; pp. 281–297.
40. Li, Y.; Hung, E.; Chung, K.; Huang, J. Building a decision cluster classification model for high dimensional data by a variable weighting k-means method. In Proceedings of the Australasian Joint Conference on Artificial Intelligence, Auckland, New Zealand, 1–5 December 2008; pp. 337–347.
41. Quinlan, J.R.; Compton, P.J.; Horn, K.A.; Lazarus, L. Inductive knowledge acquisition: A case study. In *Proceedings of the Second Australian Conference on Applications of Expert Systems, 1987*; Addison-Wesley Longman Publishing Co., Inc.: Redwood City, CA, USA, 1987; pp. 137–156.
42. Buntine, W. Learning classification trees. *Stat. Comput.* 1992, *2*, 63–73.
43. Dempster, A.P.; Laird, N.M.; Rubin, D.B. Maximum Likelihood from Incomplete Data Via the EM Algorithm. *J. R. Stat. Soc. Ser. B* 1977, *39*, 1–22.
44. Hurvich, C.M.; Tsai, C.L. Regression and time series model selection in small samples. *Biometrika* 1989, *76*, 297–307.
45. Jing, L.; Tian, K.; Huang, J.Z. Stratified feature sampling method for ensemble clustering of high dimensional data. *Pattern*

Recognit. 2015, *48*, 3688–3702.

46. Qiu, W.; Joe, H. clusterGeneration: Random cluster generation (with specified degree of separation). *R Packag. Vers.* 2009, *1*, 70122–75275.
47. Dua, D.; Graff, C. UCI Machine Learning Repository. Available online: http://archive.ics.uci.edu/ml (accessed on 14 September 2019).
48. Alcalá-Fdez, J.; Fernández, A.; Luengo, J.; Derrac, J.; García, S. KEEL Data-Mining Software Tool: Data Set Repository, Integration of Algorithms and Experimental Analysis Framework. *Mult.-Valued Log. Soft Comput.* 2011, *17*, 255–287.
49. Geusebroek, J.M.; Burghouts, G.J.; Smeulders, A.W. The Amsterdam Library of Object Images. *Int. J. Comput. Vis.* 2005, *61*, 103–112.
50. Benaglia, T.; Chauveau, D.; Hunter, D.R.; Young, D. mixtools: An R Package for Analyzing Finite Mixture Models. *J. Stat. Softw.* 2009, *32*.
51. Hubert, L.; Arabie, P. Comparing partitions. *J. Classif.* 1985, *2*, 193–218.
52. Manning, C.; Raghavan, P.; Schütze, H. Introduction to information retrieval. *Nat. Lang. Eng.* 2010, *16*, 100–103.

CHAPTER

12

Multi-Partitions Subspace Clustering

Vincent Vandewalle [1,2]

[1]Biostatistics Department, Univ. Lille, CHU Lille, ULR 2694—METRICS: Évaluation des Technologies de Santé et des Pratiques MéDicales, F-59000 Lille, France

[2]Inria Lille—Nord Europe, 59650 Villeneuve d'Ascq, France

ABSTRACT

In model based clustering, it is often supposed that only one clustering latent variable explains the heterogeneity of the whole dataset. However, in many cases several latent variables could explain the heterogeneity of the data at hand. Finding such class variables could result in a richer interpretation

Citation: (APA): Vandewalle, V. (2020). Multi-Partitions Subspace Clustering. Mathematics, 8(4), 597. (18 pages) DOI: https://doi.org/10.3390/math8040597 URL: https://www.mdpi.com/2227-7390/8/4/597

of the data. In the continuous data setting, a multi-partition model based clustering is proposed. It assumes the existence of several latent clustering variables, each one explaining the heterogeneity of the data with respect to some clustering subspace. It allows to simultaneously find the multi-partitions and the related subspaces. Parameters of the model are estimated through an EM algorithm relying on a probabilistic reinterpretation of the factorial discriminant analysis. A model choice strategy relying on the BIC criterion is proposed to select to number of subspaces and the number of clusters by subspace. The obtained results are thus several projections of the data, each one conveying its own clustering of the data. Model's behavior is illustrated on simulated and real data.

INTRODUCTION

In exploratory data analysis, the statistician often uses clustering and visualization in order to improve his knowledge on the data. In visualization he looks for some principal components explaining some major characteristics of the data. For example in principal component analysis (PCA) the goal is to find a linear combination of the variables explaining the major variability of the data. In cluster analysis, the goal is to find some clusters explaining the major heterogeneity of the data. In this article we suppose that the data can contain several clustering latent variables, that is, we are in the multiple partition setting, and we are simultaneously looking for clustering subspaces, that is, linear projections of the data each one related to some clustering latent variable thus the developed model is later called multi-partitions subspace clustering. A solution to perform multi-partition subspace clustering is to use a probabilistic model on the data such as a mixture model [**1**], it allows to perform the parameters estimation, and model selection such as the choice of the number of subspaces and the number of clusters per subspace using standard model choice criteria such as BIC [**2**]. Thus the main fields related to our work are model based subspace clustering and multi-partitions clustering.

In the model based subspace clustering framework, let first notice that PCA can be re-interpreted in a probabilistic way by considering a parsimonious version of a multivariate Gaussian distribution [**3**] and that the *k*-means algorithm can be re-interpreted as a particular parsimonious Gaussian mixture model estimated using a classification EM algorithm [**4**]. A re-interpretation of the probabilistic PCA has also been used in clustering by Bouveyron et al. [**5**] in order to cluster high-dimensional data. Although the proposed high dimensional mixture does not performs dimension reduction, it rather operates a class per class dimension reduction which does not allow to have a global model-based data visualization. Thus Bouveyron and Brunet [**6**] proposed the so called Fisher-EM algorithm which simultaneously performs clustering and dimension reduction. This is performed through a modified version of the EM algorithm [**7**] by including a Fisher step between the E and the M step. This approach allows the same projection to be applied to all data, but does not guarantee the increasing of the likelihood at each iteration of the algorithm.

In the context of multi-partitions clustering, Galimberti and Soffritti [**8**] assumed that the variables can be partitioned into several independent blocks, each one following a full-covariance Gaussian mixture model. The model selection was done by maximizing the BIC criterion by a forward/backward approach. Then, Galimberti et al. [**9**] generalized thier previous work by relaxing the assumption of block independence. The proposed extension takes into account three types of variables, classifying variables, redundant variables and non-classifying variables. In this context, the choice of the model is difficult because several roles have to be taken into account for each variable, which requires a lot of calculations, even for the reallocation of only one variable. Poon et al. [**10**] also took into account the multi-partition setting, called as facet determination in their article. The model considered is similar to that of Galimberti and Soffritti [**8**], but it also allows tree dependency between latent class variables, resulting in the Pouch Latent Tree Models (PLTM). Model selection is performed by a greed search to maximize the BIC criterion. The resulting model allows a broad understanding of the data, but the tree structure search makes estimation even more difficult as the number of variables increases. More recently, Marbac and Vandewalle [**11**] proposed a tractable muti-partition clustering algorithm not limited to continuous data; in the Gaussian setting it can be seen as particular case of Galimberti and Soffritti [**8**] where they assume a diagonal covariance matrix allowing a particularly efficient search of the partition of the variables in sub-vectors.

In this article we suppose that the data can contain *several* clustering latent variables that is we are in the multiple partition setting. But contrary to Marbac and Vandewalle [**11**] where it is assumed that variables are divided into blocks each one related to some clustering of the data, we are looking for *clustering subspaces*, i.e., linear projections of the data each one related to some particular clustering latent variable thus replacing the combinatorial question of finding the partition of the variables in independent sub-vectors by the question of finding the coefficients of the linear combinations. The proposed approach can be related to the independent factor analysis [**12**] where the author deals with source separation, in our framework a source can be interpreted as some specific clustering subspace; however, their approach becomes intractable as the numbers of sources increases and does not allow to consider multivariate subspaces. Moreover, it is not invariant up to a rotation and rescaling of the data, where our proposed methodology is.

The organisation of the paper is the following, in Section 2 we present a reinterpretation of the factorial discriminant analysis as a search of discriminant components and of independent non-discriminant components. In Section 3, the multi-partitions subspace clustering model and the EM algorithm to estimate the parameters of the model will be presented. In Section 4, results on simulated and real data will show the interest of the method in practice. In Section 5, a conclusion and discussion of future extension of the paper will be made.

PROBABILISTIC INTERPRETATION OF THE FACTORIAL DISCRIMINANT ANALYSIS

Linear Discriminant Analysis (LDA)

It is supposed that n quantitative data in dimension d are available, the data number i will be denoted by $x_i=(x_{i1},\ldots,x_{id})^T$, where x_{ij} is the value of variable j of data i. The whole dataset will be denoted by $x=(x_1,\ldots,x_n)^T$. Let assume that the data is clustered in K clusters, the class label of data i will be denoted by $z_i=(z_{i1},\ldots,z_{iK})^T$, with z_{ik} equals to 1 if data i belongs to cluster k and 0 otherwise. Let also denote by $z=(z_1,\ldots,z_n)$ the partition of x. In this section z is supposed to be known. For sake of simplicity the random variables and their realisations will be denoted in lower case, and p will be used as a generic notation to denote a probability distribution function (p.d.f.) which will be interpreted according to its arguments.

In the context of linear discriminant analysis [**13**], it is supposed that the distribution xi given the cluster follows a d-variate Gaussian distribution with common covariance matrices:

$$\forall k \in \{1,\ldots,K\},\ \ x_i|z_{ik}=1 \sim \mathcal{N}_d(\mu_k,\Sigma),$$

with μ_k the vector of means in cluster k and Σ the common class conditional covariance matrix. Let also denote by $\pi_k=p(z_{ik}=1)$ the prior weights of each cluster.

The posterior cluster membership probabilities can be computed using the Bayes formula:

$$p(z_{ik}=1|\boldsymbol{x}_i)=\frac{\pi_k\phi_d(\boldsymbol{x}_i;\boldsymbol{\mu}_k,\boldsymbol{\Sigma})}{\sum_{k'=1}^{K}\pi_{k'}\phi_d(\boldsymbol{x}_i;\boldsymbol{\mu}_{k'},\boldsymbol{\Sigma})}, \tag{1}$$

where $\phi_d(\cdot;\mu_k,\Sigma)$ stands for the p.d.f. of the d-variable Gaussian distribution with expectation μ_k and covariance matrix Σ.

Let $\bar{\boldsymbol{x}}_k$ denote the class conditional mean in cluster k:

$$\bar{\boldsymbol{x}}_k=\frac{1}{n_k}\sum_{i=1}^{n} z_{ik}\boldsymbol{x}_i,$$

with $n_k=\sum_{i=1}^{n} z_{ik}$ the number of data in cluster k, and by $\bar{\boldsymbol{x}}$ the unconditional mean. Let also denote by W the empirical intra-class covariance matrix:

$$\mathbf{W}=\frac{1}{n}\sum_{k=1}^{K}\sum_{i=1}^{n} z_{ik}(\boldsymbol{x}_i-\bar{\boldsymbol{x}}_k)(\boldsymbol{x}_i-\bar{\boldsymbol{x}}_k)^{\mathrm{T}},$$

and by B the empirical between class covariance matrix:

$$\boldsymbol{B}=\frac{1}{n}\sum_{k=1}^{K} n_k(\bar{\boldsymbol{x}}_k-\bar{\boldsymbol{x}})(\bar{\boldsymbol{x}}_k-\bar{\boldsymbol{x}})^{\mathrm{T}}.$$

If the data are supposed to be independent, the likelihood can simply be written as:

$$\begin{aligned}\ell(\pi_1,\ldots,\pi_K,\mu_1,\ldots,\mu_K,\Sigma;\mathbf{x},\mathbf{z}) &= -\frac{n}{2}\log(\det(\Sigma))-\frac{1}{2}\sum_{i=1}^{n}\sum_{k=1}^{K} z_{ik}\|x_i-\mu_k\|^2_{\Sigma^{-1}}\\ &+\sum_{i=1}^{n}\sum_{k=1}^{K} z_{ik}\log(\pi_k)-\frac{n}{2}\log(2\pi).\end{aligned}$$

The maximum likelihood estimators of the parameters of π_k, μ_k and Σ are $\hat{\pi}_k=\frac{n_k}{n}$, $\hat{\mu}_k=\bar{x}_k$ and $\hat{\boldsymbol{\Sigma}}=\mathbf{W}$. A new data point can be then classified by plugin the estimated values of the parameters in Equation (**1**):

$$\hat{z}_i=\underset{k\in\{1,\ldots,K\}}{\operatorname{argmax}}\ \hat{\mu}_k^{\mathrm{T}}\hat{\boldsymbol{\Sigma}}^{-1}x_i-\frac{1}{2}\mu_k^{\mathrm{T}}\hat{\boldsymbol{\Sigma}}^{-1}\mu_k+\log(\hat{\pi}_k),$$

the resulting classification boundary being linear in this case.

Factorial Discriminant Analysis (FDA)

Let us note that, from a descriptive viewpoint, one can be interested in dimension reduction in order to visualize the data. This could be done by using PCA, but from a classification perspective the component explaining the largest variability in the data are often not the same that the components providing the best separation between the clusters.

The goal of factorial discriminant analysis (FDA) is to find the component maximizing the variance explained by the cluster above the intra-class variance. The coefficients of the first discriminant component $v_1 \in \mathbb{R}^d$ is defined by

$$v_1 = \arg\max_{v \in \mathbb{R}^d} \frac{v^{\mathrm{T}} B v}{v^{\mathrm{T}} W v}.$$

It is well known that v_1 is the eigen vector associated with the highest eigen value λ_1 of $W^{-1}B$ [**14**]. The remaining discriminant components are obtained through the remaining eigen vectors of $W^{-1}B$. Let denote by $\lambda_1,\ldots,\lambda_{K-1}$ the eigen values of $W^{-1}B$ sorted in decreasing order and by $v_1,\ldots,v_{K-1}$ the associated eigen vectors. Moreover, if each component is constrained to have an intra-class variance equal to one (i.e., $v_k{}^{\mathrm{T}} W v_k = 1, \forall k \in \{1,\ldots,K-1\}$), the classification obtained using the Mahalonobis distance can simply be obtained by using the Euclidean distance on the data projected on the discriminant components.

Equivalence between LDA and FDA

As proved in Campbell [**15**] and detailed in Trevor Hastie [**16**], the FDA can be interpreted in a probabilistic way as an LDA where the rank of $\{\mu_1,\ldots,\mu_K\}$ is constrained to be equal to p with $p \leq K-1$ under the common class covariance matrix assumption. This allows us to reparametrize the probabilistic model in the following way:

$$x_i | z_{ik} = 1 \sim \mathcal{N}_d\left(A\begin{pmatrix}\nu_k\\ \gamma\end{pmatrix}, AA^{\mathrm{T}}\right),$$

where $\nu_k \in \mathbb{R}^p, \gamma \in \mathbb{R}^{d-p}$ and $A \in \mathcal{M}_{d,d}(\mathbb{R})$. Let notice that this new parametrization at this step is not unique but the model can easily be made identifiable by imposing some constraints on the parameters.

Let $y_i \in \mathbb{R}^p$ and $u_i \in \mathbb{R}^{d-p}$ two random variables, the new parametrization can be reinterpreted in the following generative framework:

- Draw $z_i : z_i \sim \mathcal{M}(1; \pi_1, \ldots, \pi_K)$ where M stands for the multinomial distribution
- Draw $y_i|z_i : y_i|z_{ik} = 1 \sim \mathcal{N}_p(v_k, I_p)$
- Draw $u_i : u_i \sim \mathcal{N}_{d-p}(\gamma, I_{d-p})$
- Compute x_i based on y_i and $u_i : x_i = A \begin{pmatrix} y_i \\ u_i \end{pmatrix}$.

Thus the p.d.f. of x_i can be factorized in the following way:

$$p(x_i) = \frac{1}{|A|} p(u_i) p(y_i).$$

From a graphical angle the model can be reinterpreted as in **Figure 1**, where u_i and y_i are latent random variables.

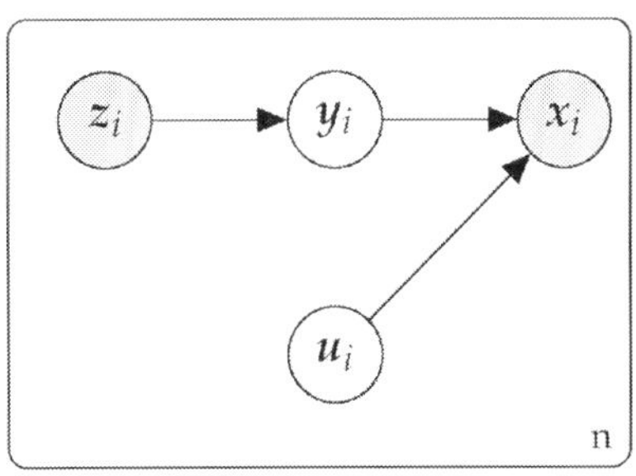

Figure 1. Bayesian dependency graph for factorial discriminant analysis.

In practice we are interested in finding y_i and u_i from x_i. We will denote by $V \in \mathcal{M}_{p,d}(\mathbb{R})$ and $R \in \mathcal{M}_{d-p,d}(\mathbb{R})$ the matrices which allow this computation: $y_i = Vx_i$, $u_i = Rx_i$. It is obvious that $\begin{pmatrix} V \\ R \end{pmatrix} = A^{-1}$, for the rest of the article only the parametrization in terms of V and R will be used.

The main interest of this parametrization is that

$$p(z_{ik} = 1|x_i) = p(z_{ik} = 1|y_i, u_i) = p(z_{ik} = 1|y_i) = p(z_{ik} = 1|Vx_i).$$

It means that only Vx_i is required to compute the posterior class membership probabilities:

$$p(z_{ik} = 1|\boldsymbol{x}_i) = \frac{\pi_k \phi_p(\boldsymbol{V}\boldsymbol{x}_i; \nu_k, \boldsymbol{I}_p)}{\sum_{k'=1}^{K} \pi_{k'} \phi_p(\boldsymbol{V}\boldsymbol{x}_i; \nu_{k'}, \boldsymbol{I}_p)}. \tag{2}$$

Parameters are estimated by maximum likelihood, using the variable change formulae the likelihood can be written:

$$\begin{aligned}\ell(\pi_1, \ldots, \pi_K, \boldsymbol{V}, \boldsymbol{R}, \gamma, \nu_1, \ldots, \nu_K; \mathbf{x}, \mathbf{z}) &= n \log \left| \det \begin{pmatrix} \boldsymbol{V} \\ \boldsymbol{R} \end{pmatrix} \right| - \frac{1}{2} \sum_{i=1}^{n} \sum_{k=1}^{K} z_{ik} \|\boldsymbol{V}\boldsymbol{x}_i - \nu_k\|^2 \\ &+ \sum_{i=1}^{n} \sum_{k=1}^{K} z_{ik} \log(\pi_k) - \frac{1}{2} \sum_{i=1}^{n} \|\boldsymbol{R}\boldsymbol{x}_i - \gamma\|^2 - \frac{n}{2} \log(2\pi).\end{aligned}$$

The first term is related to the variable change, the second term is related to the discriminant components and the third term is related to prior class membership probabilities and the fourth term is related to the non-discriminant components. In practice, this decomposition is the corner stone of the proposed approach since it separates the clustering part from the non-clustering part.

As stated in Campbell [**15**] the maximum likelihood estimator of V are the first p eigen vectors $v_1, \ldots, v_p$ of $W^{-1}B$ in rows:

$$\hat{\boldsymbol{V}} = \begin{pmatrix} v_1^{\mathrm{T}} \\ \vdots \\ v_p^{\mathrm{T}} \end{pmatrix},$$

renormalized such that:

$$\hat{\boldsymbol{V}}\boldsymbol{W}\hat{\boldsymbol{V}}^{\mathrm{T}} = \boldsymbol{I}_p.$$

The maximum likelihood estimator of R is obtained such that:

$$\hat{\boldsymbol{R}}\boldsymbol{W}\hat{\boldsymbol{V}}^{\mathrm{T}} = 0$$

and that

$$\hat{\boldsymbol{R}}\boldsymbol{W}\hat{\boldsymbol{R}}^{\mathrm{T}} = \boldsymbol{I}_{d-p}, \tag{3}$$

with D_{d-p} the diagonal matrix of the d−p last eigen-values of $W^{-1}B$ (with the d−K last eigen-values which are null). $\hat{\boldsymbol{R}}^{\mathrm{T}}$ can simply be obtained by multiplying the last d−p egien-vectors of $\boldsymbol{W}^{-1/2}\boldsymbol{B}\boldsymbol{W}^{-1/2}$ by $\boldsymbol{W}^{-1/2}$ then renormalize them such that Equation (**3**) is satisfied. Such renormalization makes the parameters *V* et *R* identifiable up to a sign.

Moreover ν_k and γ are estimated by

$$\hat{\nu}_k = \frac{1}{n_k}\sum_{i=1}^{n} z_{ik}\hat{V}x_i,$$

$$\hat{\gamma} = \frac{1}{n}\sum_{i=1}^{n} \hat{R}x_i.$$

As stated in Trevor Hastie [**16**] the link between the two parametrizations is as follows:

$$\hat{\mu}_k = W\hat{V}^{\mathrm{T}}\hat{R}(\bar{x}_k - \bar{x}) + \bar{x},$$

and

$$\hat{\Sigma} = W + W\hat{R}^{\mathrm{T}}\hat{R}B\hat{R}^{\mathrm{T}}\hat{R}W.$$

From this formula we can clearly see that the reduced rank constraint operates some regularization on the parameters estimation. Moreover, from a practical perspective the reparametrization Equation (**2**) is more efficient for computing the posterior class membership probabilities.

Application in the Clustering Setting

As in Trevor Hastie [**16**], the model can easily be used in the clustering setting using the EM algorithm to maximise the likelihood. Thus W, B, $\bar{x}_k$ are recomputed at each iteration by using their version weighted by their posterior membership probabilities.

The EM algorithm is now presented. The algorithm is first initialized with some starting value of the parameters or of the partition. Then the E step and the M step are iterated until convergence. Let (r) denote the value of the parameters at iteration *r*, the *E* and the *M* steps are the followings:

- E step: compute the posterior class membership probabilities.

$$t_{ik}^{(r+1)} = \frac{\pi_k \phi_d(x_i;\mu_k^{(r)},\Sigma^{(r)})}{\sum_{k'=1}^{K}\pi_{k'}\phi_d(x_i;\mu_{k'}^{(r)},\Sigma^{(r)})} = \frac{\pi_k \phi_p(y_i^{(r)};\nu_k^{(r)},I_p)}{\sum_{k'=1}^{K}\pi_{k'}\phi_p(y_i^{(r)};\nu_{k'}^{(r)},I_p)},$$

with $y_i^{(r)} = V^{(r)}x_i$.

- M step: Compute

$$\bar{x}_k^{(r+1)} = \frac{1}{n_k^{(r+1)}} \sum_{i=1}^{n} t_{ik}^{(r+1)} x_i,$$

$$W^{(r+1)} = \frac{1}{n} \sum_{k=1}^{K} \sum_{i=1}^{n} t_{ik}^{(r+1)} (x_i - \bar{x}_k^{(r+1)})(x_i - \bar{x}_k^{(r+1)})^{\mathrm{T}},$$

$$B^{(r+1)} = \frac{1}{n} \sum_{k=1}^{K} n_k^{(r+1)} (\bar{x}_k^{(r+1)} - \bar{x})(\bar{x}_k^{(r+1)} - \bar{x})^{\mathrm{T}}.$$

Then deduce $V^{(r+1)}, R^{(r+1)}, v_1^{(r+1)}, \ldots, v_K^{(r+1)}$ and $\gamma^{(r+1)}$ as in the previous section using the eigenvalue decomposition of $W^{(r+1)^{-1}} B^{(r+1)}$.

As noticed in Trevor Hastie [**16**], this approach is not equivalent to performing a standard EM algorithm and then performing FDA at the end of the EM algorithm. FDA must be computed at each iteration of the EM algorithm since the posterior membership probabilities are only computed based on the p first clustering projections.

Let us also notice that the M step can be interpreted in a Fisher-M step since FDA is required. In this sense it can be interpreted as a particular version of the Fisher-EM algorithm of Bouveyron and Brunet [**6**]. Although the homoscedasticity could be seen as particularly constraining, it is the best framework for introducing our model in the next section, since it is easily interpretable and allows for efficient computation owing to the closed form of FDA. This limitation could be easily overcome by using rigorous extensions of the FDA in the heteroscedastic setting as in Kumar and Andreou [**17**]; however, the computation would be much more intensive, since in this case no closed form formula is available and an iterative algorithm would be required even for finding the best projection.

MULTI-PARTITION SUBSPACE MIXTURE MODEL

Presentation of the Model

Let us now suppose that instead of having only one class variable z_i for data i, we now have H class variables $z_i^1, \ldots, z_i^H$ with $K_1, \ldots, K_H$ modalities. It is assumed that $z_i^1, \ldots, z_i^H$ are independent, with $p(z_{ik}^h = 1)$ denoted by π_k^h. Let also denote by y_i^h the variables related to the clustering variable z_i^h such that:

$$y_i^h | z_{ik}^h = 1 \sim \mathcal{N}_{p_h}(v_k^h, I_{p_h})$$

and that we will denote by $p_\bullet = \sum_{h=1}^{H} p_h$.

Let us still denote by u_i the non clustering variables

$$u_i \sim \mathcal{N}_{d-p_\bullet}(\gamma, I_{d-p_\bullet}).$$

Let us also define x_i by:

$$x_i = \begin{pmatrix} V_1 \\ \vdots \\ V_H \\ R \end{pmatrix}^{-1} \begin{pmatrix} y_i^1 \\ \vdots \\ y_i^H \\ u_i \end{pmatrix}.$$

Thus,

$$p(x) = \left| \det \begin{pmatrix} V_1 \\ \vdots \\ V_H \\ R \end{pmatrix} \right| p(u) \prod_{h=1}^{H} p(y^h).$$

The **Figure 2** illustrates the model in the case of H=2.

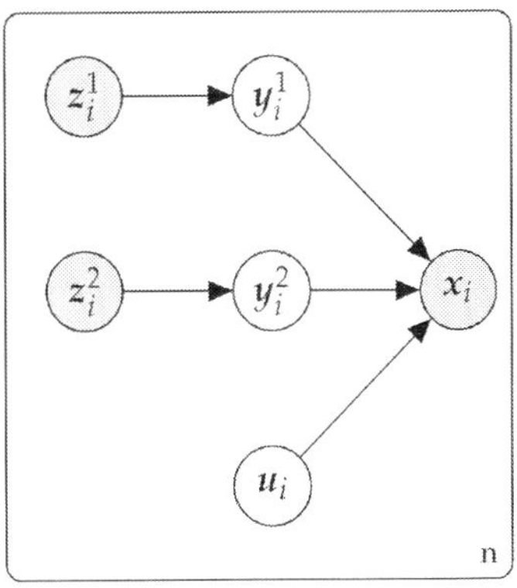

Figure 2. Adapted Bayesian dependency graph to the multi-partition setting, for H=2 clustering variables.

Let us notice that this model allows us to visualize many clustering viewpoints in a low dimensional space since xi can be summarized by $y_i^1, \ldots, y_i^H$. For instance, one can suppose that $p_1 = \ldots = p_H = 1$. In this case each clustering variable can be visualized on one component. We will denote by

$\theta = (V_1, \ldots, V_H, R, \gamma, v_1^1, \ldots, v_{K_H}^H)$ the parameters of the model to be estimated.

Discussion about the Model

The Cartesian product of cluster spaces results in $\prod_{h=1}^{H} K_h$ clusters, which can be very large without needing many parameters. Thus the proposed model can be interpreted as being a very sparse Gaussian mixture model allowing the possibility to deal with a very large number of clusters, the resulting conditional means and covariances matrices are given in the following formulas:

$$\mathbb{E}(x_i | z_{ik_1}^1 = 1, z_{ik_1}^2 = 1, \ldots, z_{ik_H}^H = 1) = \begin{pmatrix} V_1 \\ \vdots \\ V_H \\ R \end{pmatrix}^{-1} \begin{pmatrix} v_{k_1}^1 \\ \vdots \\ v_{k_H}^H \\ \gamma \end{pmatrix},$$

and

$$\mathbb{V}(x_i | z_{ik_1}^1 = 1, z_{ik_1}^2 = 2, \ldots, z_{ik_H}^H = 1) = (V_1^{\mathrm{T}} V_1 + \cdots + V_H^{\mathrm{T}} V_H + R^{\mathrm{T}} R)^{-1}.$$

Thus, the expectation of x_i given in all the clusters is a linear combination of the cluster specific means which can be referred to as a multiple-way MANOVA setting. On the one hand, as a particular homoscedastic Gaussian mixture, our model is more prone to model bias than free homoscedastic Gaussian mixture, and in the case when our model would be well-specified the homoscedastic Gaussian mixture would give a similar clustering for a large sample size (i.e., the same partitions with respect to the partition resulting from the product space of our multi-partitions model). On the other hand, our approach produces a factorised version of the partition space as well as the related clustering subspaces which is not a standard output of clustering methods, and it can deal with a large number of clusters in a sparse way which can be particularly useful for a moderated sample size. In practice, the choice between our model and an other mixture model can simply be performed through the BIC criterion.

In some sense our model can be linked with the mixture of factor analyzers [**18**]. In mixture of factor analyzers the model is of the type:

$$x_i = Ay_i + u_i,$$

where A is a low rank matrix. But here we have chosen a model of the type

$$x_i = A \begin{pmatrix} y_i \\ u_i \end{pmatrix},$$

which allows us to deal with the noise in a different way. Actually, our model is invariant up to a bijective linear transformation of the data which is not the case for the mixtures of factor analyzers. On the other hand, our model can only deal with data with moderated dimension with respect to the number of statistical units; it assumes that the sources y_i can be recovered from the observed data xi.

Estimation of the Parameters of the Model in the Supervised Setting

The likelihood of the model can be written:

$$\ell(\theta; \mathbf{x}, \mathbf{z}) = n \log \left| \det \begin{pmatrix} V_1 \\ \vdots \\ V_H \\ R \end{pmatrix} \right| - \sum_{i=1}^{n} \sum_{h=1}^{H} \sum_{k=1}^{K_h} z_{ik}^h \| V_h^T x_i - \nu_k^h \|^2$$
$$+ \sum_{i=1}^{n} \sum_{h=1}^{H} \sum_{k=1}^{K_h} z_{ik}^h \log(\pi_k^h) - \sum_{i=1}^{n} \| R^T x_i - \gamma \|^2 - \frac{n}{2} \log(2\pi).$$

The likelihood cannot be maximized directly. However, in the case of H=1, it reduces to the problem of **Section 2**. Let notice that if all the parameters are fixed except V_h and R, ν_k^h and γ, the optimisation can be easily performed by constraining $V_h^{(r+1)}$ and $R^{(r+1)}$ to be linear combinations of $V_h^{(r)}$ and $R^{(r)}$. Thus the likelihood will be optimized by using an alternate optimization algorithm. Let $M \in \mathcal{M}_{d-p_\bullet+p_h, d-p_\bullet+p_h}(\mathbb{R})$ the matrix which allow to compute $V_h^{(r+1)}$ and $R^{(r+1)}$ based on $V_h^{(r)}$ and $R^{(r)}$:

$$\begin{pmatrix} V_h^{(r+1)} \\ R^{(r+1)} \end{pmatrix} = M \begin{pmatrix} V_h^{(r)} \\ R^{(r)} \end{pmatrix} = \begin{pmatrix} M_1 \\ M_2 \end{pmatrix} \begin{pmatrix} V_h^{(r)} \\ R^{(r)} \end{pmatrix},$$

where M_1 is the sub-matrix containing the p_h first rows of M and M_2 the matrix containing the last d−p· rows of M.

Thus, the increase of the likelihood when all the parameters are fixed except V_h, R, ν_k^h and γ becomes:

$$\begin{aligned} C(M_1, M_2, \nu_k, \gamma) &= n \log|\det(M)| - \frac{1}{2}\sum_{i=1}^{n}\sum_{k=1}^{K_h} z_{ik}^h \left\| M_1 \begin{pmatrix} V_h^{(r)} \\ R^{(r)} \end{pmatrix} x_i - \nu_k^h \right\|^2 \\ &\quad -\frac{1}{2}\sum_{i=1}^{n} \left\| M_2 \begin{pmatrix} V_h^{(r)} \\ R^{(r)\mathrm{T}} \end{pmatrix} x_i - \gamma \right\|^2 . \end{aligned}$$

By denoting

$$\begin{pmatrix} y_i^{h\,(r)} \\ u_i^{(r)} \end{pmatrix} = \begin{pmatrix} V_h^{(r)} \\ R^{(r)} \end{pmatrix} x_i,$$

we have to maximize:

$$\begin{aligned} C(M_1, M_2, \nu_k, \gamma) &= n \log|\det(M)| - \frac{1}{2}\sum_{i=1}^{n}\sum_{k=1}^{K_h} z_{ik}^h \left\| M_1 \begin{pmatrix} y_i^{h\,(r)} \\ u_i^{(r)} \end{pmatrix} - \nu_k^h \right\|^2 \\ &\quad -\frac{1}{2}\sum_{i=1}^{n} \left\| M_2 \begin{pmatrix} y_i^{h\,(r)} \\ u_i^{(r)} \end{pmatrix} - \gamma \right\|^2 . \end{aligned}$$

Consequently, M and the others parameters can be obtained by applying a simple FDA on the data $(y_i^{h\,(r)\mathrm{T}}, u_i^{(r)\mathrm{T}})$. In order to optimize over all the parameters, we can loop over all the clustering dimensions.

Thus, in the case of mixed continuous and categorical data, this model can be used to visualize the clustering behavior of the categorical variables with respect to the quantitative ones.

Estimation of the Parameters of the Model in the Clustering Setting

Here our main goal is to consider the clustering setting, that is, when $z_i^1, \ldots, z_i^H$ are unknown. Consequently we will use an EM algorithm to "reconstitute the missing label" in order to maximize the likelihood. Thus the algorithm stays the same as in the supervised setting, except that the data at each iteration are now weighted by $t_{ik}^{h\,(r+1)}$ instead of z_{ik}^h.

The algorithm is the following:

- Until convergence, for h∈{1,…,H} iterates the following steps:
- E step: compute

$$t_{ik}^{h\,(r+1)} = \frac{\pi_k p(y_i^{h\,(r)}; \nu_k^{h\,(r)}, I_p)}{\sum_{k'=1}^{K} \pi_{k'} p(y_i^{h\,(r)}; \nu_{k'}^{h\,(r)}, I_p)}.$$

- M step: compute $\pi_1^{h(r+1)}, \ldots, \pi_{K_h}^{h\ (r+1)}, V_h^{(r+1)}, R^{(r+1)}, \gamma^{(r+1)}$ and $\nu_k^{h(r+1)}$ based on formulas given in the supervised setting.

Parsimonious Models and Model Choice

The proposed model needs the user to define the number of clustering subspaces *H*, the number of cluster in each clustering subspace $K_1, \ldots, K_H$, and the dimensionality $p_1, \ldots, p_H$ of each subspace. The constraints are that $H<d$, that $p_h \leq K_{h-1}$ and $p\cdot=p_1+\cdots+p_H<d$. It is clear that the number of possible models can become very high. To limit the combinatorial aspect, one can impose $K_1=\cdots=K_H=K$ and/or $p_1=\cdots=p_h=p$. In practice the choice of $p=1$ enforces to find clustering which could be visualized in one dimension, which can help the practitioner. Moreover, choosing $K=2$ is the minimal requirement in order to investigate a clustering structure. However, if possible we recommend to explore the largest possible number of models and choosing the best one with the BIC. Let us define the following parsimonious models and their related number of parameters:

- $[K_h p_h]$ the general form of the proposed model, the index *h* will be removed if values are the same for each clustering subspace.

$$\sum_{h=1}^{H}(K_h-1)+\sum_{h=1}^{H}\frac{p_h(2K_h+2d-p_h+1)}{2}+\frac{(d-p_\bullet)(d+p_\bullet+3)}{2},$$

- $[K_{ph}]$ where the number of clusters is the same for each subspace

$$H(K-1)+\sum_{h=1}^{H}\frac{p_h(2K+2d-p_h+1)}{2}+\frac{(d-p_\bullet)(d+p_\bullet+3)}{2},$$

- $[K_{hp}]$ where the dimensionalities are the same for each subspaces

$$\sum_{h=1}^{H}(K_h-1)+\sum_{h=1}^{H}\frac{p(2K_h+2d-p+1)}{2}+\frac{(d-Hp)(d+Hp+3)}{2},$$

- $[K_{h1}]$ where the dimensionalities are equals to one for each subspaces

$$\sum_{h=1}^{H}(K_h-1)+\sum_{h=1}^{H}(K_h+d)+\frac{(d-H)(d+H+3)}{2},$$

- $[K_p]$ where the number of clusters is the same for each subspace and the dimensionalities are the same for each subspace

$$H(K-1)+\frac{Hp(2K+2d-p+1)}{2}+\frac{(d-Hp)(d+Hp+3)}{2},$$

- $[K_1]$ where the number of clusters is the same for each subspace and the dimensionalities are equals to one for each subspaces

$$H(K-1)+H(K+d)+\frac{(d-H)(d+H+3)}{2}.$$

For a given model *m* the BIC is computed as:

$$BIC(m)=\ell(\hat{\theta}_m;\mathbf{x})-\frac{\nu_m}{2}\log n,$$

where ν_m is the number of parameters of the model detailed above. Thus the model choice consists of choosing the model maximising the BIC. BIC enjoys good theoretical consistency properties, thus providing a guarantee to select the true model as the number of data increases. The ICL criterion [**19**] could also be used to enforce the choice of well separated clusters, since from a classification perspective BIC is known to over-estimate the number of clusters if model assumption are violated. Let us however notice that in practice the user could be mainly interested by a low value of *H*, since even H=2 can provide him with new insights about his data, focusing on finding several clustering view points.

EXPERIMENTS

Experiments on Simulated Data

We now present a tutorial example. Let us consider n=100 data, with H=2 clustering subspaces each one of dimension one (p_1=p_2=1) and containing each of two clusters (K_1=K_2=2). Let us draw y_i^1, the first clustering variable, according to a mixture of two Gaussian univariate distributions: y_i^1 ~0.5N(0,1)+0.5N(4,1) and draw independently y_i^2, the second clustering variable, according to the same mixture distribution: y_i^2 ~0.5N(0,1)+0.5N(4,1), then draw u_i the non-classifying components according to a multivariate Gaussian distribution in $\mathbb{R}^4, \boldsymbol{u}_i \sim \mathcal{N}_4(0, \boldsymbol{I}_4)$.

Finally compute x_i based on the formula: $x_i = A\begin{pmatrix} y_i^1 \\ y_i^2 \\ u_i \end{pmatrix}$, where the 36 entries of the matrix A have been drawn according to independent N(0,1) for sake of simplicity.

Thus the proposed model, based on the observed data x, aims at recovering the clustering variables y^1 and y^2 as well as the associated cluster variables z^1 and z^2. The initial data x are presented on **Figure 3**, we see that the clustering structure of the data is not apparent from these scatter plots. The underlying clustering variables are presented on **Figure 4**, where we see the separation of the colors on the first clustering variable Y_1 and separation of the shapes on the second clustering variable Y_2. Let us notice that such factorization gives a more synthetic view of the clustering than seeing these clusters as four clusters. Using standard dimension reduction techniques such as PCA does not succeed in recovering the clustering subspace see **Figure 5**. Performing a factorial discriminant analysis considering the four clusters in the supervised setting we get **Figure 6**, it finds good separation between the clusters, however we do not obtain the factorised interpretation of the clustering as the Cartesian product of two independent clusterings. Finally by performing the estimation of the model parameters in an unsupervised setting based on the data x we get **Figure 7**. We see that $\hat{Y}_1$ succeeds in recovering Y_1 up to a sign and that $\hat{Y}_2$ succeeds in recovering Y_2.

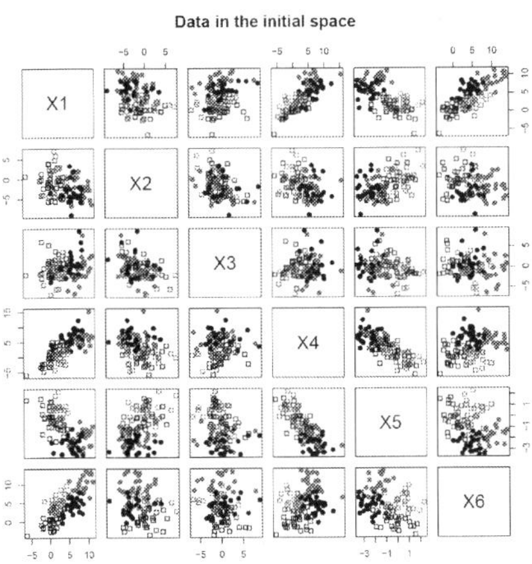

Figure 3. Scatter plots of the initial data on the illustrative example. The color depends on the first cluster variable, and the shape depends on the second cluster variable.

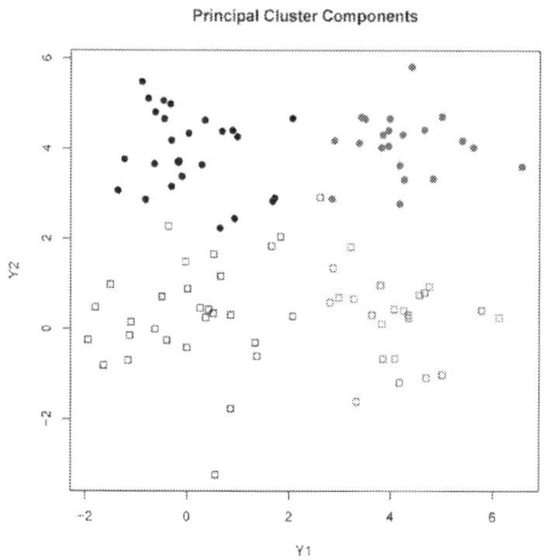

Figure 4. Scatter plot of the illustrative data on the two original clustering subspaces.

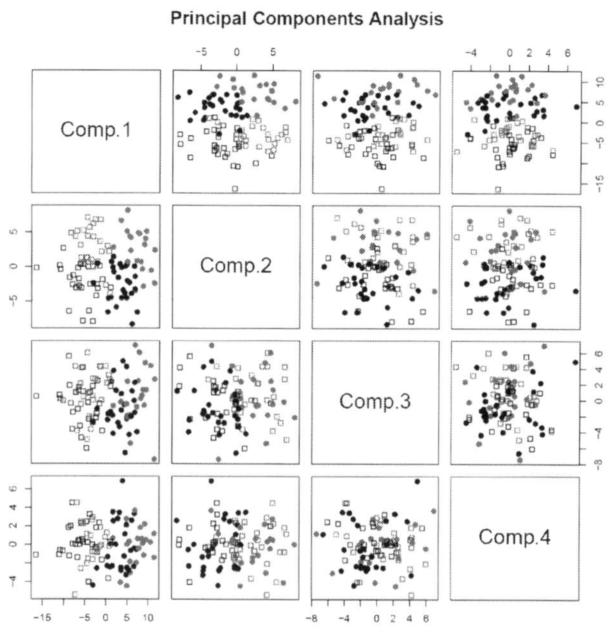

Figure 5. Scatter plot of the illustrative data on components one to four of the proimcipal component analysis (PCA).

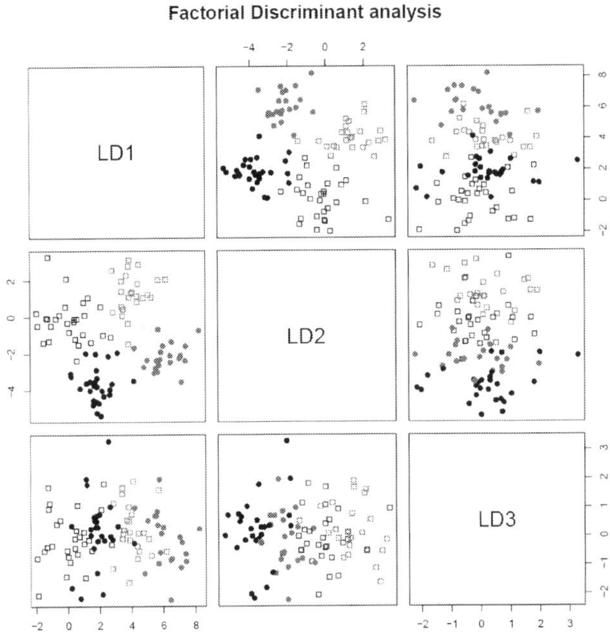

Figure 6. Scatter plot of the component of the factorial discriminant analysis for the illustrative example.

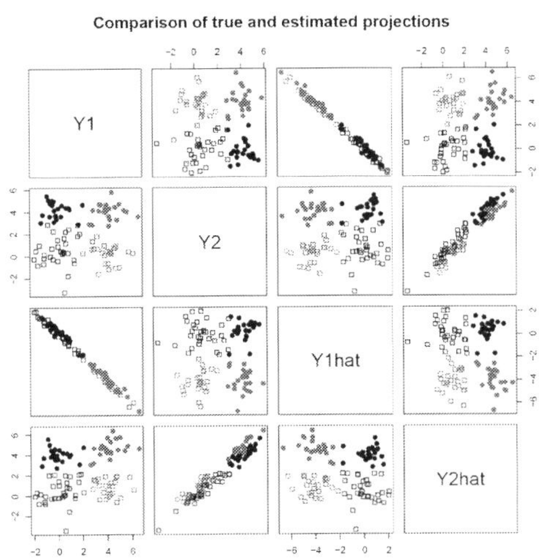

Figure 7. Comparison of the true and of the estimated clustering subspaces on the illustrative example, points are marked according to the estimated clusters.

Moreover, supposing p_1=p_2=1 we can choose K_1 and K_2 according to the BIC criterion. Values of the BIC criterion are presented in **Table 1**, where we show that the selected model is the true one with K_1=K_2=2. The lower diagonal of the table is not presented for symmetry reasons, and we limited ourselves to K_1,K_2∈{1,…,5}.

Table 1. Value of the BIC criterion according to K_1 and K_2, for the choice of the number of clusters on the illustrative example, best value in bold.

K_1\K_2	**1**	**2**	**3**	**4**	**5**
1	−1318.40	−1301.15	−1305.20	−1305.47	−1310.08
2		**−1291.80**	−1293.22	−1292.28	−1307.07
3			−1296.70	−1303.68	−1310.95
4				−1306.09	−1320.29
5					−1319.22

Experiments on Real Data

Let us consider the crabs dataset [**20**]. It consists of 200 crabs morphological data, each crab has two categorical (cluster) attributes—the species, orange or blue, and the sex, male or female. The dataset is composed of 50 males orange, 50 males blue, 50 females orange, 50 females blue for which 5 numerical attributes have been measured: the frontal lobe size, the rear width, the carapace length, the carapace width and the body depth. We can see the PCA of the data in **Figure 8**. We see that component two separates males and females well, whereas component three separates orange and blue subspecies. However, we will see that by applying our model we obtained a better separation of the clusters.

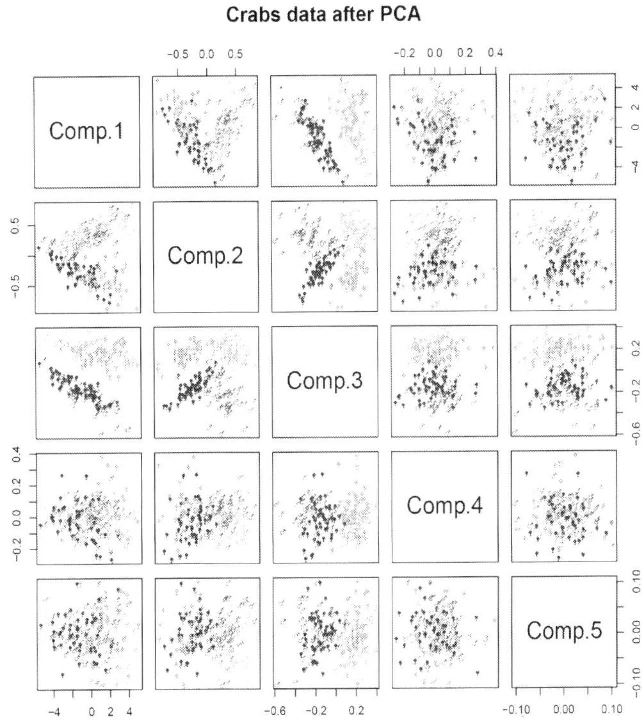

Figure 8. Scatter plots of the crabs data after PCA. Subspecies are represented according to their color, and sex is represented according to its symbol.

Like in the tutorial example we will take $p_1=p_2=1$ and $K_1,K_2\in\{1,\ldots,5\}$. The resulting BIC tabular is given in **Table 2**, it suggests the choice of $K_1=3$ and $K_2=4$. The resulting visualization of the clustering variables in given **Figure 9**. Let us notice that Y_2 is divided in four clusters however, however we only see three since two of them have the same mean. We can see that even if the numbers of clusters do not correspond, the first clustering subspace finds the subspecies, whereas the second clustering subspace finds the sex. We could also look at the solution provided by $K_1=K_2=2$ on **Figure 10**, this one has a lower BIC but seems more natural for the problem at hand. We see that the obtained map is in fact quite similar the map obtained **Figure 9**; however, we notice that from a density approximation point of view we obtain a lower fit. In fact, if we look at the correlations between Y_1 **Figure 9** and Y_2 **Figure 10** we have a correlation of −0.97, and a similar correlation is obtained between Y_2 **Figure 9** and Y_1 **Figure 10**. Thus, the produced subspace are finally quite similar.

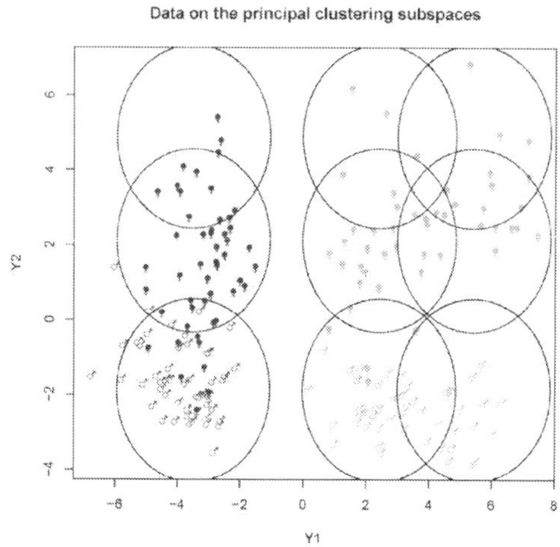

Figure 9. Scatter plots of the clustering subspace on the crabs data for K_1=3 and K_2=4, 95% isodensity is given for each component resulting of the Cartesian product.

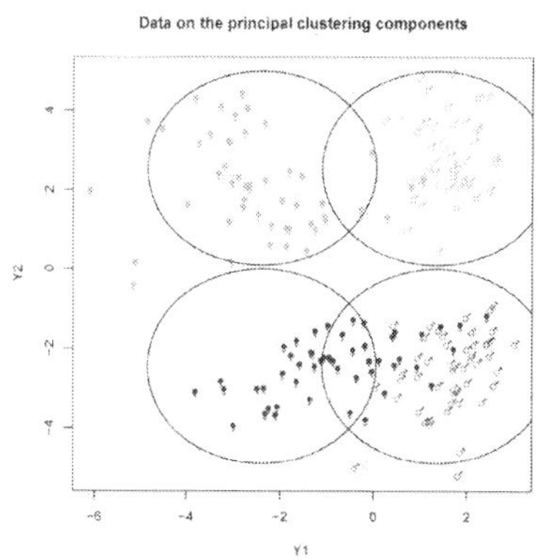

Figure 10. Scatter plots of the clustering subspace on the crabs data for K_1=K_2=2, 95% isodensity is given for each component resulting of the Cartesian product.

Table 2. Value of the BIC criterion according to K_1 and K_2, for the choice of the number of clusters on the crabs dataset, best value in bold.

K_1\K_2	**1**	**2**	**3**	**4**	**5**
1	−62.66	0.41	10.40	5.11	0.80
2		17.82	16.57	18.88	0.49

3			3.75	**22.52**	17.44
4				−26.65	−26.64
5					12.06

CONCLUSIONS AND PERSPECTIVES

We have proposed a model which allows us to combine visualization and clustering with many clustering viewpoints. Moreover, we have shown the possibility of performing model choice by using the BIC criterion. The proposed model can provide new information on the structure present in the data by trying to reinterpret the cluster as a result of the Cartesian product of several clustering variables.

The proposed model is limited to the homoscedatic setting, which could be seen as a limitation; however, from our point of view this is more robust than the heteroscedastic setting, which is known to be jeopardized by the degeneracy issue [**21**]. However, the extension of our work on the heteroscedastic setting can easily be performed from the modeling point of view; the main issue in this case would be the parameters estimation where an extension of the FDA to the heteroscedastic setting would be needed, as presented in Kumar and Andreou [**17**]. Another difficult issue is the choice of *H*, $K_1,\dots,K_H$ and $p_1,\dots,p_H$, which is very combinatorial. Here we have proposed an estimation strategy for all these tuning parameters being fixed, and then performed a selection of the best tuning according to BIC. However, in future work, a model selection strategy to perform the model selection through a modified version of the EM algorithm will also be investigated as in Green [**22**]; it would thus limit the combinatorial aspect of the global model search through EM-wise local model searches.

Funding

This research received no external funding.

ACKNOWLEDGMENTS

I would like to thank Rohit BHAGWAT for his preliminary work on the experimental part of the topic during his internship. I also would like to thank anonymous reviewers for their comments which contributed to improve the quality of the article.

Conflicts of Interest

The author declares no conflict of interest.

REFERENCES

1. McLachlan, G.; Peel, D. *Finite Mixture Models*; John Wiley & Sons: Hoboken, NJ, USA, 2004.
2. Schwarz, G. Estimating the Dimension of a Model. *Ann. Stat.* **1978**, *6*, 461–464.
3. Tipping, M.E.; Bishop, C.M. Probabilistic principal component analysis. *J. R. Stat. Soc. Ser. B Stat. Methodol.* **1999**, *61*, 611–622.
4. Celeux, G.; Govaert, G. Gaussian parsimonious clustering models. *Pattern Recognit.* **1995**, *28*, 781–793.
5. Bouveyron, C.; Girard, S.; Schmid, C. High-dimensional data clustering. *Comput. Stat. Data Anal.* **2007**, *52*, 502–519.
6. Bouveyron, C.; Brunet, C. Simultaneous model-based clustering and visualization in the Fisher discriminative subspace. *Stat. Comput.* **2012**, *22*, 301–324.
7. Dempster, A.; Laird, N.; Rubin, D. Maximum likelihood from incomplete data via the EM algorithm. *J. R. Stat. Society. Ser. B Methodol.* **1977**, *39*, 1–38.
8. Galimberti, G.; Soffritti, G. Model-based methods to identify multiple cluster structures in a data set. *Comput. Stat. Data Anal.* **2007**, *52*, 520–536.
9. Galimberti, G.; Manisi, A.; Soffritti, G. Modelling the role of variables in model-based cluster analysis. *Stat. Comput.* **2018**, *28*, 145–169.
10. Poon, L.K.; Zhang, N.L.; Liu, T.; Liu, A.H. Model-based clustering of high-dimensional data: Variable selection versus facet determination. *Int. J. Approx. Reason.* **2013**, *54*, 196–215.
11. Marbac, M.; Vandewalle, V. A tractable multi-partitions clustering. *Comput. Stat. Data Anal.* **2019**, *132*, 167–179.
12. Attias, H. Independent factor analysis. *Neural Comput.* **1999**, *11*, 803–851.
13. Fisher, R.A. The use of multiple measurements in taxonomic problems. *Ann. Eugen.* **1936**, *7*, 179–188.
14. Friedman, J.; Hastie, T.; Tibshirani, R. *The Elements of Statistical Learning*; Springer Series in Statistics; Springer: Berlin, Germany, 2001; Volume 1.
15. Campbell, N.A. Canonical variate analysis—A general model formulation. *Aust. J. Stat.* **1984**, *26*, 86–96.

16. Hastie, T.; Tibshirani, R. Discriminant Analysis by Gaussian Mixtures. *J. R. Stat. Society. Ser. B Methodol.* **1996**, *58*, 155–176.
17. Kumar, N.; Andreou, A.G. Heteroscedastic discriminant analysis and reduced rank HMMs for improved speech recognition. *Speech Commun.* **1998**, *26*, 283–297.
18. Ghahramani, Z.; Hinton, G.E. *The EM Algorithm for Mixtures of Factor Analyzers*; Technical Report, Technical Report CRG-TR-96-1; University of Toronto: Toronto, ON, Canada, 1996.
19. Biernacki, C.; Celeux, G.; Govaert, G. Assessing a mixture model for clustering with the integrated completed likelihood. *IEEE Trans. Pattern Anal. Mach. Intell.* **2000**, *22*, 719–725.
20. Campbell, N.; Mahon, R. A multivariate study of variation in two species of rock crab of the genus Leptograpsus. *Aust. J. Zool.* **1974**, *22*, 417–425.
21. Biernacki, C.; Chrétien, S. Degeneracy in the maximum likelihood estimation of univariate Gaussian mixtures with EM. *Stat. Probab. Lett.* **2003**, *61*, 373–382.
22. Green, P.J. On use of the EM for penalized likelihood estimation. *J. R. Stat. Soc. Ser. Methodol.* **1990**, *52*, 443–452.

CHAPTER

13

Shrinkage Simplex-Centroid Designs for a Quadratic Mixture Model

Taha Hasan[1], Sajid Ali[2], Munir Ahmed[3]

[1] Department of Statistics, Islamabad Model College for Boys, Sector F-10/4, Islamabad, Pakistan

[2] School of Business and Economics, Lahore, Pakistan

[3] Department of Mathematics, Islamabad Model College for Boys, Sector F-10/4, Islamabad, Pakistan

ABSTRACT

A simplex-centroid design for q mixture components comprises of all possible subsets of the q components, present in equal proportions. The design does not contain full mixture blends except the overall centroid.

Citation: (APA): Hasan, T., Ali, S., & Ahmed, M. (2018). Shrinkage simplex-centroid designs for a quadratic mixture model. Journal of Industrial Engineering International, 14(1), 87-93. (7 pages) DOI: https://doi.org/10.1007/s40092-017-0210-1 URL: https://link.springer.com/article/10.1007/s40092-017-0210-1

In real-life situations, all mixture blends comprise of at least a minimum proportion of each component. Here, we introduce simplex-centroid designs which contain complete blends but with some loss in *D*-efficiency and stability in *G*-efficiency. We call such designs as shrinkage simplex-centroid designs. Furthermore, we use the proposed designs to generate component-amount designs by their projection.

INTRODUCTION

In a mixture experiment with q components, the proportion of ingredients may be denoted by $x_1, x_2, \ldots, x_q$, where $x_i \geq 0$ for $i = 1, 2, \ldots, q$ and $\sum_{i=1}^{q} x_i = 1$. The response depends only on the mixture and not on the total amount of the mixture. The factor space is a (q−1)-dimensional regular simplex S_{q-1}:

$$S_{q-1} = \left\{ x : (x_1, x_2, \ldots, x_q) \middle| \sum_{i=1}^{q} x_i = 1, x_i \geq 0 \right\}.$$

There are various mixture models available in the literature, and for the estimation of their parameters, many mixture designs have been proposed. Simplex-centroid design is the simplest and widely used mixture design by the practitioners. The main feature of this article is the construction of simplex-centroid mixture design with the real mixture blends, though with less efficiency. The design points in a simplex-centroid design do not completely explore the whole mixture region, whereas in our proposed designs, the design points fall uniformly inside the mixture space and explore it in true sense. Prescott (1998) shrunk the co-ordinates of design points towards the centroid after their re-parameterization, using orthogonally blocked mixture designs. The resulting designs were composed of full mixture blends, falling inside the simplex. The designs were less efficient as compared to the original design.

The article is organized as follows. First simplex-centroid design is reviewed. Next, the concept of shrinkage design and the re-parameterization of co-ordinate system for three and fours components are discussed. Then, the shrinkage simplex-centroid designs for three and four components are constructed and their *D*- and *G*-efficiencies are compared. The proposed designs are further used to develop component-amount designs by their projection. Finally, the application of shrinkage design is given and conclusions are made.

REVIEW OF SIMPLEX-CENTROID DESIGN

Scheffé (1963) introduced the simplex-centroid design, where only such mixtures are considered in which the components presented have equal proportions. It has 2^q-1 design points. These design points have $\binom{q}{1}$ permutations of (1,0,...,0), the $\binom{q}{2}$ permutations of (12,12,0,...,0), the $\binom{q}{3}$ permutations of $\left(\frac{1}{3},\frac{1}{3},\frac{1}{3},0,\ldots,0\right)$, and so on, and the overall centroid $\left(\frac{1}{q},\frac{1}{q},\frac{1}{q},\ldots,\frac{1}{q}\right)$. Such mixtures are located at the centroid of the $(q-1)$-dimensional simplex and the centroids of all the lower dimensional simplexes within the $(q-1)$-dimensional simplex. Responses are collected at the design points and a polynomial is fitted. The design points in the simplex-centroid design will support the polynomial equation. The general form of the quadratic Scheffé polynomial equation is

$$E(Y)=\sum_{i=1}^{q}\beta_i X_i+\sum_{i=1}^{q-1}\sum_{j=i+1}^{q}\beta_{ij}X_iX_j.$$

Simplex-centroid designs for three and four components are given in Tables 1 and 2, and are depicted by Fig. 1a and b.

Table 1 Simplex-centroid design for three components

Run	x_1	x_2	x_3
1	1	0	0
2	0	1	0
3	0	0	1
4	1/2	1/2	0
5	1/2	0	1/2
6	0	1/2	1/2
7	1/3	1/3	1/3

Table 2 Simplex-centroid design for four components

Run	x_1	x_2	x_3	x_4
1	1	0	0	0
2	0	1	0	0
3	0	0	1	0
4	0	0	0	1
5	1/2	1/2	0	0
6	1/2	0	1/2	0
7	1/2	0	0	1/2
8	0	1/2	1/2	0
9	0	1/2	0	1/2
10	0	0	1/2	1/2
11	1/3	1/3	1/3	0
12	1/3	1/3	0	1/3
13	1/3	0	1/3	1/3
14	0	1/3	1/3	1/3
15	1/4	1/4	1/4	1/4

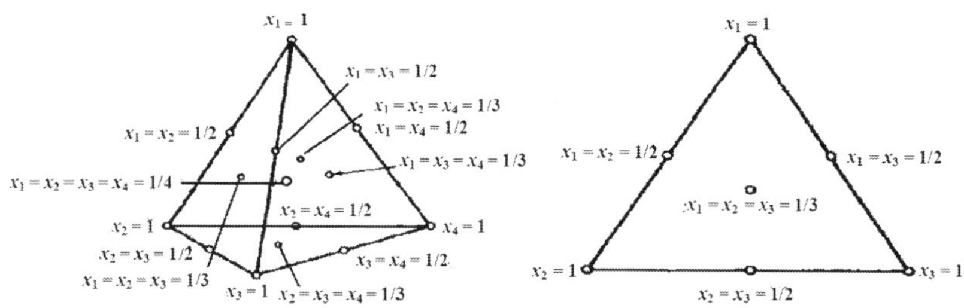

Figure. 1: Simplex-centroid designs for three and four components

In real-life situations, we always use a mixture that has at least a minimum proportion of all the ingredients present. Therefore, we need to construct the optimal designs with complete mixture blends. The standard simplex design is a boundary point design with the exception of the overall centroid that is all the other points are on the boundary of the simplex. We propose the construction of three and four components' simplex-centroid designs by the shrinkage of design points towards its centroid and call them shrinkage simplex-centroid designs.

SHRINKAGE DESIGN

The pioneering work on this issue was done by Prescott (1998). He constructed nearly optimal designs for Scheffé's quadratic mixture model with three and four components, using Latin square-based orthogonal blocking scheme. Such optimal designs although have complete mixture compositions, but are less efficient. Aggarwal et al. (2011) proposed nearly D-, A-, and E-optimal designs for Scheffé's, Kronecker, Becker's, and Darroch and Waller's quadratic mixture models in four components, using F-square-based orthogonal blocking scheme. Hasan and Khan (2011, 2012) proposed nearly D-, A-, and E-optimal orthogonally blocked designs in three components using Scheffé's and Kronecker mixture models. All the above references deal with the construction of shrinkage designs in orthogonal blocks. We have adapted the same methodology for the construction of shrinkage optimal design using simplex-centroid design, where the mixture blends are not in orthogonal blocks.

Re-parameterization of the co-ordinate system

Consider a mixture region in $(q-1)$ dimensions and denote the co-ordinates in the region by the symbols a,b,c,..., such that a+b+c+⋯=1. The orthogonal blocks containing pairs of Latin squares provide mathematical expression for the information matrix X′X in terms of the symbols a,b,c,... Using any optimal criteria, we can determine the optimal values of a,b,c,... Prescott (1998) discussed the re-parameterization of co-ordinates of points in a mixture region and constructed nearly D-optimal designs. We review this re-parameterization method for the construction of Shrinkage simplex-centroid design in three and four components.

1. Three components' mixture

Consider a two-dimensional simplex formed by three components, as given in Fig. 2. Take any design point P (a, b, c) in the simplex, such that $a \geq b \geq c$. Express the co-ordinates a, b, and c in terms of s and f, where s is the shrinkage parameter with f be a co-ordinate of the point on the edge of simplex, and (1/3, 1/3, 1/3) is the centroid of the simplex:

$$a = (1-s)f + \frac{s}{3}$$

$$b = (1-s)(1-f) + \frac{s}{3}$$

$$c = \frac{s}{3}. \qquad (1)$$

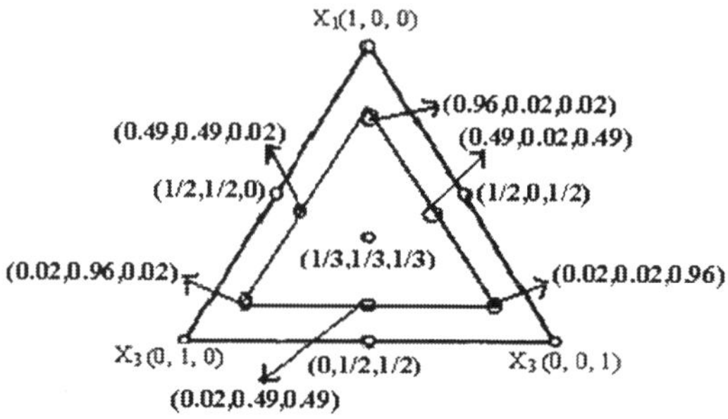

Figure. 2: Simplex-centroid design and shrinkage simplex-centroid design with $s = 0.05$

For instance, when s = 0, the point P (a, b, c) reduces to P (f, $1 - f$, 0, 0), falling on the edge of the simplex.

2. Four components' mixture

We re-parameterize a point P with the co-ordinates (a,b,c,d) in a three-dimensional simplex, where a≥b≥c≥d. As a special case when c=dc=d, the co-ordinates (a,b,c,d) of the point P, expressed as a function of (f,s), are given as follows:

$$a = (1 - s)f + s/4$$
$$b = (1 - s)(1 - f) + s/4$$
$$c = d = s/4. \qquad (2)$$

The extension of re-parameterization to mixture experiments with more than four components is very simple.

SHRINKAGE SIMPLEX-CENTROID DESIGNS FOR THREE AND FOUR COMPONENTS

We construct shrinkage simplex-centroid design by considering quadratic Scheffé's mixture models in three and four components. The D-efficiency of shrinkage designs is obtained with regard to the D-criterion, where $D = |X'X|^{1/P}$ and P denotes the number of parameters in the model:

$$\text{D-efficiency} = |X'X|^{1/P} / |X'X|_o^{1/P} \times 100. \qquad (3)$$

Here, $|X'X|_o^{1/P}$ is the value of D for non-shrinkage design. Furthermore, the prediction capability of a design can be assessed using G-optimality criterion, which searches for the design that minimizes the maximum prediction variance over the experimental region. A design criterion related to G-optimality is *G*-efficiency, defined as

$$\text{G-efficiency} = 100 \times \left(\frac{p/n}{\text{MPV}}\right), \tag{4}$$

where p/n is the average prediction variance (APV) and MPV is the maximum prediction variance. We evaluate *D*- and *G*-efficiencies of shrinkage designs for several values of parameter *s*, for a quadratic Scheffé's model. The three components' quadratic Scheffé's model is

$$E(y) = \beta_1 x_1 + \beta_2 x_2 + \beta_3 x_3 + \beta_{12} x_1 x_2 + \beta_{13} x_1 x_3 + \beta_{23} x_2 x_3. \tag{5}$$

The value of *D* for simplex-centroid design, given in Table 1, for three component quadratic Scheffé's model is 0.27049. Now, consider that the shrinkage parameter $s = 0.05$. Using re-parameterized co-ordinates given in Eq. (1), the shrinkage simplex-centroid design is obtained as follows.

The value of *D* for shrinkage simplex-centroid design given in Table 3 for quadratic Scheffé's model is 0.2293. Using *D*-efficiency criterion given in Eq. (3), the efficiency of the above design is 85%. Comparing the design in Table 3 with the simplex-centroid design in Table 1, it is clear that the constructed design is a true mixture design with full mixture blends, though less efficient. Table 4 gives *D*-efficiency of shrinkage simplex-centroid designs for several values of *s*. In Table 5, we have *G*-efficiency of the designs for different values of shrinkage parameter *s* (Fig. 3).

Table 3 Shrinkage simplex-centroid design with the parameter $s = \mathbf{0.05}$

Run	x_1	x_2	x_3
1	0.96	0.02	0.02
2	0.02	0.96	0.02
3	0.02	0.02	0.96
4	0.49	0.49	0.02
5	0.49	0.02	0.49
6	0.02	0.49	0.49
7	1/3	1/3	1/3

Table 4 *D*-efficiency of shrinkage simplex-centroid designs for $q = 3$

s	$\lvert X'X\rvert^{1/p}$	$\lvert X'X\rvert_0^{1/p}$	*D*-efficiency
0.00	0.270492	0.270492	100
0.05	0.229272	0.270492	85
0.10	0.201085	0.270492	74

Table 5 *G*-efficiency of shrinkage simplex-centroid designs for $q = 3$

s	APV	MPV	*G*-efficiency
0.00	0.857	0.992	86.4
0.05	0.857	0.992	86.4
0.10	0.857	0.992	86.4

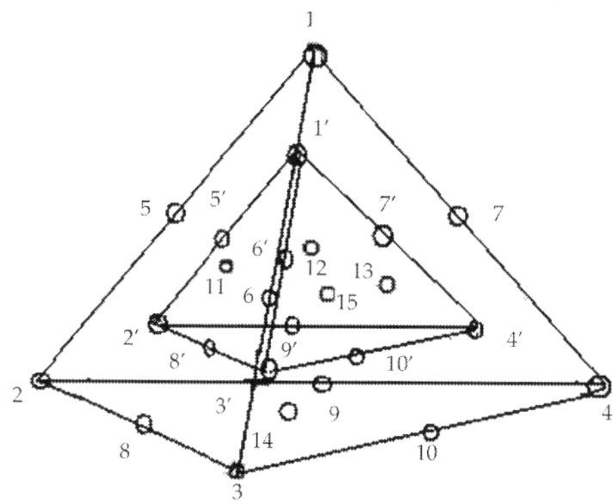

Figure. 3: Simplex-centroid design and shrinkage simplex-centroid design for $q = 4$ with $s = 0.05$ (simple number shows runs in simplex-centroid design and the numbers with primes show runs in shrinkage simplex-centroid design)

Next, we consider four components' quadratic Scheffé mixture model:

$$E(y) = \beta_1 x_1 + \beta_2 x_2 + \beta_3 x_3 + \beta_4 x_4 + \beta_{12} x_1 x_2 + \beta_{13} x_1 x_3 \\ + \beta_{14} x_1 x_4 + \beta_{23} x_2 x_3 + \beta_{24} x_2 x_4 + \beta_{34} x_3 x_4. \quad (6)$$

The value of *D* for simplex-centroid design, given in Table 2, for four components' quadratic Scheffé's model is 0.2322. Now, shrink design in Table 2 is towards the centroid, with the shrinkage parameter $s = 0.05$ and re-parameterizes the co-ordinates using Eq. (2). The resulting shrinkage simplex-centroid design is given as follows.

The value of D for shrinkage simplex-centroid design, given in Table 6, for quadratic Scheffé's model is 0.1795. Using D-efficiency criterion given in Eq. (3), the efficiency of the above design is 77.3%. Tables 7 and 8 provide D- and G-efficiencies, respectively, for shrinkage simplex-centroid designs with several values of s.

Table 6 Shrinkage simplex-centroid design for four components with s = 0.05

Run	x_1	x_2	x_3	x_4
1	0.9625	0.0125	0.0125	0.0125
2	0.0125	0.9625	0.0125	0.0125
3	0.0125	0.0125	0.9625	0.0125
4	0.0125	0.0125	0.0125	0.9625
5	0.4875	0.4875	0.0125	0.0125
6	0.4875	0.0125	0.4875	0.0125
7	0.4875	0.0125	0.0125	0.4875
8	0.0125	0.4875	0.4875	0.0125
9	0.0125	0.4875	0.0125	0.4875
10	0.0125	0.0125	0.4875	0.4875
11	0.32915	0.32915	0.3215	0.0125
12	0.32915	0.32915	0.0125	0.32915
13	0.32915	0.0125	0.32915	0.32915
14	0.0125	0.32915	0.32915	0.32915
15	0.2500	0.2500	0.2500	0.2500

Table 7 *D*-efficiency of shrinkage simplex-centroid designs for $q = 4$

s	$\|X'X\|^{1/p}$	$\|X'X\|_0^{1/p}$	D-efficiency
0.00	0.232169	0.232169	100
0.05	0.179471	0.232169	77.3
0.10	0.169251	0.232169	73.0

Table 8 *G*-efficiency of shrinkage simplex-centroid designs for $q = 4$

s	APV	MPV	G-efficiency
0.00	0.667	0.978	68.2
0.05	0.667	0.977	68.2
0.10	0.667	0.977	68.2

Hence, comparing the design in Table 6 with the simplex-centroid design in Table 2, the constructed design is a true mixture design, containing full mixture blends with some compromise on its efficiency.

CONSTRUCTION OF COMPONENT-AMOUNT DESIGN VIA PROJECTION OF SHRINKAGE SIMPLEX-CENTROID DESIGN

A mixture-amount experiment is a type of mixture experiment that is performed at two or more levels of total amount. The response is assumed to be dependent upon the individual proportions of components in the blend and also on its amount. The effect on the response after varying mixture component proportions and the total amount of the mixture is measured by fitting a mixture-amount model to the design. The design for fitting mixture-amount model is called mixture-amount design, developed by Piepel and Cornell (1985). Piepel (1988) modified that model to accommodate zero-amount condition. The alternative model uses the actual amounts of the ingredients denoted by $a_1, a_2, \ldots, a_q$, such that $a_1 + a_2 + \cdots + a_q = A$. The proportions x_i are related to the amount a_i through $x_i = a_i/A$. This is called component-amount model:

$$E(y) = \alpha_0 + \sum_{i=1}^{q} \alpha_i a_i + \sum_{i=1}^{q} \left(\alpha_{ii} a_i^2 + \sum_{i(j}^{q} \alpha_{ij} a_i a_j \right). \quad (7)$$

Prescott and Draper (2004, 2008) discussed the construction of designs for component-amount models by projecting standard symmetric mixture designs (simplex-lattice and simplex-centroid designs) in lower dimensions. The method involved the collapsing of the standard symmetric mixture designs by the removal of their one or more columns. It leads to a set of symmetric designs for each of the several levels of the amount A. We construct component-amount designs by the projection of shrinkage simplex-centroid designs, as given in "Shrinkage design".

The component-amount design formed by Prescott and Draper (2004, 2008), through projection of simplex-centroid design, given in Table 2, is provided in Table 9.

The above component-amount design has the value $D = 0.21825$ and its G-efficiency is 68.2%. The design is composed of incomplete mixture blends with different levels of amounts except the centroid point as a complete mixture blend. For real-life situations, we need component-amount designs which have full mixture blends with at least a minimum amount of each component in the mixture. Such designs can be obtained by the projection of shrinkage simplex-centroid designs.

Table 9 Simplex-centroid design for four components

Run	a_1	a_2	a_3	A
1	1	0	0	1
2	0	1	0	1
3	0	0	1	1
4	0	0	0	0
5	1/2	1/2	0	1
6	1/2	0	1/2	1
7	1/2	0	0	1/2
8	0	1/2	1/2	1
9	0	1/2	0	1/2
10	0	0	1/2	1/2
11	1/3	1/3	1/3	1
12	1/3	1/3	0	2/3
13	1/3	0	1/3	2/3
14	0	1/3	1/3	2/3
15	1/4	1/4	1/4	3/4

We project the design, given in Table 6, to lower dimension by deleting any column. Let the column for x_4 in Table 6 is deleted.

Table 10 Shrinkage simplex-centroid component-amount design for three components

Run	a_1	a_2	a_3	A
1	0.9625	0.0125	0.0125	0.9875
2	0.0125	0.9625	0.0125	0.9875
3	0.0125	0.0125	0.9625	0.9875
4	0.0125	0.0125	0.0125	0.0375
5	0.4875	0.4875	0.0125	0.9875
6	0.4875	0.0125	0.4875	0.9875
7	0.4875	0.0125	0.0125	0.5125
8	0.0125	0.4875	0.4875	0.9875
9	0.0125	0.4875	0.0125	0.5125
10	0.0125	0.0125	0.4875	0.5125
11	0.32915	0.32915	0.32915	0.9875
12	0.32915	0.32915	0.0125	0.6708
13	0.32915	0.0125	0.32915	0.6708
14	0.01250	0.32915	0.32915	0.6708
15	0.2500	0.2500	0.2500	0.7500

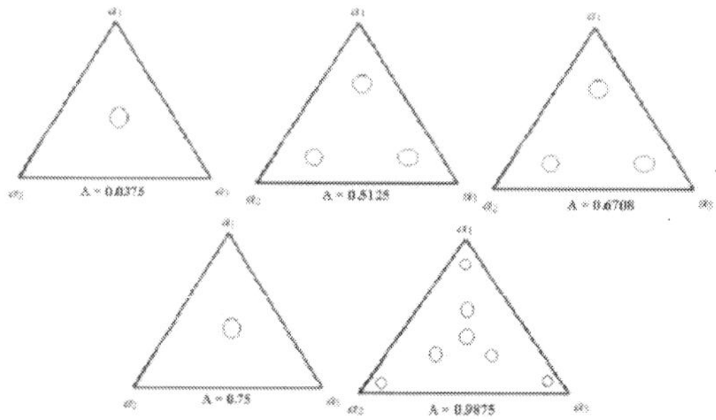

Figure. 4: Structure for the component-amount design in a_1, a_2, and a_3 simplex, formed by the projection of shrinkage simplex-centroid design with the shrinkage parameter $s = 0.05$

The resulting component-amount design in three components has the levels of amount A = 0.0375, 0.5125, 0.6708, 0.9875, and 0.7500 with replicates 1, 3, 3, 7, and 1 and is given in Table 10. This shrinkage component-amount design has the D-efficiency 91.2% and the G-efficiency design 68.2%. On comparison of this design with the design in Table 9, it can be noted that the constructed design is composed of full mixture blends, although it has less efficiency (Fig. 4).

EXAMPLE: DIAZEPAM SOLUBILITY EXPERIMENT

We take an example, given in Smith (Smith 2005, p. 54). Belloto et al. (1985) were interested in studying the solubility of drug diazepam in mixtures of ethanol, propylene glycol, and water. It was desired to develop a mixture model that would predict the diazepam in any mixture of solvents in a three-dimensional simplex. We first choose simplex-centroid design given in Table 1 and augment it with two replicates at each of the pure blends and with three axial check blends.

The G-efficiency of the design is 63.14%. We shrink augmented simplex-centroid design, given in Table 11, using the shrinkage parameter $s = 0.1$. The G-efficiency of the design is again 63.14%, while its D-efficiency is 74.4%. Therefore, mixture design with full mixture blends can be used to develop a mixture model with the stable G-efficiency that would predict the diazepam in any mixture of solvents in a three-dimensional simplex (Fig. 5).

Table 11 Augmented simplex-centroid design for three components

Run	x_1 Ethanol	x_2 Glycol	x_3 Water	y Solubility (mg/ml)
1	1	0	0	27.8
2	0	1	0	7.42
3	0	0	1	0.048
4	1/2	1/2	0	27.0
5	1/2	0	1/2	6.02
6	0	1/2	1/2	0.61
7	4/6	1/6	1/6	28.0
8	1/6	4/6	1/6	13.0
9	1/6	1/6	4/6	0.408
10	1/3	1/3	1/3	9.52
11	0	1	0	7.42
12	1	0	0	27.8
13	0	0	1	0.048

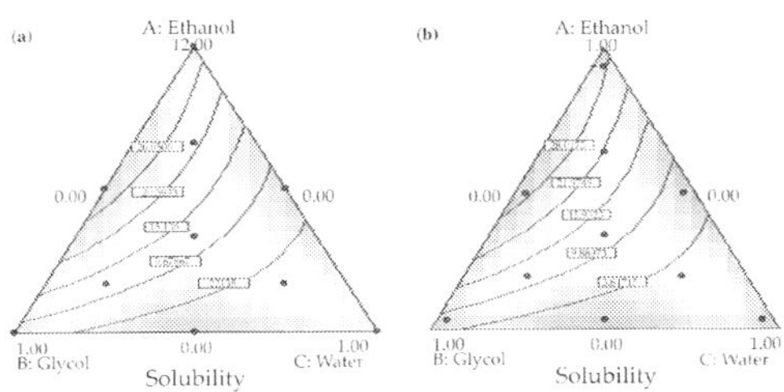

Figure. 5: **a** Contour plot of predicted solubility using the design in Table 11 and **b** contour plot of predicted solubility using shrinkage design with parameter $s = 0.1$

CONCLUSION

Simplex-centroid design is a widely used mixture design for quadratic mixture model. This classical design does not contain full mixture blends except centroid. In real-life situation, a mixture always has a least proportion of every component. In the current study, shrinkage designs are constructed from simplex-centroid designs. The proposed shrinkage simplex-centroid designs are the real-life mixture designs, although they have less efficiency as compared to the simplex-centroid designs available in the literature. This stability in *G*-efficiency of these designs shows that the prediction capability of the model does not change by shrinking conventional simplex-centroid design towards the centroid. Therefore, compromising on the loss in D-optimality of simplex-centroid design, mixture designs with full mixture blends and with the stable G-optimality can be constructed.

REFERENCES

1. Aggarwal ML, Singh P, Sarin V, Husain B (2011) Nearly optimal orthogonally blocked designs for four mixture components in three and four components based on F-squares. Commun Stat Simul 40(1):165–183
2. Belloto RJ, Dean AM, Moustafa MA, Molokhia AM, Gouda MW, Sokoloski TD (1985) Statistical techniques applied to solubility predictions and pharmaceutical formulations: an approach to problem solving using mixture response surface methodology. Int J Pharm 23:195–207
3. Hasan T, Khan MF (2011) Nearly A- and E-optimal orthogonally blocked designs for Scheffé's quadratic mixture model with three components. Proc Pak Acad Sci 48(2):101–106
4. Hasan T, Khan MF (2012) Nearly optimal orthogonally blocked designs for second degree K-model in three components. Pak J Stat 28(2):177–187
5. Piepel GF (1988) A note on models for mixture-amount experiments when the total amount takes a zero value. Technometrics 30:449–450
6. Piepel GF, Cornell JA (1985) Models for mixture experiments when the response depends on the total amount. Technometrics 27:219–227
7. Prescott P (1998) Nearly optimal orthogonally blocked designs for a quadratic mixture model with *q* components. Commun Stat Theory Methods 27(10):2259–2580
8. Prescott P, Draper NR (2004) Mixture component-amount designs via projections, including orthogonally blocked designs. J Qual Technol 36:413–431
9. Prescott P, Draper NR (2008) D-optimal component-amount designs for quadratic and cubic models. J Appl Stat 35(7):739–749
10. Scheffé H (1963) Simplex-centroid designs for experiments with mixtures. J R Stat Soc B 25:235–263
11. Smith WF (2005) Experimental design for formulation. ASA-SIAM Series on statistics and applied probability. SIAM/American Statistical Association, Philadelphia/Alexandria

CHAPTER

14

Determining Genetic Causal Variants Through Multivariate Regression Using Mixture Model Penalty

V. S. Sundar,[1,2] Chun-Chieh Fan,[1,3] Dominic Holland,[1,4] and Anders M. Dale[1,2,4,5]

[1]Center for Multimodal Imaging and Genetics, University of California, San Diego, La Jolla, CA, United States

[2]Department of Radiology, University of California, San Diego, La Jolla, CA, United States

[3]Department of Cognitive Sciences, University of California, San Diego, La Jolla, CA, United States

[4]Department of Neuroscience, University of California, San Diego, La Jolla, CA, United States

Citation: (APA): Sundar, V. S., Fan, C. C., Holland, D., & Dale, A. M. (2018). Determining genetic causal variants through multivariate regression using mixture model penalty. Frontiers in genetics, 9, 77. (8 pages) DOI:https://doi.org/10.3389/fgene.2018.00077, URL: https://www.frontiersin.org/articles/10.3389/fgene.2018.00077/full

[5]Department of Psychiatry, University of California, San Diego, La Jolla, CA, United States

ABSTRACT

With the availability of high-throughput sequencing data, identification of genetic causal variants accurately requires the efficient incorporation of function annotation data into the optimization routine. This motivates the need for development of novel methods for genome wide association studies with special focus on fine-mapping capabilities. A penalty function method that is simple to implement and capable of integrating functional annotation information into the estimation procedure, is proposed in this work. The idea is to use the prior distribution of the effect sizes explicitly as a penalty function. The estimates obtained are shown to be better correlated with the true effect sizes (in comparison with a few existing techniques). An increase in the positive and negative predictive value is demonstrated using Hapgen2 simulated data.

INTRODUCTION

Detection and estimation of the genetic causal variants associated with a particular phenotypic trait is one of the most challenging problems in modern day statistical genetics. Mathematical techniques are formulated with primary focus on fine-mapping studies, phenotype prediction, and heritability estimation (Servin and Stephens, 2007; Lee et al., 2009; Gaffney et al., 2012; Maller et al., 2012; Valdar et al., 2012; Zuber et al., 2012; de los Campos et al., 2013; International Multiple Sclerosis Genetics Consortium et al., 2013; Zhou et al., 2013; Mahajan et al., 2014; Pickrell, 2014; Spain and Barrett, 2015; Schweiger et al., 2016). Algorithms that integrate functional annotation data into the estimation procedure (Schork et al., 2013; Zhou et al., 2013; Kichaev et al., 2014; Zablocki et al., 2014; Vilhjálmsson et al., 2015) are being continually developed with the understanding that Linkage Disequilibrium (LD) and polygenicity reduces the likelihood of the identified genetic variant being biologically causal (Visscher et al., 2012). The resulting procedures have better fine-mapping and effect size estimation capabilities (Kichaev et al., 2014; Kichaev and Pasaniuc, 2015).

While fine-mapping studies focuses on detecting causal variants, regression, or Bayesian optimization methods integrate these fine-mapping results into the estimation procedure to accurately determine the effect sizes. Currently, fine-mapping studies either use summary statistics or raw genotype data to arrive at quantitative assessment of causal nature of the SNPs. For example, CAVIAR (Hormozdiari et al., 2014, 2015), PAINTOR (Kichaev et al., 2014), and RiVIERA (Li and Kellis, 2016) use summary statistics, and DAP (Wen et al., 2016), CavMeN (Brown et al., 2017) use the raw genotype data. End results of these analyses, typically probability of the SNP being causal, are used in target gene identification studies.

With the understanding that GWAS significant SNPs harbor more than one causal variant, few researchers have attempted to utilize multivariate methods to detect additional association signals (Newcombe et al., 2016; Ning et al., 2017) from summary statistics. Newcombe et al. (2016) utilized the correlation structure of the variants from the reference panel to develop a Bayesian regression framework that accounts for various models with respect to the number of causal SNPs per region. Ning et al. (2017) used the covariance structure between the variants, and between the variant and phenotype vector to obtain LASSO results for a series of λ›s (regularization parameter). These studies demonstrate the improvement achieved through additional analysis on already identified potential causal SNPs. The current line of work follows a similar strategy wherein prior information regarding the causal nature of the SNPs (in terms of p-values or posterior probabilities) are used to localize additional causal variants (using raw genotype data) that might have been gone undetected due to their small effect size, lower posterior probability, or possibly due to small sample size.

Frequently, a two-mixture model, one each to represent the causal and null SNPs, is used to model the effect size distribution obtained using Genome Wide Association Studies (GWAS) (Meuwissen et al., 2001; Wray et al., 2007; Bukszár et al., 2009; Logsdon et al., 2010; Park et al., 2010, 2011; Yang et al., 2010, 2011; Guan and Stephens, 2011; Habier et al., 2011; Xu et al., 2011; Speed et al., 2012; Zhou et al., 2013; Holland et al., 2016). Accurate identification of causal and null SNPs helps in understanding the underlying biological pathway regulating a disease (Sun et al., 2006; Yoo et al., 2009). Integrating functional annotation data into the estimation procedure is one way of improving the identifiability of causal SNPs (Schork et al., 2013; Kichaev et al., 2014; Zablocki et al., 2014; Kichaev and Pasaniuc, 2015; Vilhjálmsson et al., 2015).

The currently available methods for variable selection and estimating the effect sizes can be broadly categorized into Bayes' theorem based or penalty function based. Bayesian methods proceed by assigning a prior probability density function (pdf) to effect sizes and use either maximum likelihood estimation method or Markov Chain Monte Carlo (MCMC) simulations to determine the posterior effect sizes (effect sizes conditioned on the measured phenotypic data). The various methods that fall under this category differ in the specification of the prior pdf (Meuwissen et al., 2001; Habier et al., 2011; Zhou et al., 2013). Some of them are the Bayesian alphabet models (BayesA, BayesB, BayesC, BayesCπ, BayesR, etc.), and Bayesian Sparse Linear Mixture Model (BSLMM). Regression methods, on the other hand, aim to minimize an objective function with a penalty term, which is chosen to impart sparse characteristics to the effect size estimates. For example, the least angle absolute shrinkage operator (LASSO) uses a L_1 penalty (Tibshirani, 1996), and the Ridge Regression (RR) uses the L_2 penalty (Hoerl and Kennard, 1970). Methods that are a combination of either Bayesian and regression methods (Bayesian LASSO; Park and Casella, 2008; Li et al., 2010) or two regression based methods (Elastic Net; Zou and Hastie, 2005) have also been developed.

While in the Bayesian methods, the prior probabilities aid in variable selection, the shrinkage constraints does the equivalent job in regression based methods. Both the Bayesian and regression methods are geared toward accurate identification of the causal variants and phenotypic prediction. A review of the currently available methods can be found in Zhou et al. (2013) and de los Campos et al. (2013). The Bayesian methods though are mostly independent of tuning parameters, suffer from practical applicability to large datasets (in terms of efficient effect size estimation).

In this work, we formulate a simple and efficient optimization routine which combines the flexibility of Bayesian methods and simplicity of penalty function methods into a single framework. The idea is to use the prior pdf of the effect sizes explicitly as a penalty function. The motivation of the paper is to introduce the method, provide details regarding the implementation of the procedure, and demonstrate its various capabilities. At this stage, theoretically, we do not claim superiority over existing methods developed for effect size estimation and phenotype prediction.

PROBLEM STATEMENT

Considering N individuals, n genetic markers, and a linear model; the $N \times 1$ phenotype vector $\boldsymbol{y}$ is related to $N \times n$ genotype matrix $\mathbf{X}$ through the $n \times 1$ vector of effect sizes β as (Meuwissen et al., 2001):

$$y = X\beta + \varepsilon \qquad (1)$$

Here ε is the vector of noise terms modeled as $N(0, \Sigma_\varepsilon)$. Elements of **X** are typically coded as 0, 1, or 2 (prior to normalization). We aspire to select the causal variants and determine their effect sizes, β^ such that $\varepsilon = (\mathbf{X}\beta - y)^T (\mathbf{X}\beta - y)$ is minimun. Due to the correlated and sparse nature of the SNPs, the univariate results often end up being erroneous estimates (Kim et al., 2009; de los Campos et al., 2013; Zhou et al., 2013).

This resulted in the evolution of multivariate methods for determining the causal variants (see, for example, Hoerl and Kennard, 1970; Tibshirani, 1996; Zou and Hastie, 2005; Kim et al., 2009; de los Campos et al., 2013; Zhou et al., 2013).

METHODS

Regression methods

The Elastic Net (EN) (Zou and Hastie, 2005) provides the most generalized representation of the commonly used objective functions in penalty methods, and is given as:

$$F_{EN} = (\mathbf{X}\beta - y)^T(\mathbf{X}\beta - y) + \lambda \sum_{j=1}^{n} \left[\frac{1}{2}(1-\alpha)\beta_j^2 + \alpha|\beta_j| \right];$$
$$\lambda > 0; 0 < \alpha \leq 1 \quad (2)$$

In the above expression, α = 1 corresponds to the LASSO (Tibshirani, 1996), and α = 0 results in the Ridge regression (RR) (Hoerl and Kennard, 1970). There exists several algorithms that minimizes the above objective function while efficiently determining the regularization parameters (Tibshirani, 1996; Fu, 1998; Efron et al., 2004; Zou and Hastie, 2005; Park and Casella, 2008; Wu and Lange, 2008). A variant of the LASSO, termed group LASSO (Meier et al., 2008) employs multiple regularization parameters to different groups of SNPs.

Bayesian methods

Modeling β as a mixture pdf, $\beta \sim \pi_1 p_1(\beta) + (1 - \pi_1)p_0(\beta)$ (Habier et al., 2011; de los Campos et al., 2013; Zhou et al., 2013), with $p_1(\beta)$ denoting pdf of the causal SNPs and $p_0(\beta)$ denoting the pdf of the null SNPs, the posterior pdf of β is given as

$$p(\beta|y) = Kp(y|\beta)p(\beta) \quad (3)$$

with $K^{-1} = \int_{-\infty}^{\infty} p(y|\beta)p(\beta)d\beta$. The estimate of the effect sizes is determined as $\langle\beta|\boldsymbol{y}\rangle$, where $\langle\bullet\rangle$ denotes the mathematical expectation operator. Irrespective of the distribution of β, the likelihood function, $p(\boldsymbol{y}|\beta)$ can be shown to be Normal with mean $\mathbf{X}\beta$, and variance Σ_ε (Robert, 2004). Existing variants of the Bayesian methods could be obtained by changing the pdfs $p_1(\beta)$ and $p_0(\beta)$ (de los Campos et al., 2013; Zhou et al., 2013). Supplementary Material provides information on the equivalence between Bayesian and regression based methods for a few priors.

Mixture model penalty method

We design a penalty function that intuitively accomplishes shrinking the regression coefficients while incorporating any prior information about the causal nature of the SNPs. The motivation for this penalty function stems from the understanding that the effect sizes can be realistically represented using a multimodal pdf (Meuwissen et al., 2001; Bukszár et al., 2009; Logsdon et al., 2010; Guan and Stephens, 2011; Habier et al., 2011; Yang et al., 2011; Zhou et al., 2013; Holland et al., 2016) and functional annotations help in classifying the SNPs as either being causal or not (Schork et al., 2013). In our formulation, the main error minimizing term is the negative log-likelihood function, and a mixture prior cost function imparts the necessary sparsity to the effect size estimates. Several researchers in the genetics community have used the Spike and Slab pdf (Ishwaran and Rao, 2005) as prior pdf of effect sizes (de los Campos et al., 2013; Zhou et al., 2013). However, using this pdf explicitly as a penalty function has not been attempted in genetic association studies. This also sidesteps the computationally expensive Markov Chain Monte Carlo (MCMC) method used for obtaining posterior effect size estimates.

The likelihood function, $p(\boldsymbol{y}|\beta) \sim N(\mathbf{X}\beta, \Sigma_\varepsilon)$ is expressed as

$$p(y|\beta) = \frac{1}{(2\pi)^{n/2}|\Sigma_\varepsilon|^{1/2}} e^{-\frac{1}{2}(y-\mathbf{X}\beta)^T \Sigma_\varepsilon^{-1}(y-\mathbf{X}\beta)} \tag{4}$$

We construct a cost function, C that has the ability to capture the causal nature of SNPs:

$$C = \sum_{j=1}^{n} c_j(\hat{\beta}_j) \tag{5}$$

where the cost associated with the jth SNP is given as

$$c_j(\beta_j) = -\log\left[\tilde{\pi}_{1j}p_{1j}(\beta_j) + (1-\tilde{\pi}_{1j})p_{0j}(\beta_j)\right] \tag{6}$$

Here $\tilde{\pi}_1 = [\tilde{\pi}_{11}, \tilde{\pi}_{12}, \cdots, \tilde{\pi}_{1n}]^T$ is the $n \times 1$ vector of non-null prior probabilities associated with the functional annotation of the SNPs. That is, if the jth SNP is highly likely to be causal, then a higher value (say 0.5) is specified to that SNP. $p_{1j}(\bullet)$ and $p_{0j}(\bullet)$ denote the pdf of causal and null SNPs, respectively. The cost function, thus, acts as a medium to incorporate the enrichment details of individual SNPs. Typically, we use a normal pdf, $\phi_{1j}(\bullet; 0; \tilde{\sigma}_{1j})$, to model the causal effects. Note that $\tilde{\pi}_1$ denotes the assumed prior probability and π_1 denotes the true unknown probability. Denoting by L the negative log-likelihood function, the function to be minimized is written as

$$F = L + C \tag{7}$$

with

$$L = -\log\left[(2\pi)^{-n/2}|\tilde{\Sigma}_\varepsilon|^{-1/2}\right] + \frac{1}{2}(y - \mathbf{X}\beta)^T\tilde{\Sigma}_\varepsilon^{-1}(y - \mathbf{X}\beta) \tag{8}$$

The nonlinear conjugate gradient method (NCG) (Hestenes and Stiefel, 1952; Fletcher and Reeves, 1964; Polak and Ribiere, 1969; Shewchuk, 1994; Dai and Yuan, 1999; Hager and Zhang, 2006) with Newton-Raphson line search algorithm is used to minimize F. A step-wise implementation of the optimization procedure is given in Supplementary Material. We show that when the NCG method is used for optimization, the evaluation of whole $n \times n$ Hessian matrix can be avoided. This significantly reduces the computational cost whilst not compromising the accuracy of the solution (Equation S12).

Remarks

1. The cost function shown in Equation (6) is conceptually similar to the penalty function proposed by Ročková and George (2016). The authors, using a mixture Laplace pdf, estimate the prior probability of the effects using a coordinate-wise optimization routine. We, however, specify different prior probabilities for each variant, so as to incorporate any information on LD or functional annotations. Furthermore, our motivation to use a mixture model stems from our understanding of the genetic architecture of the human genome.

2. The likelihood ratio and the cost function are weighed equally, so that the minimum error solution is sparse. Unequal weights can be specified, say higher for L if it is known that the genetic architecture is highly polygenic, and low if only a few genetic causal variants influence the phenotype under consideration.
3. Variants of the proposed method could be obtained by changing the pdfs used in constructing the mixture model—for example, Laplace or non-local pdfs (Johnson and Rossell, 2010) could be used instead of two normal pdfs. These however, are minor modifications, and our main contribution lies in proposing an explicit mixture model pdf as a penalty function.
4. Instead of considering cost associate with individual SNPs, the SNPs can be clustered through specification of suitable correlation. This could possibly capture the underlying LD information. However, this requires incorporation of LD metrics such as r^2 into covariance structure of the clustered SNPs. Such studies have not been pursued in this present work, and will likely be a part of future efforts.
5. Higher prior probabilities can be specified to cluster of SNPs in a given LD block that is envisioned to contain the causal signal. SNPs not in this LD block may be provided lower or zero prior probability.
6. SNPs that belong to certain functional annotation category have higher likelihood of being causal. Hence, SNPs in these regions are deemed to be enriched, i.e., have higher probability of influencing a particular phenotype. Typically SNPs tagging regulatory and coding regions are considered to be enriched in comparison with introns and intergenic SNPs (Schork et al., 2013). SNPs in the MHC region can be considered to be enriched when studying immune related diseases (Ellinghaus et al., 2016).
7. Using existing packages such as CAVIAR, DAP, PAINTOR, RiVIERA, and S-LDSC (Finucane et al., 2015), one could obtain a quantitative assessment of the causal nature of individual SNPs. These results can be directly used as prior probabilities ($\pi\tilde{}1$) in the proposed optimization routine. Probabilities could also be based on GWAS p-values. However, these values tend to alter with increase in power.

8. As mentioned earlier, for each regression based method, there exists a Bayesian equivalent. In the Bayesian methods, assuming a prior pdf, samples are drawn from the posterior distribution using MCMC. The proposed method avoids sampling from the prior and posterior pdf of the effect sizes by specifying the prior information explicitly as a penalty function. This distinguishes the method from the Bayesian LASSO and BSLMM.
9. Fine-mapping methods typically require data from dense genotyping arrays, which are further imputed using reference panels, such as 1,000 Genomes (1000 Genomes Project Consortium et al., 2012). The mixture-model method, on the other hand, uses whole genome wide data to locate the causal signal. In this aspect, genotype data preferred for fine-mapping studies, may be unsuitable for the proposed method.

Simulation studies

Hapgen2 (Su et al., 2011) and 1,000 Genomes (1000 Genomes Project Consortium et al., 2012) is used for simulating realistic genotypes for an European population of size 100,000 considering all the 22 chromosomes (80378054 SNPs). True effect sizes are simulated based on the understanding that a proportion of the SNPs are causal with effect sizes distributed as $N(0, 1)$.

Whole genome analysis

For whole genome analysis, due to computational issues, the analysis can be carried out using SNP windows such that no two potential causal SNPs in LD are separated (Berisa and Pickrell, 2016). In this work, we consider SNPs associated with individual chromosomes in each sliding window. For each chromosome, the first 20,000 SNPs with minor allele frequency >0.01 are considered in the analysis, resulting in a total of 440,000 SNPs. The number of causal variants are taken to be 50% of the SNPs in the functional annotation category—Exon, 3′UTR, 5′UTR. This gives rise to 4,233 causal SNPs—approximately 1% of the total SNPs considered. Thus, SNPs belonging to these categories are considered to be enriched, i.e., have higher likelihood of being causal. Three different genotype matrices and three different true effect sizes are considered for the analysis, resulting in a total of 18 cases for estimating the normalized mean squared error (NMSE). The phenotype is simulated using Equation (1) with a heritability of 0.5.

Specifying prior probability to individual SNPs requires functional annotation information for the genotyped SNPs. Assuming such significant information is unavailable, we initially specify equal $\tilde{\pi}_1$ and $\tilde{\sigma}_1$ values for all the SNPs, i.e., $\tilde{\pi}_{1j} = \tilde{\pi}_1, \tilde{\sigma}_{1j} = 1, \forall j$—Mixture Model with Constant Priors (MM-CP). An estimate of Σ_ε for determining the likelihood function is obtained by assuming a heritability of 0.0227 per SNP window. For comparison, results are obtained using the Regularized Pseudo Inverse (RegPI)—analytical solution with $\tilde{\pi}_1 = \pi_1$ and no mixture model (Supplementary Material), LASSO, and univariate regression method. The enrichment factors used in simulating the data are specified as prior probabilities in the optimization routine, and the resulting estimates are denoted as Mixture-Model with Enriched Priors (MM-EP), that is, $\tilde{\pi}_1 = \pi_1$. The RegPI method and MM-EP methods differ only in the procedure followed to obtain the effect size estimates, and in principle, are equivalent. While the RegPI method provides a closed form solution for the effect sizes, the MM-EP method utilizes an optimization algorithm to achieve the same goal. As mentioned in section 3.3.1, results from a prior analysis (typically fine-mapping studies) could be used to improve the detection and estimation capabilities of the method. We use end results of DAP as prior probabilities and the resulting estimates are denoted as MM-DAP. It is to be noted that the MM-DAP method utilizes the genotype information twice, once for estimating the prior probability of causality (DAP), and once in our optimization routine (for effect size estimation and detection of additional causal variants). However, the context in which the information is used slightly differs. An adaptive/iterative method for estimating the causal probabilities could avoid this. We are working toward achieving this. Matlab implementation of the proposed method is included along with this paper. The implementation has provision for gradual increment or decrement of $\pi\tilde{}1$ and σ_0 values.

RESULTS

The correlation between the true and estimated effect sizes, the percentage of positive and negative predictive values (PPV, NPV) are used to measure the accuracy of different methods (Figure 1). PPV is defined as the ratio of number of true variants identified to the total number of variants identified. Similarly, NPV is defined as the ratio of number of true null SNPs identified to the total number of null SNPs identified. For the RegPI and MM-EP methods, $\tilde{\pi}_1 = \pi_1$, hence these methods provide an upper bound for both the effect size correlation, PPV and NPV. This can be considered as a case where complete genetic architecture of the effect size distribution is known. For the MM-CP method, $\tilde{\pi}_{1j}=0.01 \forall j$, and the null pdf is taken to be Laplacian. Specifying $\tilde{\pi}_{1j}=1$ is the special case of the infinitesimal model, where all the SNPs are assumed to be causal with Normal effect size distribution. The MM-DAP method used DAP results as prior probabilities for the genetic variants. This constitutes partial knowledge about the distribution of causal SNPs in the genome.

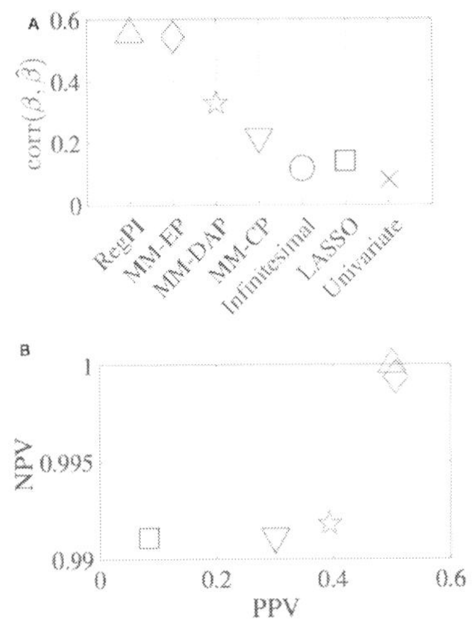

Figure 1: Estimates obtained using various methods. **(A)** correlation between the estimated and true effect sizes, **(B)** PPV and NPV. RegPI: Regularized pseudo inverse (green triangle); MM-EP: Mixture model with enriched priors (magenta diamond); MM-DAP: Mixture model with DAP priors (orange star); MM-CP: Mixture model with constant priors (red inverted triangle); Infinitesimal: Normal prior (no mixture) (blue circle); LASSO (black square); Univariate (brown cross).

The MM-CP, Infinitesimal, LASSO, and Univariate methods do not use functional annotation information. Thus any improvement in the effect size estimates obtained using MM-CP method, in comparison with the other three, is deemed significant. Though a sparse structure is imposed on the penalty function in the MM-CP method, the method essentially does not incorporate any enrichment factors. A slight improvement in the effect size correlation can be observed in Figure 1. The figure illustrates the advantage of the proposed formulation in terms of locating the causal variants accurately. The positive and negative predictive value for the Infinitesimal and Univariate methods are not shown in the figure. In obtaining the LASSO estimates, initially, a two-fold cross validation has been carried out for SNP window 1 (i.e., chromosome 1), resulting in λ = 0.508. NMSE estimates are obtained for a grid of values between 0.45 and 0.70 for all the other chromosomes, and the estimates corresponding to the minimum NMSE value is reported in Figure 1. The MM-DAP estimates lie between the MM-CP and MM-EP estimates, as the prior probabilities are based on a previous analysis which identifies few significant SNPs.

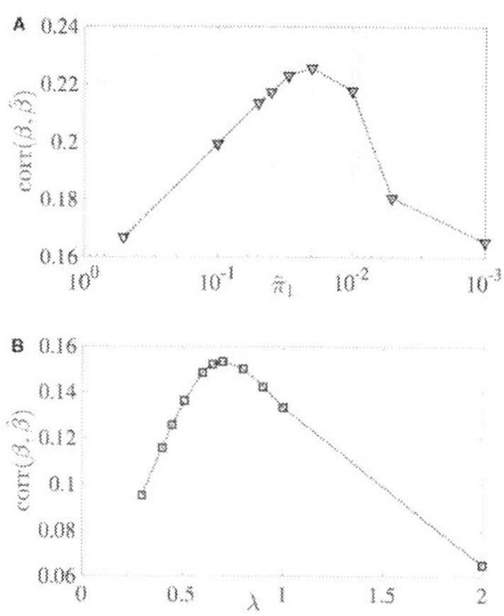

Figure 2: Variation in the correlation between estimated and true effect sizes. **(A)** MM-CP, **(B)** LASSO.

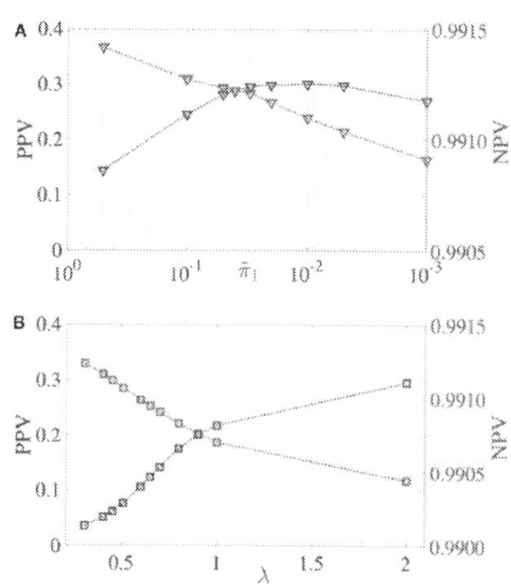

Figure 3: Variation of PPV and NPV values. **(A)** MM-CP, **(B)** LASSO.

The correlation between estimated and true effect sizes, PPV and NPV values have been obtained for a grid of $\tilde{\pi}_1$ values and plotted in Figures 2, 3, respectively. Similar study is carried out for LASSO (with respect to the regularization parameter λ). For the MM-CP method, the x-axis is plotted in the reverse direction so that moving along the x-axis toward right implies increase in sparseness (consistent with the x-axis of LASSO).

DISCUSSION

From Figures 2, 3, it can be observed that the MM-CP and LASSO estimates follow a similar trend with increase in sparseness. Enforcing a sparse structure with arbitrary prior probabilities for the effect sizes result in estimates that are better at localizing the genetic causal variants. A good understanding of the distribution of the SNPs across the genome helps in incorporating the functional annotations, thereby further improving the effect size estimates—RegPI and MM-EP methods. A sensible choice of the enrichment factors require prior knowledge about the phenotype under study. For example, SNPs in the MHC region are shown to have a larger impact on Ankylosing Spondylitis than the non MHC SNPs (Ellinghaus et al., 2016). In this case, one could provide a higher π_1 value for the SNPs in the MHC region (differential enrichment). The MM-DAP results are obtained using this strategy, i.e. use prior information to improve the performance of the estimation procedure.

The correlation plotted in Figure 2 reflects the accuracy of the estimation procedures, that is, how close the estimated effect size is to the true effect size (which is unknown). Here $\tilde{\pi}_1=1$ implies that all the SNPs (regression coefficients) contribute to the phenotype $\mathbf{y}$. This leads to the underestimation of the effect sizes due to the distribution of the true signal among several SNPs. The sparse representation, say $\tilde{\pi}_1=10^{-2}$ on the other hand has the advantage of distributing the total signal among few selected non-zero SNPs. Depending on the selected SNPs, the correlation between true and estimated SNPs may vary. For the infinitesimal case, though all the causal SNPs have been identified, will result in low correlation value, because the effect sizes of these causal variants have been underestimated. The same phenomenon can be observed when the infinitesimal model ($\tilde{\pi}_1=1$) is compared with LASSO in Figure 1.

It is straightforward to note that the amount of shrinkage achieved depends on the characteristics of the penalty function used. Using the mixture model pdf explicitly as a penalty function places a probabilistic sparse constraint on the effect sizes, as opposed to the distance based constraints used typical in penalty function based method. Specifying a sparse structure without any knowledge about the underlying genetic architecture (i.e., specifying an arbitrary π_1) is shown to equip the optimization routine with better fine-mapping capabilities, and result in estimates that are at least as good as the LASSO and Univariate methods.

The non-linear conjugate gradient method is used to solve the optimization problem efficiently by harnessing the structure of the objective function's Hessian matrix (Supplementary Material). Thus, the method in its current form can be applied to whole genome analysis without any difficulty. Application of the method to specialized chip sequenced data, say the Immunochip or Oncochip, requires a careful approach in specifying the $\tilde{\pi}_1$ values to various SNPs. We are working toward developing methodologies to automatically determine the probability of SNP association, and SNP correlation with in the optimization framework. Alternately, end results from existing fine-mapping studies such as DAP or PAINTOR can be used as prior probabilities. Therefore, the method needs to be viewed as an efficient optimization algorithm capable of integrating functional annotation data (if available). Interpretation of the framework as a means to incorporate functional annotation and LD information, while at the same time achieving good variable selection and effect size estimation capabilities are some of the features we believe are important to the genetics community.

Author contributions

VS: Wrote manuscript, performed analyses, contributed to study design, interpretation of results, and critical revision of manuscript; C-CF and DH: Contributed to data preparation, interpretation of results, and critical revision of manuscript; AD: Performed analyses, contributed to study design, interpretation of results, and critical revision of manuscript.

Conflict of interest statement

The authors declare that the research was conducted in the absence of any commercial or financial relationships that could be construed as a potential conflict of interest.

Footnotes

Funding. National Institute of Health NIDA-ABCD-USA Consortium (5U24DA041123).

REFERENCES

1. 1000 Genomes Project Consortium. Abecasis G. R., Auton A., Brooks L. D., DePristo M. A., Durbin R. M., et al. . (2012). An integrated map of genetic variation from 1,092 human genomes. Nature 491, 56–65. 10.1038/nature11632
2. Berisa T., Pickrell J. K. (2016). Approximately independent linkage disequilibrium blocks in human populations. Bioinformatics 32, 283–285. 10.1093/bioinformatics/btv546
3. Brown A. A., Viñuela A., Delaneau O., Spector T. D., Small K. S., Dermitzakis E. T. (2017). Predicting causal variants affecting expression by using whole-genome sequencing and rna-seq from multiple human tissues. Nat. Genet. 49:1747. 10.1038/ng.3979
4. Bukszár J., McClay J. L., van den Oord E. J. (2009). Estimating the posterior probability that genome-wide association findings are true or false. Bioinformatics 25, 1807–1813. 10.1093/bioinformatics/btp305
5. Dai Y.-H., Yuan Y. (1999). A nonlinear conjugate gradient method with a strong global convergence property. SIAM J. Optimiz. 10, 177–182.
6. de Los Campos G., Hickey J. M., Pong-Wong R., Daetwyler H. D., Calus M. P. (2013). Whole-genome regression and prediction methods applied to plant and animal breeding. Genetics 193, 327–345. 10.1534/genetics.112.143313
7. Efron B., Hastie T., Johnstone I., Tibshirani R. (2004). Least angle regression. Ann. Stat. 32, 407–499. 10.1214/009053604000000067
8. Ellinghaus D., Jostins L., Spain S. L., Cortes A., Bethune J., Han B., et al. . (2016). Analysis of five chronic inflammatory diseases identifies 27 new associations and highlights disease-specific patterns at shared loci. Nat. Genet. 48, 510–518. 10.1038/ng.3528
9. Finucane H. K., Bulik-Sullivan B., Gusev A., Trynka G., Reshef Y., Loh P.-R., et al. . (2015). Partitioning heritability by functional annotation using genome-wide association summary statistics. Nat. Genet. 47, 1228–1235. 10.1038/ng.3404
10. Fletcher R., Reeves C. M. (1964). Function minimization by conjugate gradients. Comput. J. 7, 149–154. 10.1093/comjnl/7.2.149
11. Fu W. J. (1998). Penalized regressions: the bridge versus the lasso. J. Comput. Graph. Stat. 7, 397–416.
12. Gaffney D. J., Veyrieras J.-B., Degner J. F., Pique-Regi R., Pai A. A., Crawford G. E., et al. . (2012). Dissecting the regulatory architecture

of gene expression QTLs. Genome Biol. 13:R7. 10.1186/gb-2012-13-1-r7

13. Guan Y., Stephens M. (2011). Bayesian variable selection regression for genome-wide association studies and other large-scale problems. Ann. Appl. Stat. 5, 1780–1815. 10.1214/11-AOAS455
14. Habier D., Fernando R. L., Kizilkaya K., Garrick D. J. (2011). Extension of the bayesian alphabet for genomic selection. BMC Bioinformatics 12:186. 10.1186/1471-2105-12-186
15. Hager W. W., Zhang H. (2006). A survey of nonlinear conjugate gradient methods. Pac. J. Optim. 2, 35–58.
16. Hestenes M. R., Stiefel E. (1952). Methods of conjugate gradients for solving linear systems. J. Res. Natl. Bureau Stand. 49, 409–436.
17. Hoerl A. E., Kennard R. W. (1970). Ridge regression: biased estimation for nonorthogonal problems. Technometrics 12, 55–67.
18. Holland D., Wang Y., Thompson W. K., Schork A., Chen C.-H., Lo M.-T., et al. . (2016). Estimating effect sizes and expected replication probabilities from gwas summary statistics. Front. Genet. 7:15. 10.3389/fgene.2016.00015
19. Hormozdiari F., Kichaev G., Yang W.-Y., Pasaniuc B., Eskin E. (2015). Identification of causal genes for complex traits. Bioinformatics 31, i206–i213. 10.1093/bioinformatics/btv240
20. Hormozdiari F., Kostem E., Kang E. Y., Pasaniuc B., Eskin E. (2014). Identifying causal variants at loci with multiple signals of association. Genetics 198, 497–508. 10.1534/genetics.114.167908
21. International Multiple Sclerosis Genetics Consortium (IMSGC) Beecham A. H., Patsopoulos N. A., Xifara D. K., Davis M. F., Kemppinen A., et al. . (2013). Analysis of immune-related loci identifies 48 new susceptibility variants for multiple sclerosis. Nat. Genet. 45, 1353–1360. 10.1038/ng.2770
22. Ishwaran H., Rao J. S. (2005). Spike and slab variable selection: frequentist and bayesian strategies. Ann. Stat. 33, 730–773. 10.1214/009053604000001147
23. Johnson V. E., Rossell D. (2010). On the use of non-local prior densities in bayesian hypothesis tests. J. R. Stat. Soc. Ser. B 72, 143–170. 10.1111/j.1467-9868.2009.00730.x
24. Kichaev G., Pasaniuc B. (2015). Leveraging functional-annotation data in trans-ethnic fine-mapping studies. Am. J. Hum. Genet. 97, 260–271.

10.1016/j.ajhg.2015.06.007

25. Kichaev G., Yang W.-Y., Lindstrom S., Hormozdiari F., Eskin E., Price A. L., et al. . (2014). Integrating functional data to prioritize causal variants in statistical fine-mapping studies. PLoS Genet. 10:e1004722. 10.1371/journal.pgen.1004722
26. Kim S., Sohn K.-A., Xing E. P. (2009). A multivariate regression approach to association analysis of a quantitative trait network. Bioinformatics 25, i204–i212. 10.1093/bioinformatics/btp218
27. Lee S.-I., Dudley A. M., Drubin D., Silver P. A., Krogan N. J., Pe'er D., et al. . (2009). Learning a prior on regulatory potential from eQTL data. PLoS Genet. 5:e1000358. 10.1371/journal.pgen.1000358
28. Li J., Das K., Fu G., Li R., Wu R. (2010). The bayesian lasso for genome-wide association studies. Bioinformatics 27, 516–523. 10.1093/bioinformatics/btq688
29. Li Y., Kellis M. (2016). RiVIERA-MT: a bayesian model to infer risk variants in related traits using summary statistics and functional genomic annotations. bioRxiv 059345. 10.1101/059345
30. Logsdon B. A., Hoffman G. E., Mezey J. G. (2010). A variational bayes algorithm for fast and accurate multiple locus genome-wide association analysis. BMC Bioinformatics 11:58. 10.1186/1471-2105-11-58
31. Mahajan A., Go M. J., Zhang W., Below J. E., Gaulton K. J., Ferreira T., et al. . (2014). Genome-wide trans-ancestry meta-analysis provides insight into the genetic architecture of type 2 diabetes susceptibility. Nat. Genet. 46, 234–244. 10.1038/ng.2897
32. Maller J. B., McVean G., Byrnes J., Vukcevic D., Palin K., Su Z., et al. . (2012). Bayesian refinement of association signals for 14 loci in 3 common diseases. Nat. Genet. 44, 1294–1301. 10.1038/ng.2435
33. Meier L., Van De Geer S., Bühlmann P. (2008). The group lasso for logistic regression. J. R. Stat. Soc. Ser. B 70, 53–71. 10.1111/j.1467-9868.2007.00627.x
34. Meuwissen T. H., Hayes B. J., Goddard M. E. (2001). Prediction of total genetic value using genome-wide dense marker maps. Genetics 157, 1819–1829. [PMC free article] [PubMed]
35. Newcombe P. J., Conti D. V., Richardson S. (2016). JAM: a scalable bayesian framework for joint analysis of marginal snp effects. Genet. Epidemiol. 40, 188–201. 10.1002/gepi.21953
36. Ning Z., Lee Y., Joshi P. K., Wilson J. F., Pawitan Y., Shen X. (2017). A

selection operator for summary association statistics reveals allelic heterogeneity of complex traits. Am. J. Hum. Genet. 101, 903–912. 10.1016/j.ajhg.2017.09.027

37. Park J.-H., Gail M. H., Weinberg C. R., Carroll R. J., Chung C. C., Wang Z., et al. . (2011). Distribution of allele frequencies and effect sizes and their interrelationships for common genetic susceptibility variants. Proc. Natl. Acad. Sci. U.S.A. 108, 18026–18031. 10.1073/pnas.1114759108
38. Park J.-H., Wacholder S., Gail M. H., Peters U., Jacobs K. B., Chanock S. J., et al. . (2010). Estimation of effect size distribution from genome-wide association studies and implications for future discoveries. Nat. Genet. 42, 570–575. 10.1038/ng.610
39. Park T., Casella G. (2008). The bayesian lasso. J. Am. Stat. Assoc. 103, 681–686. 10.1198/016214508000000337
40. Pickrell J. K. (2014). Joint analysis of functional genomic data and genome-wide association studies of 18 human traits. Am. J. Hum. Genet. 94, 559–573. 10.1016/j.ajhg.2014.03.004
41. Polak E., Ribiere G. (1969). Note sur la convergence de méthodes de directions conjuguées. Revue Française d'informatique et de Recherche Opérationnelle. Série Rouge 3, 35–43.
42. Robert C. P. (2004). Monte Carlo Methods. Wiley Online Library.
43. Ročková V., George E. I. (2016). The spike-and-slab lasso. J. Am. Stat. Assoc. 10.1080/01621459.2016.1260469
44. Schork A. J., Thompson W. K., Pham P., Torkamani A., Roddey J. C., Sullivan P. F., et al. (2013). All SNPs are not created equal: genome-wide association studies reveal a consistent pattern of enrichment among functionally annotated SNPs. PLoS Genet. 9:e1003449 10.1371/journal.pgen.1003449
45. Schweiger R., Kaufman S., Laaksonen R., Kleber M. E., März W., Eskin E., et al. . (2016). Fast and accurate construction of confidence intervals for heritability. Am. J. Hum. Genet. 98, 1181–1192. 10.1016/j.ajhg.2016.04.016
46. Servin B., Stephens M. (2007). Imputation-based analysis of association studies: candidate regions and quantitative traits. PLoS Genet. 3:e114. 10.1371/journal.pgen.0030114
47. Shewchuk J. R. (1994). An Introduction to the Conjugate Gradient Method Without the Agonizing Pain. Carnegie Mellon University.

48. Spain S. L., Barrett J. C. (2015). Strategies for fine-mapping complex traits. Hum. Mol. Genet. 24, R111–R119. 10.1093/hmg/ddv260
49. Speed D., Hemani G., Johnson M. R., Balding D. J. (2012). Improved heritability estimation from genome-wide SNPs. Am. J. Hum. Genet. 91, 1011–1021. 10.1016/j.ajhg.2012.10.010
50. Su Z., Marchini J., Donnelly P. (2011). Hapgen2: simulation of multiple disease SNPs. Bioinformatics 27, 2304–2305. 10.1093/bioinformatics/btr341
51. Sun L., Craiu R. V., Paterson A. D., Bull S. B. (2006). Stratified false discovery control for large-scale hypothesis testing with application to genome-wide association studies. Genet. Epidemiol. 30, 519–530. 10.1002/gepi.20164
52. Tibshirani R. (1996). Regression shrinkage and selection via the lasso. J. R. Stat. Soc. Ser. B 58, 267–288.
53. Valdar W., Sabourin J., Nobel A., Holmes C. C. (2012). Reprioritizing genetic associations in hit regions using lasso-based resample model averaging. Genet. Epidemiol. 36, 451–462. 10.1002/gepi.21639
54. Vilhjálmsson B. J., Yang J., Finucane H. K., Gusev A., Lindström S., Ripke S., et al. . (2015). Modeling linkage disequilibrium increases accuracy of polygenic risk scores. Am. J. Hum. Genet. 97, 576–592. 10.1016/j.ajhg.2015.09.001
55. Visscher P. M., Brown M. A., McCarthy M. I., Yang J. (2012). Five years of GWAS discovery. Am. J. Hum. Genet. 90, 7–24. 10.1016/j.ajhg.2011.11.029
56. Wen X., Lee Y., Luca F., Pique-Regi R. (2016). Efficient integrative multi-snp association analysis via deterministic approximation of posteriors. Am. J. Hum. Genet. 98, 1114–1129. 10.1016/j.ajhg.2016.03.029
57. Wray N. R., Goddard M. E., Visscher P. M. (2007). Prediction of individual genetic risk to disease from genome-wide association studies. Genome Res. 17, 1520–1528. 10.1101/gr.6665407
58. Wu T. T., Lange K. (2008). Coordinate descent algorithms for lasso penalized regression. Ann. Appl. Stat. 2, 224–244. 10.1214/07-AOAS147
59. Xu L., Craiu R. V., Sun L. (2011). Bayesian methods to overcome the winner's curse in genetic studies. Ann. Appl. Stat. 5, 201–231. 10.1214/10-AOAS373

60. Yang J., Benyamin B., McEvoy B. P., Gordon S., Henders A. K., Nyholt D. R., et al. . (2010). Common snps explain a large proportion of the heritability for human height. Nat. Genet. 42, 565–569. 10.1038/ng.608
61. Yang J., Weedon M. N., Purcell S., Lettre G., Estrada K., Willer C. J., et al. . (2011). Genomic inflation factors under polygenic inheritance. Eur. J. Hum. Genet. 19, 807–812. 10.1038/ejhg.2011.39
62. Yoo Y. J., Pinnaduwage D., Waggott D., Bull S. B., Sun L. (2009). Genome-wide association analyses of north american rheumatoid arthritis consortium and framingham heart study data utilizing genome-wide linkage results. BMC Proc. 3:S103. 10.1186/1753-6561-3-S7-S103
63. Zablocki R. W., Schork A. J., Levine R. A., Andreassen O. A., Dale A. M., Thompson W. K. (2014). Covariate-modulated local false discovery rate for genome-wide association studies. Bioinformatics 30, 2098–2104. 10.1093/bioinformatics/btu145
64. Zhou X., Carbonetto P., Stephens M. (2013). Polygenic modeling with bayesian sparse linear mixed models. PLoS Genet. 9:e1003264. 10.1371/journal.pgen.1003264
65. Zou H., Hastie T. (2005). Regularization and variable selection via the elastic net. J. R. Stat. Soc. Ser. B 67, 301–320. 10.1111/j.1467-9868.2005.00503.x
66. Zuber V., Silva A. P. D., Strimmer K. (2012). A novel algorithm for simultaneous SNP selection in high-dimensional genome-wide association studies. BMC Bioinformatics 13:284. 10.1186/1471-2105-13-284

CHAPTER

15

Mixture Models for Analyzing Product Reliability Data: A Case Study

S. Ruhi[1], S. Sarker[2] and M. R. Karim[2]

[1] Department of Mathematics, Pabna University of Science & Technology, Pabna, Bangladesh.

[2] Department of Statistics, University of Rajshahi, Rajshahi 6205, Bangladesh.

ABSTRACT

In the case of manufactured products, there are situations where some components of a product are produced over a period of time by collecting items from different vendors, using different raw materials, machines, and manpower. The physical characteristics and the reliabilities of such

Citation: (APA): Ruhi, S., Sarker, S., & Karim, M. R. (2015). Mixture models for analyzing product reliability data: a case study. SpringerPlus, 4(1), 634. (14 pages) DOI: https://doi.org/10.1186/s40064-015-1420-x URL: https://springerplus.springeropen.com/articles/10.1186/s40064-015-1420-x

components may be different, but sometimes it is difficult to distinguish them clearly. In such situations, mixtures of distributions are often used in the analysis of reliability data for these components. Here a twofold Weibull–Weibull mixture model is applied to analyze product reliability data that consist of both failure and censored lifetimes. The Expectation–Maximization (EM) algorithm is used to find the maximum likelihood estimates of the model parameters. As a case study, it analyses an Aircraft component (Windshield) failure data and various characteristics of the mixture model, such as the reliability function, B10 life, mean time to failure, etc., are estimated to assess the reliability of the component. Simulation studies are performed to investigate the properties and uses of the proposed method.

Keywords: Case study, Data analysis, EM algorithm, Mixture model, Reliability

BACKGROUND

Reliability of a product is defined by the probability that the product will perform well it is intended function for a specified time period or usage limit under normal operating condition (Meeker and Escobar 1998; Blischke and Murthy 2000). Because of rapid advances in manufacturing technology, customers expect to purchase products that will be highly sophisticated, reliable and safe. In recent years many manufacturers are collecting and analyzing field failure data to enhance the reliability of their products and to improve goodwill and customer satisfaction (Blischke et al. 2011).

The paper analyses an Aircraft component (Windshield of a particular model) failure data that contain both failure and censored times. The same data are analyzed by Murthy et al. (2004) and Ruhi (2015). Murthy et al. (2004) have fitted a twofold Weibull mixture model for the data and estimated the parameters of the model by applying a graphical method based on Weibull probability paper (WPP) plot. The graphical method has been used widely to estimate the parameters of the Weibull mixture model. The bulk of the existing literature deals with the well-separated subpopulations cases and uses various approximations for plotting and the characterizations of the asymptotes. As mentioned in Murthy et al. (2004), there are two serious drawbacks in the graphical method. These are as follows:

- The graphical method yields very crude estimates unless applied repeated iteration and evaluated by vision. As such, they can be used as starting points for more sophisticated statistical methods.

- They do not provide any statistical confidence limits for the estimated parameters.

To overcome these drawbacks, here we apply the Expectation–Maximization (EM) algorithm to find the maximum likelihood estimates of the twofold Weibull mixture model and investigate the performance of the proposed method over the method of Murthy et al. (2004). The performance of the method will be evaluated by numerical simulation studies.

The outline of the paper is as follows. "Mixture models" describes the assumed mixture model. "Maximum-likelihood estimation of model parameters" explains the parameter estimation method. "Case study: analysis of aircraft Windshield failure" data presents a case study based on aircraft Windshield failure data. "Simulation study" presents a simulation study to investigate the performance of the method and "Conclusion" concludes the paper. Finally Appendix provides R codes that used in the paper for estimating the parameters of the model.

MIXTURE MODELS

Various types of statistical models have been applied extensively in the analysis of failure data for manufactured products. However, there are situations where some components of a product are produced over a period of time by collecting items from different vendors, using different raw materials, machines, and manpower. In such situations, mixtures of distributions are often used in the analysis of reliability data as the physical characteristics and the reliabilities of such components may be different and difficult to distinguish easily and clearly.

The cumulative distribution function (cdf), probability density function (pdf) and hazard function (hf) of a general *K*-fold mixture model involving *K* subpopulations have presented in Murthy et al. (2004) and Blischke et al. (2011). The cumulative distribution function (cdf) of a general *K*-fold mixture model can be written as

$$G(t) = \sum_{j=1}^{K} p_j F_j(t) \tag{1}$$

where $F_j(t)$ is the cdf of the *j*-th sub-population, p_j is the mixing probability of the *j*-th sub-population, $p_j > 0$ and $p_j > 0$ and $\sum_{j=1}^{K} p_j = 1$. The probability density function (pdf) is given by

$$g(t) = \sum_{j=1}^{K} p_j f_j(t) \tag{2}$$

where $f_j(t)$ is the pdf associated with $F_j(t)$. The hazard function $h(t)$ is

$$h(t) = \sum_{j=1}^{K} w_j(t)\, h_j(t) \tag{3}$$

where $h_j(t)$ is associated with subpopulation j, and

$$w_j(t) = \frac{p_j R_j(t)}{\sum_{j=1}^{K} p_j R_j(t)}, \quad j = 1, 2, \ldots, K, \tag{4}$$

where $\sum_{j=1}^{K} w_j(t) = 1$ with

$$R_j(t) = 1 - F_j(t), \quad j = 1, 2, \ldots, K. \tag{5}$$

From (3), we see that the failure rate for the model is a weighted mean of the failure rate for the subpopulations with the weights varying with t, $t \geq 0$.

Special case: twofold Weibull mixture model

The cdf of the twofold mixture model (putting $K = 2$ in (1)) for the random variable T is given by

$$G(t) = pF_1(t) + (1 - p)F_2(t), \quad t \geq 0 \tag{6}$$

If $F_1(t)$ follows Weibull (α_1, β_1) and $F_2(t)$ follows Weibull (α_2, β_2) distributions, the cdf for twofold Weibull–Weibull mixture model from Eq. (6) becomes

$$G(t) = \left[1 - \exp\left[-\left(\frac{t}{\alpha_2}\right)^{\beta_2}\right]\right] + p\left[\exp\left\{-\left(\frac{t}{\alpha_2}\right)^{\beta_2}\right\} - \exp\left\{-\left(\frac{t}{\alpha_1}\right)^{\beta_1}\right\}\right], \quad \{\beta_1, \alpha_1, \beta_2, \alpha_2\} > 0, \; t \geq 0 \tag{7}$$

The corresponding probability density function (pdf) is

$$g(t) = p\left[\frac{\beta_1}{\alpha_1}\left(\frac{t}{\alpha_1}\right)^{\beta_1 - 1} \exp\left(-\frac{t}{\alpha_1}\right)^{\beta_1}\right] + (1 - p)\left[\frac{\beta_2}{\alpha_2}\left(\frac{t}{\alpha_2}\right)^{\beta_2 - 1} \exp\left(-\frac{t}{\alpha_2}\right)^{\beta_2}\right], \quad t \geq 0 \tag{8}$$

The reliability function is

$$R(t) = p\left[\exp\left[-\left(\frac{t}{\alpha_1}\right)^{\beta_1}\right]\right] + (1-p)\left[\exp\left[-\left(\frac{t}{\alpha_2}\right)^{\beta 2}\right]\right], \quad t \geq 0 \tag{9}$$

And the hazard function is

$$h(t) = \frac{pf_1(t) + (1-p)f_2(t)}{p\exp\left[-\left(\frac{t}{\alpha_1}\right)^{\beta_1}\right] + (1-p)\exp\left[-\left(\frac{t}{\alpha_2}\right)^{\beta_2}\right]}, \quad t \geq 0 \tag{10}$$

Other two-fold mixture models can be derived by using different cdfs from different lifetime distributions, similarly. In the remainder of the paper we apply this twofold Weibull mixture model to analyze the data set.

MAXIMUMLIKELIHOOD ESTIMATION OF MODEL PARAMETERS

If the data contain the failure/censored random variable t_i and failure/censored indicator δ_i, (if the *i*th observation is failure then $\delta_i = 1$ and if it is censored then $\delta_i = 1$) for i=1,2,⋯ni=1,2,⋯n, the likelihood function under random censoring scheme for the data is given by

$$L(\theta) = \prod_{i=1}^{n} [f(t_i)]^{\delta_i}[1 - F(t_i)]^{1-\delta_i} \tag{11}$$

where, θ is the parameter vector for the assumed model. Taking log on both sides of Eq. (11), we get,

$$\ln L = \sum_{i=1}^{n} [\delta_i \ln f(t_i) + (1-\delta_i)\ln\{1 - F(i_i)\}] \tag{12}$$

In the case of twofold Weibull mixture model with $\theta = \{\beta_1, \alpha_1, \beta_2, \alpha_2, p\}$, the log-likelihood function (12) becomes

$$\ln L = \sum_{i=1}^{n} \delta_i \ln\left[p\left[\frac{\beta_1}{\alpha_1}\left(\frac{t_i}{\alpha_1}\right)^{\beta_1 - 1}\exp\left\{-\left(\frac{t_i}{\alpha_1}\right)^{\beta_1}\right\}\right](1-p)\left[\frac{\beta_2}{\alpha_2}\left(\frac{t_i}{\alpha_2}\right)^{\beta_2 - 1}\exp\left\{-\left(\frac{t_i}{\alpha_2}\right)^{\beta_2}\right\}\right]\right]$$
$$+ (1-\delta_i)\ln\left[\exp\left\{-\left(\frac{t_i}{\alpha_2}\right)^{\beta_2}\right\} - p\left[\exp\left\{-\left(\frac{t_i}{\alpha_2}\right)^{\beta_2}\right\} - \exp\left\{-\left(\frac{t_i}{\alpha_1}\right)^{\beta_1}\right\}\right]\right] \tag{13}$$

The maximum likelihood estimates of the parameters are obtained by taking the partial derivatives of (13) with respect to β_1, α_1, β_2, α_2 and p and setting to zero. The maximum likelihood estimating equations obtained from (13) do not give closed form solutions for the parameters $\theta = \{\beta_1, \alpha_1, \beta_2, \alpha_2, p\}$. Therefore, it requires a numerical iterative procedure for finding the MLEs of the parameters.

Estimation of Mixing Proportions Using EM Algorithm

The Expectation–Maximization (EM) algorithm is an efficient iterative procedure to compute the Maximum Likelihood Estimates (MLEs) of the parameters of the distribution in the presence of missing or hidden data (Dempster et al. 1977; McLachlan and Krishnan 2008). Bordes and Chauveau (2012) discussed several iterative methods based on EM and stochastic EM methodology to estimate parametric or semi parametric mixture models for randomly right censored lifetime data, conditioned that they are identifiable. Here we discuss the EM algorithm for finding the MLEs of the parameters of a general K-fold mixture model with parameters $\Theta = (p_1, \cdots, p_K, \theta_1, \cdots, \theta_K)$, where p_j is mixing parameters and θ_j is the parameters for the density function f_j, j=1,2,…,K_j=1,2,…,K. Let y=(t_1,…,t_n)′y=(t_1,…,t_n)′ denotes the observed random sample obtained from the mixture density. Let us introduce the unobservable or missing data vectors $z = (z_1', \ldots, z_n')'$, where z_i is a K-dimensional vector of zero–one indicator variables and where z_{ij} is one or zero according to whether t_i arose or did not arise from the j-th component of the mixture (i=1,2,…,n;j=1,2,…,K) (i=1,2,…,n;j=1,2,…,K). The EM algorithm handles the unobservable data to the problem by working with the current conditional expectation of the complete-data log likelihood given the observed data. Let us define the complete-data vector x as x=(y′,z′)′.x=(y′,z′)′.

Each iteration of the EM algorithm consists of two steps. The Expectation or E-step computes the conditional expectation of the complete-data log-likelihood for Θ given observed data, which at the $(m+1)$th iteration can be expressed as

$$Q\left(\Theta, \Theta^{(m)}\right) = \sum_{i=1}^{n}\sum_{j=1}^{K} E_{\Theta^{(m)}}\left(Z_{ij}\middle|y\right)\partial_i \ln\left[p_j^{(m)} f_j\left(t_i\middle|\theta_j^{(m)}\right)\right]$$
$$+\sum_{i=1}^{n}\sum_{j=1}^{K} E_{\Theta^{(m)}}\left(Z_{ij}\middle|y\right)(1-\partial_i) \ln\left[p_j^{(m)} R_j\left(t_i\middle|\theta_j^{(m)}\right)\right] \quad (14)$$

As (14) is linear in the unobservable data z_{ij}, the E-step (on the $(m+1)$ th iteration) simply requires the calculation of the current conditional expectation of Z_{ij} given the observed data y, where Z_{ij} is the random variable corresponding to z_{ij}. Now

$$E_{\Theta^{(m)}}\left(Z_{ij} \middle| y\right) = z_{ij}^{(m)}$$

Here $z^{(m)}_{ij}$ are the posterior probabilities which can be expressed using the Bayes's theorem as

$$z_{ij}^{(m)} = \frac{p_j^{(m)}\left[\partial_i f_j\left(t_i \middle| \theta_j^{(m)}\right) + (1-\partial_i)R_j\left(t_i \middle| \theta_j^{(m)}\right)\right]}{\sum_{j=1}^{m} p_j^{(m)}\left[\partial_i f_j\left(t_i \middle| \theta_j^{(m)}\right) + (1-\partial_i)R_j\left(t_i \middle| \theta_j^{(m)}\right)\right]} \tag{15}$$

The evaluation of this expectation is called the E-step of the algorithm. The second step (the Maximization or M-step) maximizes (14) with respect to the parameters to obtain new parameter estimations $\Theta^{(m+1)}$. To maximize (14), we can maximize the term containing p_j and the term containing θ_j independently since they are not related. To find the expression for p_j, we introduce the Lagrange multiplier λ with the constraint $\sum_{j=1}^{K} p_j = 1$. Under this constraint, taking the derivative of (14) with respect to p_j and setting equal to zero, we get

$$\sum_{i=1}^{n} \frac{1}{p_j} z_{ij}^{(m)} + \lambda = 0 \tag{16}$$

Summing both sizes over j and using $\sum_{j=1}^{m} z_{ij}^{(m)} = 1$; we get that $\lambda = -n$ resulting in

$$p_j^{(m+1)} = \frac{1}{n}\sum_{i=1}^{n} z_{ij}^{(m)} \tag{17}$$

See, Bucar et al. (2004) for more details.

For some distributions, it is possible to get closed-form analytical expressions for θ_j. However, in the case of Weibull distributions with $\theta_j = (\alpha_j, \beta_j)$, j=1,2,…,Kj=1,2,…,K, we have to apply numerical procedures to find MLEs of the parameters. Here we apply the *survreg* function with weight (weight > 0) given in the *survival* package of the R-program. The algorithm proceeds by using the newly derived parameters as the guess for the next iteration. The E- and M-steps are iterated until the algorithm converges.

Finally, the EM algorithm for estimating the parameters of a K-fold Weibull mixture distribution can be summarized in a step-by-step procedure as follows:

Step 1 Begin with initial guesses of $p_j^{(0)}$, $\alpha_j^{(0)}$ and $\beta_j^{(0)}$ for $j = 1, 2, \ldots, K$

Step 2 Using the initial values of $p_j^{(0)}$, $\alpha_j^{(0)}$ and $\beta_j^{(0)}$, at m-th iteration calculate the conditional expectation of z_{ij}, i.e., $z_{ij}^{(m)}$ using (15).

Step 3 At the$m$-th iteration, find the MLEs of $p_j^{(m+1)}$, $\alpha_j^{(m+1}$ and $\beta_j^{(m+1)}$ as follows:

The applications of the EM algorithm are broad because of its flexibility in analyzing incomplete or missing data. In any fields, when it is difficult to maximize the complicated likelihood function, various extensions and modifications of the EM algorithm have been proposed to simplify the computations, e.g., see Wei and Tanner (1990), Meng and Rubin (1993) and Liu and Rubin (1994). More detailed theory and applications of the EM algorithm can be found in McLachlan and Krishnan (2008).

CASE STUDY: ANALYSIS OF AIRCRAFT WIND-SHIELD FAILURE DATA

In this section, as a case study, we analyze a set of aircraft Windshield failure data. We apply twofold Weibull mixture model for the failure data and estimate various characteristics of the Windshield, such as the reliability function, B10 life, mean time to failure, etc. to assess the reliability of the Windshield.

Aircraft Windshield failure data

Data on failure and censored times for a particular model of Windshield given in Table 1 are taken from Murthy et al. (2004), originally given in Blischke and Murthy (2000). The data consist of 88 failure times and 65 censored times out of 153 observations. Here censored time (or service time) means that the Windshields have not failed at the time of observation. The unit for measurement of time is 1000 h.

Table 1 Windshield failure data

T	δ	T	δ	T	δ	T	δ	T	δ
0.040	1	2.154	1	3.595	1	1.183	0	3.003	0
0.301	1	2.190	1	3.699	1	1.244	0	3.102	0

0.309	1	2.194	1	3.779	1	1.249	0	3.304	0
0.557	1	2.223	1	3.924	1	1.262	0	3.483	0
0.943	1	2.224	1	4.035	1	1.360	0	3.500	0
1.070	1	2.229	1	4.121	1	1.436	0	3.622	0
1.124	1	2.300	1	4.167	1	1.492	0	3.665	0
1.248	1	2.324	1	4.240	1	1.580	0	3.695	0
1.281	1	2.349	1	4.255	1	1.719	0	4.015	0
1.281	1	2.385	1	4.278	1	1.794	0	4.628	0
1.303	1	2.481	1	4.305	1	1.915	0	4.806	0
1.432	1	2.610	1	4.376	1	1.920	0	4.881	0
1.480	1	2.625	1	4.449	1	1.963	0	5.140	0
1.505	1	2.632	1	4.485	1	1.978	0		
1.506	1	2.646	1	4.570	1	2.053	0		
1.568	1	2.661	1	4.602	1	2.065	0		
1.615	1	2.688	1	4.663	1	2.117	0		
1.619	1	2.823	1	4.694	1	2.137	0		
1.652	1	2.890	1	0.046	0	2.141	0		
1.652	1	2.902	1	0.140	0	2.163	0		
1.757	1	2.934	1	0.150	0	2.183	0		
1.795	1	2.962	1	0.248	0	2.240	0		
1.866	1	2.964	1	0.280	0	2.341	0		
1.876	1	3.000	1	0.313	0	2.435	0		
1.899	1	3.103	1	0.389	0	2.464	0		
1.911	1	3.114	1	0.487	0	2.543	0		
1.912	1	3.117	1	0.622	0	2.560	0		
1.914	1	3.166	1	0.900	0	2.592	0		
1.981	1	3.344	1	0.952	0	2.600	0		
2.010	1	3.376	1	0.996	0	2.670	0		
2.038	1	3.385	1	1.003	0	2.717	0		
2.085	1	3.443	1	1.010	0	2.819	0		
2.089	1	3.467	1	1.085	0	2.820	0		
2.097	1	3.478	1	1.092	0	2.878	0		
2.135	1	3.578	1	1.152	0	2.950	0		

Nonparametric estimate of reliability function

Figure 1 is the reliability (or survival) plot for the component. The plot appears to be reasonable; it shows the estimated MTTF is 3.03549 thousand hours or approximately 127 days. The nonparametric estimate of median lifetime is 2964 h, indicates that 50 % of the Windshield fails at 2964 h. The nonparametric estimate of cdf, known as empirical distribution function (edf) is one minus the estimated reliability function.

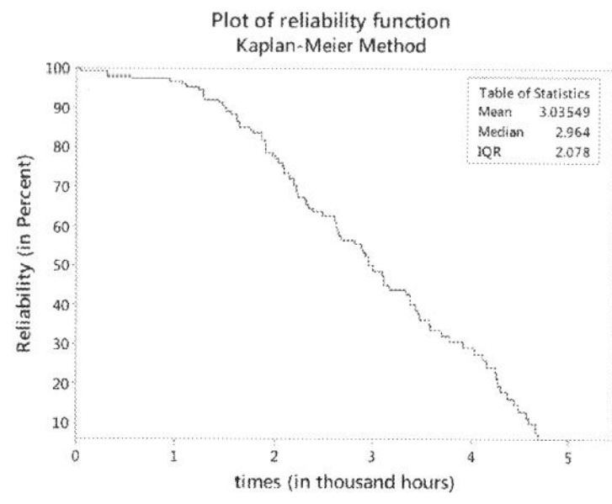

Figure. 1: Nonparametric reliability plots

Parametric estimate of reliability function

Murthy et al. (2004) assumed the twofold Weibull mixture model for this data set and estimated the model parameters based on Weibull Probability Plots (WPP) method. In this article we apply the EM algorithm, discussed in "Estimation of mixing proportions using EM algorithm", to find the maximum likelihood estimates of the parameters $\theta = \{\beta_1, \alpha_1, \beta_2, \alpha_2, p\}$ for the twofold Weibull mixture model and investigate the performance of the proposed method over the method of Murthy et al. (2004). A comparison between the estimates of the parameters obtained by two different methods is given in Table 2.

Table 2 Estimates of parameters of twofold Weibull mixture model

Parameters	Estimates based on WPP	Estimates based on EM algorithm
$\hat{\beta}_1$	0.429	1.2098
$\hat{\alpha}_1$	8.230	0.2541
$\hat{\beta}_2$	2.990	2.7802
$\hat{\alpha}_2$	3.210	3.4856
$\hat{p}$	0.136	0.0176
$(1-\hat{p})$	0.864	0.9823

We have estimated the cdfs and reliability functions of twofold Weibull mixture model based on both non-parametric (by *Kaplan–Meier* method) and parametric (by EM Algorithm) approaches. The cdf and reliability function are also estimated by using the WPP plot method [estimates are taken from Murthy et al. (2004)]. Figures 2 and 3 compare the estimated reliability functions and cdfs, respectively to find out the best approach for the data set.

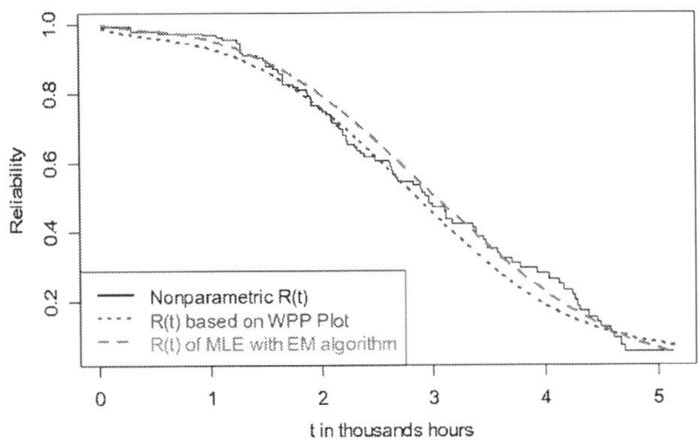

Figure. 2: Comparison of reliability functions

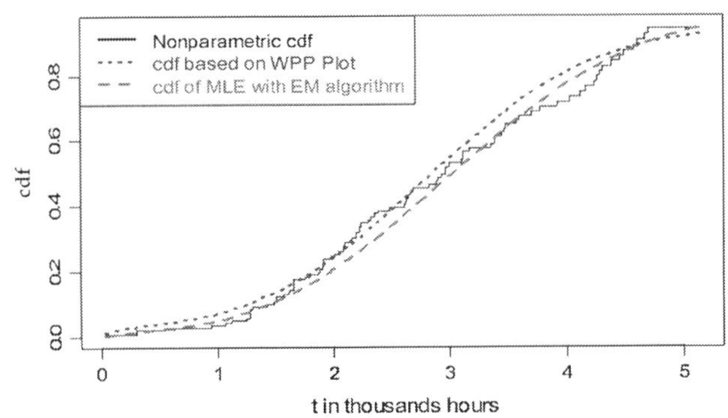

Figure. 3: Comparison of cdfs

From Fig. 2, we observe that the reliability function obtained by the EM algorithm method is much closer to the Kaplan–Meier estimate than that of the reliability function estimated by the WPP plot method. The plots of cdfs shown in Fig. 3 conclude the same. These indicate that the method of estimation with the EM algorithm procedure is better than the WPP plot procedure.

The estimates of adjusted Anderson–Darling (AD) test statistic based on WPP method and EM algorithm method are 412.5845 and 410.2851, respectively. This again indicates that the EM algorithm method provides better fit for the data set than the WPP method.

Reliability Characteristics of Windshield Data

Some of the reliability related important characteristics such as mean time to failure (MTTF), B10 lifetime, B50 (median) lifetime of the Windshield obtained by two methods are displayed in Table 3.

Table 3 Estimates of reliability characteristics of Windshield

Quantities	EM algorithm method	WPP Plot method
MTTF	3.0525	5.5782
B10-Lifetime	1.5248	1.3125
B50-Lifetime	3.0046	2.9298

Table 3 indicates that the estimates of MTTF obtained from maximum likelihood method via the EM algorithm and from WPP plot method are 3.0525 (thousand hours) and 5.5782 (thousand hours), respectively. Estimate of MTTF obtained by EM algorithm is very close to the nonparametric estimate of MTTF (3.03549 thousand hours) given in Fig. 1. The WPP method overestimates the MTTF in this case. From the estimates of B10-lifetime and B50-lifetime, we may conclude according to EM algorithm method that, 10 % of the total components fail approximately at 1524 h and 50 % fail at 3004 h.

SIMULATION STUDY

In this section, we use computer simulation to evaluate the performance of the method numerically. Numerically generated twofold mixture data are used to develop the twofold Weibull mixture model and to find the ML estimates of model parameters under right censored data. Using simulated data, the ML estimates of the model parameters, the sample means (SMs) and the mean squared errors (MSEs) of estimates are computed. Simulation programming codes are written using statistical software package R.

Steps of simulation study

Here we describe the step-by-step algorithm for simulation of twofold Weibull mixture model and estimation of model parameters via the EM algorithm.

- *Step 1* We consider a set of true value for the 5 parameters $\theta = \{\beta_1, \alpha_1, \beta_2, \alpha_2, p\}$ of twofold Weibull mixture model. Under this set of parameter, we generate $n = n_1 + n_2$ samples from the twofold Weibull mixture model using the software

R-Language (version-3.2.2). A desired percent (10, 20 and 30 %) of the largest generated sample out of 200, are considered as the right censored observations and remaining are assumed as failed lifetime.

- *Step 2* Based on the generated right censored data, we estimate the parameters via the EM algorithm assuming that the mixing sub-populations are unknown. The methodology is discussed in "Estimation of mixing proportions using EM algorithm".
- *Step 3* The above Steps 1 and 2 are repeated 1000 times under two Cases:
- *Case (i)* for a variety percent of censored observations (10, 20 and 30 %) and
- *Case (ii)* for different sample sizes ($n = 200$, 400 and 600).
- We compute the sample means (SMs) and mean squared errors (MSEs) of the estimates for the both Cases (i) and (ii).
- *Steps 4* Summarize and discuss the simulation results based on 1000 repetition.

Simulation output analysis

The simulation results are shown in Tables 4 and 5.

Table 4 Sample means of the MLEs for different percent of censored observations

Parameters	True values	Sample means of the MLEs		
		Set-01 [N = 200; 10 % cens. obs.]	Set-02 [N = 200; 20 % cens. Obs.]	Set-03 [N = 200; 30 % cens. Obs.]
$\hat{\beta}_1$	3.50	3.7945	3.9587	4.1968
$\hat{\alpha}_1$	700.00	699.3536	700.1768	705.3863
$\hat{\beta}_2$	1.20	1.1601	1.1454	1.1709
$\hat{\alpha}_2$	850.00	944.5603	1055.9218	912.0363
$\hat{p}$	0.30	0.3737	0.3692	0.3732
$(1-\hat{p})$	0.70	0.6263	0.6308	0.6268

Table 5 Sample means of the MLEs for different sample sizes

Parameters	True values	Sample means of the MLEs		
		Set-04 [N = 200; 20 % cens. obs.]	Set-05 [N = 400; 20 % cens. obs.]	Set-06 [N = 600; 20 % cens. obs.]
$\hat{\beta}_1$	3.50	3.9587	3.8619	3.8365
$\hat{\alpha}_1$	700.00	700.1768	703.0983	702.2486
$\hat{\beta}_2$	1.20	1.1454	1.1664	1.1850
$\hat{\alpha}_2$	850.00	1055.9218	891.5291	873.1364
$\hat{p}$	0.30	0.3692	0.3398	0.3192
$(1-\hat{p})$	0.70	0.6308	0.6602	0.6808

Tables 4 and 5 present the summary results of the simulations based on 1000 repetitions under the given true values. In these tables, the first column shows the parameters of the model and second column shows the true values of the parameters. Tables 4 and 5 give the sample means of the MLEs of parameters obtained by the EM algorithm. For all of the sets, the sample means of the estimated parameters are close to the corresponding true values of the parameters. If the percent of censored observations decrease (i.e., if number of failures increase), the sample means of the MLEs become more closers to the true values for all most all sets, as expected. Similarly, the sample means of the MLEs become more closers to the true values for increasing sample sizes.

The mean squared errors (MSEs) of the MLEs of parameters for different percent of censored observations and for different sample sizes are given in Tables 6 and 7, respectively. The MSEs decrease for decreasing of the percent of censored observations (i.e., for increasing the number of failures). Also the MSEs decrease for increasing of the sample sizes. These comparisons indicate that the proposed method of estimation is applicable for analyzing mixture model for censored data.

Table 6 Mean squared errors for different percent of censored observations

Parameters	True values	Mean squared errors (MSEs) of MLEs		
		Set-01 [N = 200; 10 % cens. obs.]	Set-02 [N = 200; 20 % cens. obs.]	Set-03 [N = 200; 30 % cens. obs.]
$\hat{\beta}_1$	3.50	2.17953	2.6223	4.2156
$\hat{\alpha}_1$	700.00	3232.01855	4873.0835	9777.0312
$\hat{\beta}_2$	1.20	0.15011	0.0619	0.0711
$\hat{\alpha}_2$	850.00	87452.76431	1614229.0410	269,062.7034
$\hat{p}$	0.30	0.00114	0.0391	0.0500
$(1-\hat{p})$	0.70	0.03382	0.0391	0.0500

Table 7 Mean squared errors for different sample sizes

Parameters	True values	Mean squared errors (MSEs) of MLEs		
		Set-04 [N = 200; 20 % cens. obs.]	Set-05 [N = 400; 20 % cens. obs.]	Set-06 [N = 600; 20 % cens. obs.]
$\hat{\beta}_1$	3.50	2.6223	1.9544	1.5033
$\hat{\alpha}_1$	700.00	4873.0835	1907.8766	1574.6524
$\hat{\beta}_2$	1.20	0.0619	0.0294	0.0210
$\hat{\alpha}_2$	850.00	1614229.0410	25655.13287	10622.3852
$\hat{p}$	0.30	0.0391	0.0242	0.0159
$(1-\hat{p})$	0.70	0.0391	0.0242	0.0159

CONCLUSION

There are situations where variations in product reliability can be occurred across different component vendors. In such situations, mixture of distributions can model the variability resulting from parts being bought from K different suppliers with $F_k(t)$ denoting the failure distribution for parts obtained from supplier k, k=1,2,…,Kk=1,2,…,K. This paper has applied a twofold Weibull mixture model for analyzing product reliability data with failure and censored observations. It has proposed the Expectation–Maximization (EM) algorithm to find the maximum likelihood estimates of the parameters of mixture model and compared this method with a method based on Weibull Probability Paper plots. An aircraft component (Windshield) failure data is analyzed as an example and investigated that the performance of the proposed method of estimation is impressive. The results would be useful for managerial implications in assessing and predicting the reliability of the component more accurately.

The mixture model considered here is the twofold Weibull mixture model. The proposed method is easily extendable for other mixture models also. A scope of the future research with other mixture models and with various types of censored data would be interesting.

REFERENCES

1. Blischke WR, Murthy DNP (2000) Reliability modelling, prediction, and optimization. Wiley, New York
2. Blischke WR, Karim MR, Murthy DNP (2011) Warranty data collection and analysis. Springer Verlag, London
3. Bordes L, Chauveau D (2012). EM and stochastic EM algorithms for reliability mixture models under random censoring, Hal-VousConsultezL'Archive
4. Bucar T, Nagode M, Fajdiga M (2004) Reliability approximation using finite Weibull mixture distributions. Reliab Eng Syst Saf 84:241–251
5. Dempster AP, Laird NM, Rubin DB (1977) Maximum likelihood from incomplete data via the EM algorithm (with discussion). J R Stat Soc B 39:1–38
6. Liu C, Rubin DB (1994) The ECME algorithm: a simple extension of EM and ECM with faster monotone convergence. Biometrika 81:633–648
7. McLachlan GJ, Krishnan T (2008) The EM algorithm and extensions, 2nd edn. Wiley, New York
8. Meeker WQ, Escobar LA (1998) Statistical methods for reliability data. Wiley, New York
9. Meng X-L, Rubin D (1993) Maximum likelihood estimation via the ECM algorithm: a general framework. Biometrika 80:267–278
10. Murthy DNP, Xie M, Jiang R (2004) Weibull model. Wiley, New York
11. Ruhi S (2015) Application of mixture models for analyzing reliability data: a case study. Open Access Libr J 2:e1815
12. Wei GCG, Tanner MA (1990) A Monte Carlo implementation of the EM algorithm and the poor man's data augmentation algorithms. J Am Stat Assoc 85:699–704

CHAPTER 16

Marginalized Mixture Models for Count Data from Multiple Source Populations

Habtamu K Benecha [1], Brian Neelon [2], Kimon Divaris [3], John S Preisser [4]

[1]National Agricultural Statistics Service, USDA, Washington, 20250 DC USA.

[2]Department of Public Health Sciences, Medical University of South Carolina, Charleston, 29425 SC USA.

[3]Departments of Epidemiology and Pediatric Dentistry, University of North Carolina, Chapel Hill, 27599-7450 NC USA.

[4]Department of Biostatistics, University of North Carolina, Chapel Hill, 27599-7420 NC USA.

Citation: (APA): Benecha, H. K., Neelon, B., Divaris, K., & Preisser, J. S. (2017). Marginalized mixture models for count data from multiple source populations. Journal of statistical distributions and applications, 4(1), 3. (17 pages) DOI: https://doi.org/10.1186/s40488-017-0057-4

ABSTRACT

Mixture distributions provide flexibility in modeling data collected from populations having unexplained heterogeneity. While interpretations of regression parameters from traditional finite mixture models are specific to unobserved subpopulations or latent classes, investigators are often interested in making inferences about the marginal mean of a count variable in the overall population. Recently, marginal mean regression modeling procedures for zero-inflated count outcomes have been introduced within the framework of maximum likelihood estimation of zero-inflated Poisson and negative binomial regression models. In this article, we propose marginalized mixture regression models based on two-component mixtures of non-degenerate count data distributions that provide directly interpretable estimates of exposure effects on the overall population mean of a count outcome. The models are examined using simulations and applied to two datasets, one from a double-blind dental caries incidence trial, and the other from a horticultural experiment. The finite sample performance of the proposed models are compared with each other and with marginalized zero-inflated count models, as well as ordinary Poisson and negative binomial regression

Keywords: Dental caries, Excess zeros, Marginal inference, Mixture model, Over-dispersion, Zero-inflation

INTRODUCTION

The analysis of data from populations with unexplained heterogeneity presents special challenges to researchers. When count data arise from mixtures of unobserved populations, models based on standard probability distributions are often inadequate to explain observed variability (Frühwirth-Schnatter 2005; Wedel and DeSarbo 1995). For example, in dental caries research and many other areas, proportions of observations with zero counts are often higher than expected under the Poisson or negative binomial distributions, and regression models based on these distributions may result in biased estimates and poor predictions. To account for such excess zeros, Mullahy (1986) and Lambert (1992) proposed zero-inflated Poisson (ZIP) regression. ZIP models, which employ twocomponent mixture distributions, hypothesize that observed counts arise from one of two latent classes within the source population: one class provides only zeros and the other produces both zero and non-zero values. However, the assumption of a model based on 'at-risk' and 'not-at-risk' latent classes may not be appropriate in some settings or may provide an inadequate fit. To model counts from multiple source populations, Wang et al. (1996) proposed multi-component Poisson mixture distributions, and their approach has been extended to other finite mixtures of non-degenerate count distributions. Despite the flexibility of finite mixtures for describing highly dispersed count data, parameters from standard mixture regression models are not directly applicable to making inferences about the overall effects of covariates on marginal means of count outcomes (Albert et al. 2014; Preisser et al. 2012). Even with the application of indirect methods of parameter estimation such as the use of post-modeling transformations, there are many instances where latent class model formulations fail to fully explain relationships between covariates and population-wide parameters. While the importance of the marginal mean as a target of inference in the analysis of finite mixtures of counts is well established (Albert et al. 2014; Böhning et al. 1999; Lambert 1992; Preisser et al. 2012), marginally-specified mean models for finite mixtures of count distributions have only recently been proposed. Within a ZIP likelihood framework, Long et al. (2014) proposed marginalized zero-inflated Poisson (MZIP) regression, which specifies a two-part model for counts with a set of regression coefficients for the marginal mean and, to complete model specification, a second set of regression coefficients for the latent parameter defining membership in the 'excess-zero' class. The marginalized zero-inflated negative binomial (MZINB) model (Preisser et al. 2016) extended the MZIP model to account for overdispersion in addition to excess zeros. Todem et al. (2016) described a general representation of two-part marginalized mean count models including distributions for bounded counts, e.g., the zero-inflated betabinomial distribution. In each case, the model is assumed to follow a two-component mixture consisting of a standard count distribution with a degenerate point-mass at zero. However, data-generating mechanisms based on mixtures of non-degenerate count distributions can often provide better fits when the data suggest that a simple degenerate point-mass is insufficiently flexible to capture the heterogeneity in the counts. This can arise, for example, when there is overdispersion in the counts that cannot be fully explained by standard count data distributions (e.g., Poisson, negative binomial) amended by excess zeros. In this article, we seek to expand the class of marginalized mixture models for zeroinflated and other heterogeneous count data to allow for greater model choice with maximum likelihood estimation, when there is interest in evaluating the effects of exposures on the overall mean count. For counts having unexplained heterogeneity, we extend the degenerate component of marginalized zero-inflated models to standard count distributions for more flexible modeling of the marginal mean. Our motivation comes from a randomized double-blind caries incidence trial conducted between 1988 and 1992 in Lanarkshire, Scotland, to compare the anti-caries efficacy of three toothpaste formulations in 4294 children ages 11–12 (Stephen et al. 1994). The outcome variable of interest was the number of new decayed, missing and filled surfaces (DMFS) two years following the baseline dental exam. Caries development is a complex process, which depends upon multiple biological and environmental factors; moreover, the clinical manifestation of disease is influenced by preventive care and restorative dental treatment decisions. For this reason, standard count models such as Poisson and negative binomial regression may not adequately account for heterogeneity in the DMFS counts. We consider marginalized, two-component finite mixture models to obtain direct inference about the relationship between toothpaste formulation and the marginal mean caries count in the trial population. "Methods and Results" section reviews traditional and marginalized zero-inflated count regression models, while "Models for mixtures of nondegenerate count distributions" section discusses traditional finite mixture regression models and proposes marginalized two-component count regression models involving mixtures of non-degenerate distributions. Simulation studies and two applications of the proposed models including the Lanarkshire caries trial are discussed in "Simulation study & Applications" sections, respectively. Concluding remarks follow in "Discussion and conclusions" section.

METHODS AND RESULTS

Zero-inflated Poisson and negative binomial models

Traditional zero-inflated models assume that counts arise from a two-component mixture of a standard count distribution with a distribution degenerate at zero. Under such models, counts are generated either from a 'non-susceptible' or 'perfect' state that always gives zeros, or from a 'susceptible', 'imperfect' state that produces both zero and positive counts according to a standard count data distribution. Lambert (1992) introduced the zero-inflated Poisson (ZIP) regression model and applied it for modeling defects in manufacturing processes, where defects are assumed coming from a 'perfect' state with a probability π or an 'imperfect' state with a probability $1 - \pi$. While counts from the 'perfect', 'no-defect' state are always zero, those from the 'imperfect' state follow a Poisson distribution.

The probability mass function (pmf) of a random variable having a ZIP or zero-inflated negative binomial (ZINB) distribution can be written as

$$Pr(Y_i = k) = \pi_i I(k = 0) + (1 - \pi_i) g(k|\boldsymbol{\theta}_i), k = 0, 1, 2, \ldots, \tag{1}$$

where the mixing parameter πi is interpreted as the probability of a count being from the 'non-susceptible' or 'not-at-risk' latent class, I(T) is an indicator variable taking 1 when T is true, and 0 when T is false; g is a Poisson or negative binomial mass function, and θ_i is the vector of parameters in g. When g is the Poisson mass function, θi is equal to the mean μ_i of the distribution, and for a negative binomial probability mass function g, $\theta_i = (\mu_i, \alpha)$, where μ_i is the mean of the distribution and α is the dispersion parameter. In this paper, we will use the following parameterization for the probability mass function of a negative binomial distribution with mean μ and dispersion parameter α. where the mixing parameter π_i is interpreted as the probability of a count being from the 'non-susceptible' or 'not-at-risk' latent class, I(T) is an indicator variable taking 1 when T is true, and 0 when T is false; g is a Poisson or negative binomial mass function, and θ_i is the vector of parameters in g. When g is the Poisson mass function, θi is equal to the mean μ_i of the distribution, and for a negative binomial probability mass function g, $\theta_i = (\mu_i, \alpha)$, where μ_i is the mean of the distribution and α is the dispersion parameter. In this paper, we will use the following parameterization for the probability mass function of a negative binomial distribution with mean μ and dispersion parameter α.

$$f(y|\mu, \alpha) = \frac{\Gamma(y + \alpha)}{y!\,\Gamma(\alpha)} \left(\frac{\alpha}{\alpha + \mu}\right)^{\alpha} \left(\frac{\mu}{\alpha + \mu}\right)^{y}, \text{where } y = 0, 1, \ldots.$$

In zero-inflated count models, the logit and the log link functions are typically specified for the mixing probability πi and the mean of the assumed standard distribution μ_i, respectively, as

$$logit(\pi_i) = \mathbf{w}_i'\gamma \quad \text{and} \quad log(\mu_i) = \mathbf{x}_i'\boldsymbol{\xi},$$

where w_i and x_i are $q \times 1$ and $p \times 1$ vectors of covariates for the i th subject, and $\gamma = (\gamma_1, \gamma_2, \ldots, \gamma_q)$ and $\xi = (\xi_1, \xi_2, \ldots, \xi_p)$ are regression parameters. For n independent observations, the ZIP likelihood function is

$$L(\boldsymbol{\xi}, \gamma|\mathbf{y}) = \prod_{i=1}^{n} \{1 + e^{\mathbf{w}_i'\gamma}\}^{-1} \left\{e^{\mathbf{w}_i'\gamma} + e^{-\exp(\mathbf{x}_i'\boldsymbol{\xi})}\right\}^{I(y_i=0)} \left\{\frac{e^{-\exp(\mathbf{x}_i'\boldsymbol{\xi})} e^{\mathbf{x}_i'\boldsymbol{\xi} y_i}}{y_i!}\right\}^{I(y_i>0)}$$

The corresponding likelihood function for the ZINB model can be written as

$$L(\boldsymbol{\xi},\boldsymbol{\gamma},\alpha|\mathbf{y}) = \prod_{i=1}^{n}\{1+e^{\mathbf{w}_i'\boldsymbol{\gamma}}\}^{-1}\left\{e^{\mathbf{w}_i'\boldsymbol{\gamma}} + \left(\frac{\alpha}{\alpha+e^{\mathbf{x}_i'\boldsymbol{\xi}}}\right)^{\alpha}\right\}^{I(y_i=0)}$$
$$\times \prod_{i=1}^{n}\left\{\frac{\Gamma(y_i+\alpha)}{y_i!\,\Gamma(\alpha)}\left(\frac{\alpha}{\alpha+e^{\mathbf{x}_i'\boldsymbol{\xi}}}\right)^{\alpha}\left(\frac{e^{\mathbf{x}_i'\boldsymbol{\xi}}}{\alpha+e^{\mathbf{x}_i'\boldsymbol{\xi}}}\right)^{y_i}\right\}^{I(y_i>0)}$$

Since interpretations of parameters γ and ξ in ZIP and ZINB models apply to the two latent subpopulations, they do not directly describe the overall population mean. Although the overall mean, $E(Y_i) = \nu_i$, for i th subject could be estimated from such models by

$$\nu_i = \frac{e^{\mathbf{x}_i'\boldsymbol{\xi}}}{1+e^{\mathbf{w}_i'\boldsymbol{\gamma}}}$$

and transformations such as the delta method could be applied to estimate the corresponding variance, it is not always easy to understand the behavior of ν_i. In particular, determining the overall effects of an exposure variable on incidence density ratios is challenging especially when the linear predictors from both the mixing proportions and the Poisson mean model contain the exposure variable (Long et al. 2014).

Marginalized ZIP and ZINB models To estimate the overall effects of covariates on the population mean, marginalized zeroinflated Poisson (Long et al. 2014) and marginalized zero-inflated negative binomial (Preisser et al. 2016) models specify parameters for the probability of being an excess zero (i.e., π_i) and the marginal mean $\nu_i = E(y_i) = (1 - \pi_i)\mu_i$ as

$$logit(\pi_i) = \mathbf{w}_i'\boldsymbol{\gamma} \quad \text{and} \quad log(\nu_i) = \mathbf{x}_i'\boldsymbol{\beta},$$

where $\beta = (\beta_1, \beta_2, \dots, \beta_p)$ is a vector of regression parameters for ν_i that have the same interpretations for the effects of exposures on the marginal mean as in Poisson and negative binomial regression, whereas the parameters in γ have the same latent class interpretations for zero-inflation as in ZIP and ZINB models. The MZIP and MZINB likelihood functions are obtained by replacing μ_i by $\nu_i/(1 - \pi_i)$ in the ZIP and ZINB likelihoods, respectively. The next section introduces methods of estimating regression parameters for the overall population mean of heterogeneous counts generated from non-degenerate mixture distributions. With the aim of expanding the pool of two-part marginalized models for counts, special consideration is given to data generating mechanisms based on mixtures of two Poissons and a negative binomial with a Poisson distribution.

Finite mixture models

Finite mixture distributions have been used to model counts obtained from heterogeneous populations (Wang et al. 1996; Schlattmann 2009; Morgan et al. 2014). In the general finite mixture model, the source population is assumed to be a partition of $m \geq 2$ latent subpopulations; with a probability π_{ij}, the count random variable Y_i corresponding to the ith individual takes a value from the jth subpopulation according to a distribution specific to the subpopulation. An m-component mixture distribution can be defined as (Frühwirth-Schnatter 2005; Wedel and DeSarbo 1995)

$$Pr(Y_i = y_i|\pi, \theta_i) = \sum_{j=1}^{m} \pi_{ij} f_j(y_i|\theta_{ij})$$

where the components $f_1, f_2, \ldots, f_m$ are probability mass functions of known distributions, $\theta_i = (\theta_{i1}, \ldots, \theta_{im})$ where θ_{ij} is the vector of parameters in f_j, and $\pi = (\pi_1, \pi, \ldots, \pi_m)$ is a vector of mixing probabilities with $0 \leq \pi_j \leq 1$ and $\sum_{j=1}^{m} \pi_j = 1$. While the mixture distribution for zero-inflated counts in equation (1) allows mixing probabilities to vary across individuals, conventional finite mixture models assume a constant probability, π_j, corresponding to the j th subpopulation and impose heterogeneity through $f_j(y_i|\theta_{ij})$.

The Poisson mixture distribution, where

$$f_j(y_i|\mu_{ij}) = \frac{e^{-\mu_{ij}} \mu_{ij}^{y_i}}{y_i!}$$

with μ_{ij} being the mean of the j th component distribution, is a popular finite mixture model for count data. In Poisson mixture regression, the mean μ_{ij}, $j = 1, \ldots, m$, is modeled as a function of covariates using the log link. Wang et al. (1996) discuss that such models are identifiable for full rank design matrices. While finite mixture models enable flexible modeling of counts from heterogeneous populations, their parameters have latent class interpretations. Such coefficients do not directly provide inferences regarding the effects of covariates on the overall population mean (Min and Agresti 2005; Roeder et al. 1999). For $m = 2$, the pmf of a random variable with a Poisson-Poisson mixture distribution can be written as

$$f(y_i|\pi, \mu_{1i}, \mu_{2i}) = \pi f_{P1}(y_i|\mu_{1i}) + (1 - \pi) f_{P2}(y_i|\mu_{2i})$$

where π is a mixing probability, and f_{P1} and f_{P2} are Poisson mass functions with corresponding mean parameters μ_{1i} and μ_{2i}. Similarly, a negative binomial-Poisson random variable has a pmf given by

$$f(y_i|\pi_i, \mu_{1i}, \mu_{2i}, \alpha) = \pi f_P(y_i|\mu_{1i}) + (1-\pi) f_{NB}(y_i|\mu_{2i}, \alpha). \quad (2)$$

In Eq. (2), f_p is a Poisson pmf with mean parameter μ_{1i} and f_{NB} a negative binomial pmf with mean and dispersion parameters μ_{2i} and α, respectively. The marginal mean, ν_i, of a random variable Y_i having either of the two mixture distributions can be written as

$$\nu_i = \pi \mu_{1i} + (1-\pi)\mu_{2i}. \quad (3)$$

In traditional finite mixture models, separate regression equations are specified for the mean of each component of the mixture. In general, ν_i depends upon a complicated function of the regression coefficients from the components. In the next section, new marginalized models are specified for direct inference regarding the effects of covariates on ν_i.

Marginalized finite mixture models

Solving for μ_{2i} in Eq. (3) gives Marginalized finite mixture models Solving for μ_{2i} in Eq. (3) gives

$$\mu_{2i} = \frac{\nu_i - \pi\mu_{1i}}{1-\pi}. \quad (4)$$

To estimate a model for ν_i, the likelihood functions of Poisson-Poisson and negative binomial-Poisson mixture models can be written as functions of ν_i by replacing μ_{2i} by the linear function of the marginal mean in Eq. (4). Thus, marginalized PoissonPoisson (MPois-Pois) and negative binomial-Poisson (MNB-Pois) pmfs can be written as in Eqs. (5) and (6), respectively:

$$f_{MPP}(y_i|\pi, \mu_{1i}, \nu_i) = \pi \frac{e^{-\mu_{1i}}\mu_{1i}^{y_i}}{y_i!} + (1-\pi)\frac{e^{-\frac{\nu_i - \pi\mu_{1i}}{1-\pi}}\left[\frac{\nu_i-\pi\mu_{1i}}{1-\pi}\right]^{y_i}}{y_i!} \quad (5)$$

$$f_{NBP}(y_i|\pi, \alpha, \mu_{1i}, \nu_i) = \pi \frac{e^{-\mu_{1i}}\mu_{1i}^{y_i}}{y_i!} + (1-\pi)\frac{\Gamma(y_i+\alpha)}{y_i!\,\Gamma(\alpha)}\left(\frac{\alpha}{\alpha + \frac{\nu_i-\pi\mu_{1i}}{1-\pi}}\right)^{\alpha}\left(\frac{\frac{\nu_i-\pi\mu_{1i}}{1-\pi}}{\alpha+\frac{\nu_i-\pi\mu_{1i}}{1-\pi}}\right)^{y_i} \quad (6)$$

The MPois-Pois model is defined through Eq. (5) with specification of generalized linear models in (7) for the relationship of covariates to ν_i and μ_{1i},

$$\begin{aligned} log(\nu_i) &= \mathbf{x}_i'\boldsymbol{\beta} \\ log(\mu_{1i}) &= \mathbf{z}_i'\boldsymbol{\xi} \\ logit(\pi) &= \rho \end{aligned} \quad (7)$$

where x_i and z_i are vectors of covariates and β and ξ are corresponding vectors of regression coefficients, and $-\infty < \rho < \infty$ is a constant. Although ξ and ρ are considered nuisance parameters that are not of primary interest, they need to be modeled to facilitate maximum likelihood estimation of β in the marginal mean model. The logarithm of μ_{1i} is modeled by using a linear predictor that involves covariates as in standard finite mixture Poisson models. The mixing parameter π is modeled as a constant using the logit link to guarantee that its estimate is between 0 and 1.

A common model specification is $x_i = z_i$ such that β and ξ are p×1 vectors of parameters. However, the covariates that are included in modeling νi and μ1i may be different. As the main interest is in β, a reduced set of covariates z_i may be considered when it is necessary for computational tractability. Alternatively, a shared-parameter model (Preisser et al. 2016) may be used to incorporate a large number of covariates with relatively few parameters. The MNB-Pois model defined through Eqs. (6) and (7) also requires estimation of the dispersion parameter α via a model specified as

$$log(\alpha) = -\tau. \tag{8}$$

The link functions in Eqs. (7) and (8) correspond to $v_i > 0$, $\mu_{1i} > 0$, $0 < \pi < 1$ and $\alpha > 0$. For n independent count random variables $Y_1, Y_2, \ldots, Y_n$ with corresponding realizations $y_1, y_2, \ldots, y_n$, the likelihood function for MPois-Pois models is given by Eq. (9).

$$L(\rho, \boldsymbol{\beta}, \boldsymbol{\xi}|\mathbf{y}) = \prod_{i=0}^{n} \frac{1}{(1+e^{\rho})y_i!} \left\{ e^{\rho} \exp(-e^{\mathbf{z}_i'\boldsymbol{\xi}}) e^{\mathbf{z}_i'\boldsymbol{\xi} y_i} + e^{-\eta(\rho,\boldsymbol{\beta},\boldsymbol{\xi};\mathbf{x}_i,\mathbf{z}_i)} \eta(\rho, \boldsymbol{\beta}, \boldsymbol{\xi}; \mathbf{x}_i, \mathbf{z}_i)^{y_i} \right\} \tag{9}$$

with

$$\eta(\rho, \boldsymbol{\beta}, \boldsymbol{\xi}; \mathbf{x}_i, \mathbf{z}_i) = e^{\mathbf{x}_i'\boldsymbol{\beta}}(1+e^{\rho}) - e^{\rho} e^{\mathbf{z}_i'\boldsymbol{\xi}}. \tag{10}$$

Similarly, the likelihood function for the MNB-Pois model can be specified as

$$L(\rho, \tau, \boldsymbol{\beta}, \boldsymbol{\xi}|\mathbf{y}) = \prod_{i=0}^{n} \left\{ \frac{\Gamma(y_i + e^{-\tau})}{(1+e^{\rho})\Gamma(y_i+1)\Gamma(e^{-\tau})} \left(\frac{e^{-\tau}}{e^{-\tau} + \eta(\rho, \boldsymbol{\beta}, \boldsymbol{\xi}; \mathbf{x}_i, \mathbf{z}_i)} \right)^{e^{-\tau}} \times \left(\frac{\eta(\rho, \boldsymbol{\beta}, \boldsymbol{\xi}; \mathbf{x}_i, \mathbf{z}_i)}{e^{-\tau} + \eta(\rho, \boldsymbol{\beta}, \boldsymbol{\xi}; \mathbf{x}_i, \mathbf{z}_i)} \right)^{y_i} \right\} + \prod_{i=0}^{n} \frac{e^{\rho} \exp(-e^{\mathbf{z}_i'\boldsymbol{\xi}}) e^{\mathbf{z}_i'\boldsymbol{\xi} y_i}}{(1+e^{\rho})y_i!}$$

where $\eta(\rho, \beta, \xi; x_i, z_i)$ has the same expression as in Eq. (10). With carefully chosen starting parameter values, marginalized finite mixture models can be fitted using quasiNewton optimization. Guidance for specifying starting values and use of SAS Proc NLMIXED for fitting the proposed models is presented as Additional file 1 (Benecha et al., 2017) along with further discussion of connections between the models in "Marginalized ZIP and ZINB models" and "Marginalized finite mixture models" sections. Finally, with respect to mixture Eq. (2), solving for μ_{1i} in (3) gives

$$\mu_{1i} = \frac{\nu_i - (1-\pi)\mu_{2i}}{\pi}. \quad (11)$$

Inserting this expression for μ_{1i} in the standard mixture likelihood function based on Eq. (2) gives a likelihood function for a model that is different from MNB-Pois. The alternative model, which marginalizes over the Poisson part versus MNB-Pois that marginalizes over the NB part, is not considered owing to unresolved computational issues in the applications.

Simulation study

Simulation studies were performed to examine the properties of MPois-Pois and MNBPois models for various sample sizes. Counts from these models were generated from the probability mass functions in Eqs. (5) and (6), where π, μ_{1i}, ν_i and α are determined from

$$\begin{aligned} log(\nu_i) &= \mathbf{x}_i'\boldsymbol{\beta} = \beta_0 + \beta_1 x_{1i} + \beta_2 x_{2i} + \beta_3 x_{3i} \\ log(\mu_{1i}) &= \mathbf{z}_i'\xi = \xi_0 + \xi_1 x_{1i} + \xi_2 x_{2i} + \xi_3 x_{3i} \\ logit(\pi) &= \rho, \\ log(\alpha) &= -\tau \end{aligned}$$

with $x_i = z_i$ and $x_{1i} \sim$ Poisson(2)/3, $x_{2i} \sim$ exp(1), $x_{3i} \sim$ Benoulli(0.4), $\beta_0 = 1.5$, $\beta_1 = -0.1$, $\beta_2 = -0.2$, $\beta_3 = 0.5$, $\xi_0 = 1.5$, $\xi_1 = -0.5$, $\xi_2 = -0.5$, $\xi_3 = 1$, $\rho = -0.4$ and $\tau = -0.5$. Using these specifications, samples of sizes 100, 200, 500 and 1000 were generated corresponding to MPois-Pois and MNB-Pois models. Poisson and negative binomial (NB) regression and four marginalized count models, namely MZIP, MZINB, MPois-Pois and MNB-Pois, were then fitted to the data, where each simulation was repeated 10,000 times. To estimate Type I error rates of testing $H_0 : \beta_1 = 0$ vs $H_1 : \beta_1 = 0$, all the simulations were repeated by generating data using $\beta_1 = 0$, but keeping all the remaining parameter and covariate values the same as described previously. For each of the six models, the Type I error rates were calculated among converged model fits as the proportion of p-values from two-sided Wald tests that were less than 0.05. For MPois-Pois generated data, estimates of β_1, β_2 and β_3 had low biases for all models and all sample sizes (Table 1). The MPois-Pois model had Type I error rates for β_1 close to 0.05, while the remaining models tended to over-estimate the error rates (Table 2). The MPois-Pois model estimated coverages of 95% confidence intervals for β_1, β_2 and β_3 that were close to the nominal value (Table 3). Whereas NB, MZINB and MNB-Pois models tended to have only slight undercoverage, Poisson and MZIP had coverage ranging from 88 to 92%. Convergence rates for MPois-Pois simulation scenarios ranged from 96.2 to 99.3%, while convergence rates ranged from 88.0 to 90.2% for MNB-Pois, from 75.9 to 98.4% for MZIP, and from 72.0 to 96.6% for the MZINB models. Convergence was 100% for Poisson and NB regression for all sample sizes. When the data are generated from the MNB-Pois model, the MNB-Pois model had low percent relative median biases for β_1, β_2 and β_3, and the biases appear to decrease as sample sizes increase (Table 4). The corresponding estimates from the Poisson, NB and MZINB models also have low biases, but those from MPois-Pois and MZIP models are generally higher. In addition, the performance of the true MNB-Pois model with regard to Type I error rates (for β_1) and coverages of 95% confidence intervals (for β_1, β_2 and β_3) is superior to Poisson, MZIP and MPois-Pois models at all sample sizes (Tables 5 and 6, respectively) and has better performance than NB and MZINB for the sample sizes of 500 and 1000. Over 96% of MNB-Pois models converged for sample sizes of 200 or more, with 91% convergence for sample size of 100. Coverage ranged from 97.4 to 100% for MZIP, from 92.0 to 99.4% for the MPois-Pois models, from 85.3 to 91.4% for MZINB models and rates were 100% for Poisson and NB regression. Overall, the simulation results indicate that when the true model is specified, MPois-Pois or MNB-Pois models estimate marginal mean regression parameters with small biases, Type I errors close to the assumed rate and coverages of 95% confidence intervals near 95% for sample sizes of 200 or greater.

Table 1 Percent relative median biases of estimates of β_1, β_2 and β_3 from marginalized mixture models fitted to data generated from the MPois-Pois model with 10,000 replications

Sample size	Parameter	Poisson	MZIP	MPois-Pois	NB	MZINB	MNB-Pois
100	β_1	2.03	1.40	−2.04	2.18	0.97	0.56
	β_2	0.77	−3.11	0.08	1.00	−3.45	1.54
	β_3	−0.30	−0.61	−0.70	−0.26	−0.74	−0.33
200	β_1	0.97	1.70	−0.68	1.38	1.89	1.34
	β_2	−0.02	−2.64	−0.69	0.08	−2.65	0.62
	β_3	0.15	−0.43	−0.29	−0.09	−0.41	0.06

Sample size	Parameter	Poisson	MZIP	MPois-Pois	NB	MZINB	MNB-Pois
500	β_1	−0.68	−0.36	−0.87	−0.79	−1.18	0.07
	β_2	0.04	−1.51	0.11	0.09	−1.44	0.78
	β_3	0.08	−0.16	−0.14	0.05	−0.11	0.19
1000	β_1	−0.14	−0.37	−0.40	−0.07	−0.64	0.43
	β_2	0.48	−1.43	0.27	0.50	−0.91	0.88
	β_3	0.09	−0.08	0.06	0.07	−0.07	0.22

Table 2 Type I error rates for the estimate of β_1 from marginalized models fitted to data generated from the MPois-Pois model with 10,000 replications

Sample size	Poisson	MZIP	MPois-Pois	NB	MZINB	MNB-Pois
100	0.127	0.102	0.068	0.077	0.070	0.073
200	0.131	0.106	0.067	0.077	0.072	0.069
500	0.135	0.112	0.060	0.079	0.073	0.065
1000	0.134	0.112	0.054	0.072	0.066	0.061

Table 3 Coverages of 95% confidence intervals for estimates of β_1, β_2 and β_3 from marginalized models fitted to data generated from the MPois-Pois model with 10,000 replications

Sample size	Parameter	Poisson	MZIP	MPois-Pois	NB	MZINB	MNB-Pois
100	β_1	89.9	91.3	93.7	93.3	93.4	93.8
	β_2	89.2	90.9	93.2	92.4	92.8	92.9
	β_3	91.8	92.9	95.2	94.9	94.7	95.1
200	β_1	89.4	91.2	94.1	93.6	93.8	94.1
	β_2	88.9	90.9	93.3	92.2	92.9	93.2
	β_3	91.4	92.6	95.1	95.1	94.9	95.2
500	β_1	88.9	90.7	94.1	92.9	93.5	93.9
	β_2	88.6	90.5	94.4	92.0	93.1	93.9
	β_3	91.0	92.0	94.9	95.1	94.8	94.9
1000	β_1	89.3	90.9	94.7	93.5	93.9	94.4
	β_2	88.5	90.8	94.7	92.1	93.1	93.8
	β_3	91.1	92.1	95.0	95.0	95.0	94.9

Table 4 Percent relative median biases of estimates of β_1, β_2 and β_3 from marginalized mixture models fitted to data generated from the MNB-Pois model with 10,000 replications

Sample size	Parameter	Poisson	MZIP	MPois-Pois	NB	MZINB	MNB-Pois
100	β_1	6.94	23.51	6.80	10.48	13.72	11.95
	β_2	2.98	7.95	4.00	5.26	1.89	4.44
	β_3	0.01	1.40	−4.35	0.80	0.88	−0.25
200	β_1	5.27	20.12	−14.85	5.45	7.41	4.57
	β_2	1.45	5.11	−1.12	2.49	0.07	2.02
	β_3	0.18	1.49	−5.44	0.38	0.36	0.33
500	β_1	0.57	11.79	−29.97	1.31	0.73	−0.75
	β_2	0.66	2.81	−3.90	1.18	0.14	0.62
	β_3	0.39	1.52	−7.66	0.59	0.61	0.46
1000	β_1	1.19	10.34	−34.68	1.92	2.39	0.00
	β_2	0.79	2.63	−4.75	1.00	0.39	0.87
	β_3	−0.01	0.97	−10.13	0.03	−0.01	−0.19

Table 5 Type I error rates for the estimate of β_1 from marginalized models fitted to data generated from the MNB-Pois model with 10,000 replications

Sample size	Poisson	MZIP	MPois-Pois	NB	MZINB	MNB-Pois
100	0.325	0.271	0.262	0.084	0.079	0.103
200	0.334	0.272	0.255	0.079	0.073	0.064
500	0.341	0.273	0.232	0.081	0.074	0.053
1000	0.340	0.273	0.240	0.076	0.072	0.049

Table 6 Coverages of 95% confidence intervals for estimates of β_1, β_2 and β_3 from marginalized models fitted to data generated from the MNB-Pois model with 10,000 replications

Sample size	Parameter	Poisson	MZIP	MPois-Pois	NB	MZINB	MNB-Pois
100	β_1	72.3	77.4	76.9	92.5	92.4	89.7
	β_2	74.0	79.6	77.8	90.8	91.8	89.6
	β_3	74.4	79.4	83.0	94.0	93.7	92.0
200	β_1	71.6	77.6	78.1	92.2	92.3	93.0

Sample size	Parameter	Poisson	MZIP	MPois-Pois	NB	MZINB	MNB-Pois
	β_2	72.8	79.1	78.9	91.0	91.8	92.7
	β_3	74.1	80.0	83.9	94.4	94.0	93.5
500	β_1	71.2	77.0	78.1	92.1	92.2	94.2
	β_2	72.3	78.6	80.8	90.5	91.3	94.5
	β_3	73.6	79.7	80.2	94.3	93.9	94.5
1000	β_1	71.7	77.5	76.2	92.7	93.1	95.0
	β_2	73.0	78.9	81.5	90.2	91.6	95.0
	β_3	74.1	80.7	71.6	94.6	94.6	95.3

APPLICATIONS

A caries incidence trial

The methods described in this article were applied to the Lanarkshire caries incidence trial introduced in "Introduction" section. A total of 4294 children ages 11–12 were randomized to either sodium fluoride (NaF), sodium fluoride plus sodium trimetaphosphate (NaFTMP) or sodium monofluorophosphate (SMFP) toothpaste formulations and dental exams were performed at baseline and after 1, 2 and 3 years. The analysis was based on 3412 children followed up until year 2 and the response variable of interest was the number of new decayed, missing and filled surfaces (DMFS). Let *NaF*=1 if the child was given sodium fluoride and 0 otherwise and let *NaFTMP*=1 if the child was randomized to the NaFTMP group and 0 otherwise; children in the SMFP group make up the reference treatment category (*NaF*=*NaFTMP*=0). In addition to treatment allocation, baseline caries (bc: 1= high, 0 = low) and baseline calculus (calc:1=yes, 0= no) were considered as exp lanatory variables. High baseline caries values correspond to at least one decayed, missing or filled anterior tooth or premolar, and a baseline calculus value of '1' refers to the existence of calcified deposits on the teeth formed by the continuous presence of dental plaque (Stephen et al. 1994; Preisser et al. 2014). An important feature of the data is the large number of zero counts in the outcome variable, as 658 (19.28%) of the 3412 children had zero DMFS counts (Fig. 1). Since the percentage of zeros is high, two-part marginalized models may provide less biased estimates and better model fits than one-part generalized linear models.

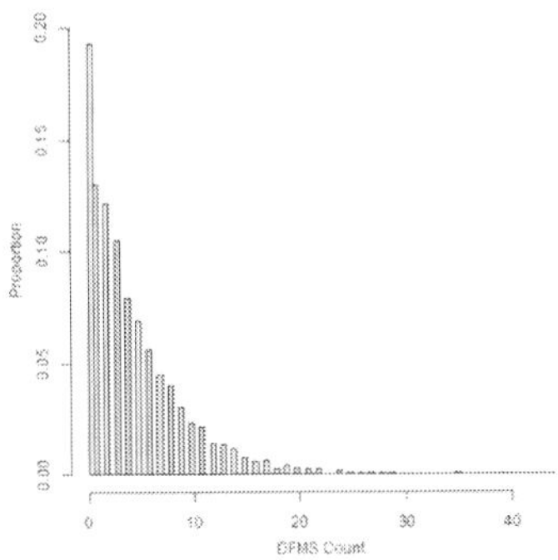

Figure. 1 Histogram of two-year DMFS increment for 3412 children ages 11–12 from a dental caries incidence trial conducted in Lanarkshire, Scotland between 1988 and 1992

Poisson, NB, MZIP, MZINB, MPois-Pois and MNB-Pois models were applied to compare the efficacy of the toothpaste formulations with respect to the marginal mean DMFS count. In the two-part models, the four binary covariates defined above were included in each model part. The three best models were NB, MZINB and MNB-Pois, which produced fitted values that best matched the observed distribution of DMFS counts (Fig. 2) and have the lowest AICs (Table 7). On the other hand, Poisson, MZIP and MPois-Pois models, which did not directly account for overdispersion, had poor fits and gave standard errors of regression coefficients for the marginal mean model that were too small. The MNB-Pois model gave the best fit (lowest AIC) while its marginal mean model parameter estimates and standard errors were similar in value to those of the next best fitting model, MZINB.

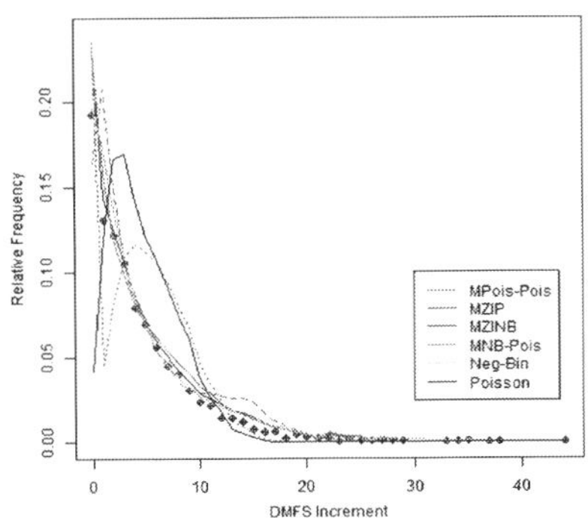

Figure. 2 Observed (*circles*) and predicted relative frequencies of two-year DMFS increment for children in the Lanarkshire caries trial

Table 7 Marginalized count regression model estimates (est) and standard errors (SE) for the Lanarkshire caries trial

	Poisson		**NB**		**MZIP**		**MZINB**		**MPois-Pois**		**MNB-Pois**	
Variable	**est**	**se**	**est**	**se**	**est**	**se**	**est**	**se**	**est**	**se**	**est**	**se**
Marginal mean model												
Intercept	1.216	0.017	1.201	0.035	1.234	0.022	1.206	0.034	1.309	0.029	1.207	0.035
bc	0.758	0.017	0.757	0.036	0.765	0.023	0.757	0.035	0.663	0.026	0.757	0.035
calc	−0.204	0.020	−0.189	0.040	−0.211	0.025	−0.195	0.039	−0.263	0.031	−0.199	0.039
NaF	−0.072	0.018	−0.056	0.039	−0.098	0.023	−0.060	0.038	−0.144	0.028	−0.060	0.038
NaFTMP	−0.052	0.022	−0.022	0.048	−0.104	0.029	−0.034	0.047	−0.062	0.035	−0.033	0.047
					Zero-inflation model				Latent class mean model			
Intercept					−1.280	0.079	−2.257	0.290	2.131	0.031	−1.425	0.513
bc					−1.309	0.113	−2.911	1.571	0.512	0.027	3.539	0.500
calc					0.129	0.095	−0.149	0.273	−0.249	0.031	0.022	0.173
NaF					0.208	0.096	0.419	0.307	−0.152	0.029	−0.013	0.188
NaFTMP					0.289	0.119	0.545	0.363	−0.081	0.037	−0.307	0.201
Mixing probability model estimates[a]												
ρ									−0.753	0.055	−1.818	0.170
π									0.320		0.140	
Dispersion estimates[b]												
τ											−0.328	0.042

	Poisson		NB		MZIP		MZINB		MPois-Pois		MNB-Pois	
Variable	**est**	**se**	**est**	**se**	**est**	**se**	**est**	**se**	**est**	**se**	**est**	**se**
ϕ			0.806	0.028			0.639	0.038			0.720	
Model fit statistics												
−2loglik	22706		17249		20410		17189		18048		17169	
AIC	22716		17261		20430		17211		18070		17193	

[a]Note that $\pi=1/(1+e^{-\rho})$

1. [b]also, $\tau=\log\phi$

	Poisson		NB[a]		MZIP		MZINB[b]		MPois-Pois	
Variable	**est**	**se**	**est**	**se**	**est**	**se**	**est**	**se**	**est**	**se**
Intercept					-4.262	0.732	-4.381	0.827	2.142	0.051
Photoperiod, 16h					4.159	0.753	4.264	0.846	-4.238	0.552
Mixing probability model[c]										
ρ									0.178	0.184
π									0.544	
Model fit statistics										
-2loglik	1566.4		1402.1		1250.2		1236.5		1236.4	
AIC	1574.4		1412.1		1262.2		1250.5		1250.4	

[a]In the NB model, $\hat{\phi} = 0.522$ (s.e. = 0.083)
[b]In the MZINB model, $\hat{\tau} = -2.662$ (s.e. 0.351) corresponding to $\hat{\phi} = e^{\hat{\tau}} = 0.070$.
[c]In the MPois-Pois model, $\pi = 1/(1 + e^{-\rho})$

Table 10 Model-predicted mean number of roots of the apple cultivar Trajan produced by the eight treatments

Treatment	No. of shoots	Observed mean	Count regression model[a]				
			Poisson	**NB**	**MZIP**	**MZINB**	**MPP**
8h + BAP 2.2	30	5.83	6.55	6.53	6.39	6.39	6.44
8h + BAP 4.4	30	7.77	6.87	6.87	6.81	6.81	6.81
8h + BAP 8.8	40	7.50	7.21	7.22	7.25	7.25	7.20
8h + BAP 17.6	40	7.15	7.56	7.60	7.73	7.72	7.60
16h + BAP 2.2	30	3.27	3.22	3.22	3.43	3.45	3.24
16h + BAP 4.4	30	2.73	2.99	2.99	3.06	3.07	2.95
16h + BAP 8.8	30	3.13	2.78	2.77	2.73	2.73	2.68

Treat-ment	No. of shoots	Ob-served mean	Count regression model[a]				
			Poisson	NB	MZIP	MZINB	MPP
16h + BAP 17.6	40	2.45	2.58	2.57	2.43	2.43	2.43

[a]MPP = MPois-Pois model

DISCUSSION AND CONCLUSIONS

In this article, marginal means of counts with unexplained heterogeneity were modeled using two-component finite mixture distributions. Regression parameters were specified in two-part marginalized models for direct estimation of exposure effects on the overall mean count using maximum likelihood methods. Specifically, the proposed MPois-Pois and MNB-Pois mixture models provide alternative model choices to MZIP and MZINB for counts that are overdispersed or have many zeros. It may not always be clear whether a zero-inflated count model or a model based on a finite mixture of two non-degenerate components is more appropriate as Poisson and negative binomial distributions with small means can generate a large amount of zeros. In the case of dental caries, zero-inflated count regression models are sometimes used (Preisser et al. 2012) even though caries researchers question whether any child can be immune to developing caries (Mwalili et al. 2008). In many fields, use of finite mixtures of non-degenerate components may have a stronger theoretical basis than assuming a mixture of at-risk and not-at-risk latent classes. While there is sometimes interest in latent classes, researchers across many fields of inquiry are frequently interested in quantifying the effects of covariates on the overall mean count while adjusting for unexplained heterogeneity. In such cases, marginal mean regression parameters in MZIP, MZINB, MPois-Pois and MNB-Pois models have straightforward interpretations in describing overall exposure effects on count outcomes.

As described in the Additional file 1 (Benecha et al. 2017), the marginalized models proposed in this article belong to a larger class of marginalized mixture models for counts. In particular, when the mixing probability component of the model is fixed either with or without covariates, the MZIP and MZINB models may be viewed as special cases of corresponding MPois-Pois and MNB-Pois models where the Poisson component of the latter two models has a mean of zero, rendering that component degenerate. In this sense, the proposed models expand the family of two-part marginalized regression models by providing alternatives to MZIP and MZINB regression. In the absence of theoretical justification, the merit of each model in the larger class of alternative marginalized models is judged based on goodness of fit considerations. Because our main interest is in modeling marginal means of counts, model parameters that are not of primary interest are allowed to depend on covariates, or none whatsoever, to complete specification of the likelihood function. This provides for model parsimony as needed while allowing all the relevant covariates to be estimated in the marginal mean model.

A simulation study indicated that when the true model is specified, each of the proposed marginalized mixture models provides low biases, Type I errors and confidence interval coverages close to the nominal levels. As shown in additional simulation studies reported in Benecha et al. (2017), model mis-specification can result in undercoverage and inflated Type I errors. Use of empirical covariance estimation as proposed by Long et al. (2014) for MZIP models would likely improve coverage and Type I errors for large samples. In any case, assessment of model goodness-of-fit is highly recommended. Unfortunately, such assessment is often hampered by computational difficulties in fitting complex models such as MNB-Pois when the data at hand do not contain sufficient information to estimate all the model parameters. Reducing the number of covariate parameters often provides an expeditious remedy for this situation. Another advance would be to develop score tests for goodness of fit, as proposed by Ridout et al. (2001) in comparing ZIP and ZINB models, that do not require fitting the model under the alternative hypothesis.

In summary, the proposed marginalized mixture modeling framework provides a wide range of alternatives to directly estimate exposure effects on marginal means of counts generated from heterogeneous populations. The methods are fairly straightforward while requiring consideration of carefully chosen starting values and can be implemented in most statistical software. Future research could extend the marginalized count regression models to mixtures of two negative binomial distributions or to those based on Eq. (11) as an alternative to MNB-Pois, to allow the mixing probabilities to depend on covariates, and to longitudinal data.

REFERENCES

1. Albert, J, Wang, W, Nelson, S: Estimating overall exposure effects for zero-inflated regression models with application to dental caries. Stat. Methods Med. Res. 23, 257–278 (2014).
2. Benecha, HK, Neelon, B, Divaris, K, Preisser, JS: Marginalized mixture models for count data from multiple source populations: supplementary material (2017). http://www.bios.unc.edu/~preisser/personal/software.html. Accessed 9 Mar 2017.
3. Böhning, D, Dietz, E, Schlattmann, P, Mendonca, L, Kirchner, U: The zero-inflated Poisson model and the decayed, missing and filled teeth index in dental epidemiology. J. R. Statist. Soc. A. 162, 195–209 (1999).
4. Frühwirth-Schnatter, S: Finite mixture and Markov switching models. Springer-Verlag, New York (2005).
5. Lambert, D: Zero-inflated Poisson regression, with application to defects in manufacturing. Technometrics. 34, 1–14 (1992).
6. Long, DL, Preisser, JS, Herring, AH: Golin CE.A marginalized zero-inflated Poisson regression model with overall exposure effects. Statist. Med. 33, 5151–5165 (2014).
7. Marin, J, Jones, O, Hadlow, W: Micropropagation of columnar apple trees. J. Hortic. Sci. 68, 289–297 (1993).
8. Min, Y, Agresti, A: Random effect models for repeated measures of zero-inflated count data. Statist. Modelling. 5, 1–19 (2005).
9. Morgan, CJ, Lenzenweger, MF, Rubin, DB, Levyd, DL: A hierarchical finite mixture model that accommodates zero-inflated counts, non-independence, and heterogeneity. Statist. Med. 33, 2238–2250 (2014).
10. Mullahy, J: Specification and testing of some modified count data models. J. Econometrics. 33, 341–365 (1986).
11. Mwalili, SM, Lesaffre, E, Declerck, D: The zero-inflated negative binomial regression model with correction for misclassification: an example in caries research. Stat. Methods Med. Res. 17, 123–139 (2008).
12. Preisser, JS, Stamm, JW, Long, DL, Kincade, M: Review and recommendations for zero-inflated count regression modeling of dental caries indices in epidemiological studies. Caries Res. 46, 413–423 (2012).

13. Preisser, JS, Das, K, Benecha, H, Stamm, JW: Logistic regression for dichotomized counts. Statist. Methods Med. Res (2014). doi:10.1177/0962280214536893.

14. Preisser, JS, Das, K, Long, DL, Divaris, K: Marginalized zero-inflated negative binomial regression with application to dental caries. Statist. Med. 35, 1722–1735 (2016).

15. Ridout, M, Demetrio, CGB, Hinde, J: Models for count data with many zeros. International Biometric Conference, Cape Town (1998). https://www.kent.ac.uk/smsas/personal/msr/webfiles/zip/ibc_fin.pdf. Accessed 27 Mar 2017.

16. Ridout, M, Hinde, J, Demetrio, CGB: A score test for testing a zero-inflated Poisson regression model against zero-inflated negative binomial alternatives. Biometrics. 57, 219–223 (2001).

17. Roeder, K, Lynch, KG, Nagin, DS: Modeling uncertainty in latent class membership: a case study in criminology. J. Am. Statist. Assoc. 94, 766–776 (1999).

18. Schlattmann, P: Medical applications of finite mixture models. Springer-Verlag, Berlin (2009).

19. Stephen, KW, Chestnutt, IG, Jacobson, APM, McCall, DR, Chesters, RK, Huntingdon, E, Schafer, F: The effect of NaF and SMFP toothpastes on three-year caries increments in adolescents. Int. Dent. J. 44, 287–295 (1994).

20. Todem, D, Kim, K, Hsu, WW: Marginal mean models for zero-inflated count data. Biometrics (2016). doi:10.1111/biom.12492.

21. Wang, P, Puterman, ML, Cockburn, I, Le, N: Mixed Poisson regression models with covariate dependent rates. Biometrics. 52, 381–400 (1996).

22. Wedel, M, DeSarbo, WS: A mixture likelihood approach for generalized linear models. J. Classif. 12, 21–55 (1995).

23. Yang, Z, Hardin, JW, Addy, CL: Testing overdispersion in the zero-inflated Poisson model. J. Stat. Plan. Infer. 139, 3340–3353 (2009).

INDEX

C

D

E

F

G

H

I

K

L

M

N

O

P